MAIZE AGROECOSYSTEM

Nutrient Dynamics and Productivity

MAIZE AGROECOSYSTEM
Nutrient Dynamics and Productivity

Dr. K.R. Krishna

Apple Academic Press

TORONTO   NEW JERSEY

© 2013 by
Apple Academic Press Inc.
3333 Mistwell Crescent
Oakville, ON L6L 0A2
Canada

Apple Academic Press Inc.
1613 Beaver Dam Road, Suite # 104
Point Pleasant, NJ 08742
USA

First issued in paperback 2021

Exclusive worldwide distribution by CRC Press, a Taylor & Francis Group

ISBN 13: 978-1-77463-196-6 (pbk)
ISBN 13: 978-1-926895-08-6 (hbk)

Library of Congress Control Number: 2012935647

Library and Archives Canada Cataloguing in Publication

Krishna, K. R. (Kowligi R.)
Maize agroecosystem: nutrient dynamics and productivity/K.R. Krishna.

Includes bibliographical references and index.
ISBN 978-1-926895-08-6
1. Corn. 2. Crop science. I. Title.

SB191.M2K75 2012   633.1>5   C2011-908701-4

Contents

List of Abbreviations

Al	Aluminum
AMF	Arbuscular mycorrhizal fungi
AN	Ammonium nitrate
APSIM	Agricultural production systems simulator
ASR	Apparent sulfur recovery
B	Boron
BBF	Broad-bed and furrows
BD	Bulk density
BMP	Best Management Practices
C	Carbon
Ca	Calcium
CEC	Cation exchange capacity
CERES	Crop Environment Research Synthesis
CI_{590}	Chlorophyll index
CIMMYT	International Maize and Wheat Improvement Center
Cl	Chloride
CMR	Chlorophyll meter readings
CNC	Critical N concentration
CP	Chisel plow
CRP	Conservation Reserve Program
CT	Conventional tillage
Cu	Copper
DMC	Direct seeded mulch-based cropping
DOC	Dissolved organic carbon
DSSAT	Decision Support Systems for Agro-Technologies
EONRs	Economically optimum nitrogen rates
FB	Flat beds
Fe	Iron
FLG	Forage legume-grass
FP	Farmer's practice
FYM	Farm yard manure
GM	Genetically modified
GNDV	Green normalized vegetation index
GPT	Gas Pressure Test
HI	Harvest index

IFDC	International Fertilizer Development Center
IHF	Integrated Horizontal Flux
INM	Integrated Nutrient Management
IPNI	International Plant Nutrition Institute
IRNT	Incubation and Residual N Test
ISNT	Illinois Soil N Test
K	Potassium
LER	Land equivalent ratio
LPM	Liquid pig manure
MBC	Microbial biomass-C
Mg	Magnesium
MLGS	Milk line growth stage
Mn	Manganese
Mo	Molybdenum
MP	Moldboard plow
N	Nitrogen
nBTPT	N-(n-Butyl) thiophosphoric triamide
$NDVI_{590}$	Vegetative Index
NNI	N Nutrition Index
NT	No-tillage
NUE	N-Use efficiency
NUTMON	Nutrient monitoring
OPOM	Occluded particulate organic matter
P	Phosphorus
PDSS	Phosphorus Decision Support System
PF	Precision farming
PGPR	Plant growth promoting rhizobacteria (PGPR)
POST-N	Post-silking N rates
PR	Phosphate rock
PRE-N	Pre-plant rate
PSNT	Pre-side Dress N Test
PUE	Precipitation use efficiency
QTLS	Quantitative trait loci
RA	Relative availability
RCM	Relative Chlorophyll Meter Reading
RF	Ridges and furrows
RFLP	Restriction fragment length polymorphism analysis
RLD	Root length density
RMW	Rice mill waste

RSN	Residual-N
RT	Reduced tillage
RY	Relative yield
RZWQM	Root Zone Water Quality Model
S	Sulfur
SADC	Southern African Development Authority
SALSA	Semi-arid Latin America Simulation Model for Agroecosystems
SEU	S use efficiency
SOM	Soil organic matter
SSNM	Site-specific Nutrient Management
SSP	Single super phosphate
UAN	Ureaammonium nitrate
UUP	Urea–urea-phosphate
VPIs	Vegetative productivity indicators
WANA	West Asia and North Africa
WSOC	Water soluble organic-C
WUE	Water use efficiency
Yat	Attainable yield
YLNS	Yield loss due to nutrient stress
YLWS	Yield loss due to water stress
Ymax	Maximum yield
Yp	Potential yield
Zn	Zinc
ZT	Zero-till

Preface

Historically seeds for a "Maize Agroecosystem" was sown several thousand years ago in Mesoamerica. Human ingenuity, migration, and necessity for diverse food crops induced its expansion. Maize cropping zones extended into North American Plains and other regions of the world, rather rapidly during ancient and medieval periods respectively. At present, Maize Agroecosystem stretches into vast plains, hills, and coastal areas of each continent. Maize belts contribute food grains to a large populace of the world. Perhaps over 4 billion humans utilize it as cereal grain. Maize silage supports a still larger number of farm animals. Maize is currently a sought after source material for Biofuel production.

Within the context of this book, "Global Maize Agroecosystem" is defined as a conglomerate of various maize belts that occur in all continents, wherever agricultural cropping is in vogue. Maize is a versatile crop. Hence, maize belts thrive in variety of geographic conditions and extend into over 145 million ha. Maize crop flourishes on several types of soil and fertility regimes. Maize belts also endure vagaries of agroclimatic conditions rather commendably. Maize is a preferred crop in dry savannas, evergreen tropics, and even in cold temperate plains.

Maize cropping zones are easily classifiable into subsistence, moderately intensive, and intensive. The genetic composition of maize belts traced in different continent varies enormously. Aspects like, geographic conditions, human preferences, and economic advantages have a share in deciding the maize cultivars that encompass or dominate the agroecosystem. Farmers often try to match crop's (cultivar) adaptability and productivity levels with economic aspirations and long term upkeep of cropping zones. Maize farming intensity is highly dependent on inherent soil fertility, nutrient, and moisture supply. Nutrient dynamics that ensue in soil and above-ground are crucial factors that decided maize productivity. Nutrients supplied through chemical fertilizers, organic manures and that derived from atmospheric deposits regulate nutrient distribution within maize belts. Water resources and nutrient supply rates have immensely influenced the maize cropping intensity and yield expectations. Several factors like erosion, seepage, runoff, and emissions have a role in regulating nutrient dynamics and upkeep of maize agroecosystems. In the intensive maize farming zones, enhanced nutrient turnover has induced high productivity, but at the same, it has also caused nutrient accumulation in the soil profile, undue loss as well as ground water pollution. During recent past, maize farming has experienced perceptible changes regarding crop genetic stocks, soil management, agronomic procedures especially nutrient, and water supply techniques. These advances have affected nutrient dynamics and productivity of maize belts all over the world. Recent knowledge accrued and discussions regarding impact of nutrient management procedures form the center piece of the book.

This book on "Maize Agroecosystem" has 8 chapters. First chapter is introductory. It includes summarizations on origin and historical aspects of maize as a crop and its spread into different continents. Descriptions on geographic settings, expanse, and productivity are also included. Maize belts around the globe are supported by wide array of soil types. Details regarding physico-chemical characteristics of soil types and their ability to support optimum productivity are described in Chapter 2. Maize farmers have adopted an assortment of tillage systems. Tillage systems that conserve soil fertility and enhance carbon sequestration have been preferred. Yet, conventional tillage systems still persist. Restricted or conservation is popular in many locations. Most recently, "No-Tillage System" has gained ground in both subsistence and intensive farming zones. Chapters 3 and 4 include detailed knowledge about major nutrients—nitrogen, phosphorus, and potassium; secondary and micronutrients. Nitrogen dynamics that ensues in the maize fields has been explained in great detail and is commensurate with large amount of knowledge accrued on this all important nutrient element. Topics like physico-chemical transformations of N, chemical tests to assess soil-N, rapid tests using chlorophyll meters and reflectance levels to ascertain plant-N, fertilizer-N supply techniques, N recovery rates, loss via erosion, seepage and emissions, extent of N transformations, and partitioning of N in plants are highlighted. Aspect like, fate of fertilizer-N supplied to maize crop, particularly its availability to roots, recovery rates, utilization patterns plus an overall picture of N cycling have been depicted. Detailed discussions about dynamics of phosphorus and potassium are other important aspects found in the same chapter. Chapter 4 includes details on distribution, supply levels, recovery, and loss of secondary and micronutrients. Within Chapter 5, Soil organic matter status and its regulation, recycling of organic residues, and green manure application have received attention. They are of relevance to preservation of soil quality, improvement of C sequestration and reduction of CO_2 emission in maize cropping zones. Chapter 6 deals with below-ground aspects of "Maize Agroecosystem". Discussions pertain to rooting pattern, architecture and traits relevant to nutrient absorption; rhizosphere and its role in nutrient acquisition; root exudations and their impact on soil nutrient availability; and soil microbes relevant to nutrient transformations, N fixation, and phosphorus acquisition.

The composition of maize belts with regard to crop genetic stock is important, since it affects nutrient dynamics. Genetic traits of maize cultivars like duration, growth pattern, nutrient acquisition rates, harvest indices, as well forage/grain yield potential have been highlighted in Chapter 7. Chapter 8 provides detailed information about crop husbandry procedures known commonly as Farmer's Practices, State Agency Recommendations, Maximum Yield Technology, Best Management Practices, and Site-specific Nutrient Management Practices and their impact on nutrient dynamics. During past 3 decades, maize farmers residing in different agroclimatic zones of the world have adopted "Integrated Nutrient Management" procedures on a preferential basis. Integrated procedures reduce burden on usage of chemical fertilizers. They aim at improving soil fertility, nutrient dynamics, and productivity of maize belts without drastically affecting various ecosystematic functions.

Over all, it is a scholarly edition within the realm of Global Agriculture. It should be a highly useful text book for students and an excellent reference book for Researchers/Professors dealing with Agriculture and Environmental Sciences.

— **Dr. Kowligi Krishna**

Acknowledgments

Several Researchers, Professors, and Maize Production Technologists have offered scientific material, unpublished reports, and publications pertaining to maize crop, its cultivation and utilization in different continents. Many have interacted constantly and few others have been a source of inspiration either directly or indirectly. Following is a list of researchers who helped me. They are:

Dr. Carina Alvarez and *Dr. Eduard Mullins* of University of Buenos Aires, Argentina, donated pictures related to maize grown on no-tillage fields in Pampas of Argentina. *Dr. Alberto Quiroga* and *Dr. Alfred Bono* of National Institute for Agricultural Technology-INTA, Anguil Experimental Agricultural Station, Anguil, Argentina provided pictures on crop rotations, no-tillage systems, mulching, drought effects, weed infestation, etc.

Dr. Kraig Roozeboom of Kansas State Agricultural University System sent me pictures of maize expanses from Rice County in West-Central Kansas, USA; *Dr. Jerry Hatfield*, Director National Agricultural and Environmental Laboratories, Iowa State University, Ames, Iowa, USA contributed pictures on soil profile from Corn Belt of USA. *Dr. Roger Elmore* and *Dr. Lori Abendroth* of Iowa State Agricultural Extension Service, Iowa State University, Ames, Iowa supplied me publications and information on maize soil fertility tests and related aspects; *Dr. Carrie Laboski*, Professor of Soil Science and Extension Specialist, University of Wisconsin, Madison, Wisconsin and *Dr. Todd Andraski* for photographs depicting nitrogen fertilizer/nytrapyrin field trials on maize grown at Arlington Experimental Station, Wisconsin, USA.

Dr. Mike Listman and his group at Communications Division, International Maize and Wheat Centre (CIMMYT), Mexico for pictures on fertilizer application, maize-based rotations, residue application, mulching, and hoeing.

Pictures on rain and sand storm in the transition zone, near Guinea-Savanna region of Southern Niger in West Africa were drawn from *International Crops Research Institute for the Semi-arid Tropics (ICRISAT)*. Pictures on sandy soil and mycorrhizas were also derived from studies by *ICRISAT* in West Africa. Photographs of red alfisols and crops in Karnataka (South India) were from Experimental Station, University of Agricultural Sciences at Bangalore, Karnataka, India.

I wish to thank my wife Dr. Uma Krishna and son Mr. Sharath Kowligi.

Chapter 1

Maize Agroecosystem: Origin, Physiography, Expanse, and Productivity

INTRODUCTION

The Beginnings of Maize Agroecosystem

Historically, seeds for a maize agroecosystem were sown 8,000 years ago, in the Pacific slopes of modern Mexican states. Obviously, human ingenuity, curiosity, and immediate necessity for food grains might have induced this event. Domestication, then selection for better traits were perhaps most crucial during initial stages of the development of a maize cropping zone around Mexican dwellings. Let us now delineate various historical facts authenticated using archeological samples, archaebotanical analysis, potteries, and physical effects found around pre-historic Mexican human settlements. Modern techniques like carbon dating of grain samples, inscriptions, and ancient literature have also contributed to better understanding of this topic. Firstly, maize is a native crop of Mesoamerica. Archeological evidences and genetic analysis suggest that maize was probably domesticated in Southern and Central Mexico some 8,000 years ago. The wild type or ancestor of domesticated maize called "Teosinte" (*Euchlaena*) is found abundantly in Central America. Primitive types of cultivated maize were derived from teosinte through repeated selection. Mexico and adjoining areas in Central America is considered the primary center of genetic diversity for maize. Carbon dating of archaeological remains such as maize cobs and grains found in Guila Naquitz caves and Oaxaca valley suggest that it was grown in Mexico some 6,250 years ago (Piperno and Flannery, 2001). Similarly, carbon isotope analyses of grains from Tehuacan in Peubla indicate that maize was cultivated during 2750 B.C. It seems by 1700 B.C. inhabitants of Tehuacan valley were regularly growing maize along with gourds and amaranth. Maize cultivation spread rapidly into several areas within Mesoamerica around 1500 B.C. It is believed that Guatemala served as conduit for adoption many of these strains of maize that entered Andean regions, Peru, Bolivia, and other parts of South America (Argentina, Chile). The initial appearance of maize in Peru has been dated to 6070 B.C. In South America, regular cultivation of maize was in vogue by 2500 B.C. Peruvians and other people in Andean highlands grew maize, potatoes, and gourds regularly during 2nd and 3rd millennium B.C. There is no doubt that maize agroecosystem that existed in its rudiments during pre-historic period was really confined to areas surrounding human settlements. Spread of maize cultivation into different locations within South America occurred gradually during Neolithic age. This is actually much after its initial domestication in Mesoamerica. Sculpted figures, folklore, and religious scripts of Mayan culture from 3rd to 9th century clearly prove the importance of maize to local population in Latin America (Jenkins, 2002). Next, the spread of maize cultivation into vast stretches of Northern Great Plain is said to have caused perceptible changes in food habits, as well as soil and environmental parameters. It is said that during 1st millennium A.D. maize cultivation diffused into most of Southwestern USA and right up to Southern Canada. It did transform the landscape perceptibly. It mostly replaced forested zones with maize. Obviously, it must have generated a sort of upheavals in biomass and nutrient turnover in the landscape. Human migratory trends had a major impact on extent of spread of maize cropping zone into North American locations. Maize cultivation spread into farther regions in north, including Canada during 1st to 5th century A.D.

Dispersal of maize in New World has been studied amply (Hudson, 1994; Gallinat, 1996; Gallinat, 1999). It seems that maize diffused into North American Plains around 700 A.D. through Rio Grande valley. Initially, Northern flint types reached the Northern plains. Later, maize spread north word along the eastern and western flanks of Rocky Mountains. Hudson (1994) suggests that maize diffused into the core of present day "Corn belt" states namely Indiana, Illinois, Ohio, and Iowa through human migration, development of infrastructure, especially trails and railways. Of course, "natural crossings" among maize genotypes must have played their part in adaptability to northern agroclimate. Maize cultivation got intensified around 800 A.D. in Mississippi (Pouketat, 2005). The spread of maize cultivation was actually induced and sustained by major rivers such as Arkansas, Mississippi, Platte, and Ohio. By 1200 A.D. maize cultivation was entrenched on the plains and undulated terrain of northern United States of America.

Maize reached European mainland through Spanish and Portuguese explorers. They introduced flint corns of Cuba to Spanish agricultural zones around 1490s (CGIAR, 2008; Harlan, 1992; Rebourg et al., 2003). It seems Spanish armadas that conquered Mexico and adjoining areas imported large quantities of maize into European ports. By mid 1500 A.D. maize had graduated from a mere curiosity to regular cereal grain in several locations within Europe. Yet, Europeans seemed to prefer maize only as a feed for farm animals. It was not immediately received as grain fit for human consumption. Europeans called it "Turkish corn" or "Turkish wheat" because it was often available with the Turkish caravans. According to Brandolini and Brandolini (2009), maize was introduced into Italy and adjoining areas immediately after voyages of Columbus. Initially, the *Everta* (popcorns) and *Indurata* (Flint corns) did not spread much due to poor adaptation. However, maize germplasm imported from Caribbean, much later during mid 1500s established well. Soon, Italy became hub for maize germplasm originating from New World. Maize diffused rapidly into other agricultural regions of Southern Europe. *Indentatas* were preferred much during medieval times. Maize, along with beans and gourds, it seems diffused into Southern European farming zones, especially Balkans during Ottoman period (Andrews, 1993).

It seems Portuguese travelers introduced maize into African continent. Maize cultivation spread into Egypt (1517 A.D.) and adjoining Arabian nations during 16th century (McCann, 2000; USDA, 2008). A few years later, maize found its way into Central and Southern African cropping zones. Maize reached interior Africa via Egypt and Arabia (Brown 1896; USDA, 2008). Maize cultivation began in West Africa during early 16th century. Introduction of maize in Sierra Leone was actually preceded by clearance of vast stretches of forests. Human migration and trade routes aided further spread of maize into interior zones, literally, spanning the breadth of the continent from Nigeria in the West to Ethiopia in the East. Chronicles by Al-Hasan Ibn Mohammed Al-Wezaz, also known as *Leo Africanus* suggests that maize was cultivated in African Coasts and along rivers Gambia and Senegal by 1526 A.D. He also reported that it was called *zaburro* along the river Congo in Central Africa. By 1550 A.D., maize was cultivated in Cape Verde islands (Brown, 1896; McCann 2000; USDA, 2008). Dutch traveler Pieter Van Broeke (1605–1612) states that "Turkish Wheat" that is maize was under cultivation all through the West Coast from Gabon to Angola and right up to Mozambique (Hair et al., 1992; La Fleur, 2000). It seems that maize reached Mozambique as early as 1495 through Spanish merchants. Maize cultivation was well entrenched in East Africa and interior regions by 1800s. Kenyans received maize in early to mid 1600s. Maize developed as a major cereal in the interior regions of this country. During early 1900s, Kenyan administration encouraged spread of maize cultivation. In Southern Africa, maize cultivation and its spread into interior areas depended vastly on mining and roadways connections. Maize served as a cheaper cereal grain to support the labor colonies around mines. Maize reached Madagascar and Mauritius during 1690s (USDA, 2008).

Maize reached West Asian agricultural zones early in 1500s. The cultivation of maize in Iraq, Arabian Peninsula, and Oman has been reported in the commentaries of Alfonso de Alboquerque dated around 1550s (Birch, 1875). Maize cultivation occurred in Iran, Baluchisthan, and Arabian coasts during early 1600s (Sykes, 1902).

Maize reached South Asian ports in 1500s via European merchants and Sea Farers. Maize was one of the commodities traded along the "silk road" and it reached China during 1500s. Maize is mentioned in Chinese botanical texts of 16th century. Details on anatomy and production are mentioned (Laufer, 1907). By mid 1700s, maize had spread into several important agricultural zones of main land China. Chinese cultivated maize in Hunan, Szechwan, Fukien, and few other provinces. There are reports that maize diffused into Philippines and Fareast from North America around 1520s. Perhaps, just during the period Magellan discovered this part of the world (CGIAR, 2008; USDA, 2008).

Maize cultivation became regular in India, Bangladesh, Pakistan, and Sri Lanka during medieval period. Portuguese introduced maize into Southern Indian peninsula during 16th century. It seems maize was one of the novelties gifted to Vijayanagara monarchs by European ambassadors who visited their courts during 16th century. The flint corns introduced into South India by Portuguese and others, it seems has undergone considerable modification to ensure local adaptation. During this past century (1900s) several types of maize and a range of germplasm material have been introduced by Americans. The genotypes were derived from Great Plains and Mesoamerica. During past 5 decades, several sweet corn genotypes introduced to South Indian farmers were originally landraces from New England zone. They were naturalized into hilly regions in Andhra Pradesh and Karnataka. Many of these introductions were impressive to South Indian farmers. Several of these maize genotypes from America have been imbibed into cropping systems practiced by South Indian farmers. Albeit, we ought to recognize that genetic diversity of maize currently available in South India is comparatively much smaller than that traced in Mesoamerica or Great plains. In addition to maize received since Medieval times (16th century), there are suggestions that maize was cultivated in South India during pre-Columbian period. For example, Hoysala sculptures (11th to 13th century A.D.) at Halebeedu and Somnathpura near Mysore in Karnataka indicate utilization of maize cobs. It is at least 2–3 centuries prior to discovery of New World by Columbus (Johannessen and Parker, 1989).

Maize was introduced into Australia during late 1780s by the British settlements around Port Jackson (Sydney). Early reports indicated that maize withstood drought and temperature prevalent in the area. In 1788, initially around 351 acres of maize were sown in the vicinity of Port Jackson (Frost, 1993; USDA, 2008).

Over all, it is reasonable to believe that maize agroecosystem, like any other cereal cropping zone expanded *gradually* initially. It is spread into Northern Great Plains and New England zone seems *rapid* during ancient period spanning 1st to 5th century A.D. Like many other Latin American crop species, a major *spurt* in its expanse occurred during medieval period. It was induced mostly by Spanish and Portuguese conquests, seafarers, and merchants. They moved maize into Europe, Asia, and Orient rather rapidly. Such a *spurt* in expansion of maize cropping must have generated a definite influence on soil, cropping combinations, and above ground vegetation in many parts of the world. Again, during mid-20th century, there was a *spurt* in expanse and intensity of maize cultivation. This event was aided by use of chemical fertilizers, high yielding genotypes, and irrigation. Maize belts in many parts of the world got intensified and productivity increased. As yet, we have no idea regarding the extent to which maize agroecosystem could be intensified, be it "Corn Belt of USA," Pampas in Argentina, European Plains, semi-arid regions of Brazil, or India. This aspect needs careful debate, decision making, and periodic revisions in strategy.

Nomenclature

The word "maize" or "mahiz" is derived from Mayan dialects. It has its origin in Araguaco in Central Mexico. It seems Columbus picked this word "maize" in the Caribbean islands and transmitted it to peoples of Old World (FAO, 2004). In most Mayan dialects it is referred as *Ixim*. In fact, in the highlands of Guatemala the word for maize is Quiche. In Mayan language it is termed *kana*. Today, maize is referred differently in many countries. In many of the Asian countries it is referred as grain brought from, or traded by or transshipped from Macca in Arabia—for example *Makka jona* or *Makka jola* (Table 1.1). Local villagers in South Asia actually bartered or purchased maize grains from traders or caravans that emanated in Macca or other Arabian locations.

Local populace in Central Africa called maize as "Ghiny-Wheat". The English and Dutch traders in Africa called it "Turkish-Wheat". It was called Turkish-wheat mainly because it was traded and supplied by Turkish travelers and businessman. Native Africans often obtained maize seeds from Turkish caravans moving through the villages. In the West Coast of Africa maize was also commonly referred as *Zaburro*. Chinese called it *Fan mai,* which meant foreign or barberian wheat. Other names were *his fan mai* (western barberian wheat) and *Jungshu* (grain of western barbarians).

Table 1.1. Maize—Its origin, distribution, botany, nomenclature, and uses: A summary.

Origin and Distribution

The primary center of origin for maize is said to be Southern Mexico. Botanically, modern maize was derived through repeated selection of wild ancestor called Teosinte (*Euchleana*). The wild type (*Zea mays* var. *Mexicana)* is well distributed in Central America. There are also suggestions that modern maize was derived from a Mesoamerican maize variety called *Chapalote*. Native Indians of Mexico and Northern plains grew a wide range of maize genotypes. Currently, we can identify at least 25 primary races of maize. Actually, over 350 races could be traced in its primary center of origin. At present, maize genotypes are worldwide in their distribution. We can demarcate several secondary and tertiary centers of diversity in different continents.

Maize was domesticated in Mesoamerica some 8000 years ago. Archeabotanical studies on grains and plant material in the caves of Guila Naquitz and Oaxaca valley suggest that maize was cultivated regularly around 6,250 years ago. The pre-historic people in Tehuacan in Peubla cultivated maize during 2750 B.C. Maize cultivation spread into Guatemala and South American locations during Neolithic age. There are innumerable references to maize in Latin American folk lore and religious scripts belonging to 3rd to 9th century A.D (Jenkins, 2002). Dispersal of maize into North American Plains, Canada, and New England zone occurred during 1st to 5th century A.D. By 1200 A.D., several maize types were cultivated by Indian tribes of North America. Maize reached European plains, Africa, and South Asia during medieval period, mainly through Spanish and Portuguese explorers. Presently, maize cropping zones extend into Americas, Africa, Europe, South Asia, and Fareast. Globally, maize agroecosystem spreads into 148 m ha. Maize belt is largest in USA followed by Brazil, China, and India.

Botany and Classification

Kingdom	Plantae
Division	Magnoliophyta
Class	Liliopsida
Order	Poales
FAmily	Poaceae
Genus	*Zea*
Tribe	*Maydae*
Species	*mays* (*Zea mays* 2n = 20)

Types of Maize

Several types of maize are utilized for food, fodder, and fuel purposes. They are:

Flour corn	*Zea mays* var. *amylaceae*;
Pop corn	*Zea mays* var. *everta*;
Dent corn	*Zea mays* var. *indentata*;
Flint corn	*Zea mays* var. *inurata*;
Waxy corn	*Zea mays* var. *ceratina*;
Striped corn	*Zea mays* var. *japonica*;
Sweet corn	*Zea mays* var. *saccharata*, or
Amylose corn	*Zea mays Zea mays* var. *rugosa*

The International Maize and Wheat Center in Mexico holds one of the largest germplasm collections of maize and its related species. In USA, USDA maintains a very large germplasm collection of maize at major Universities of Corn belt, mainly at Urbana Champaign in Illinois, Ames in Iowa, Lincoln in Nebraska, and Madison in Wisconsin, etc. Similarly, large collections of maize germplasm are available in other parts of the world such as International Institute for Tropical Agriculture (IITA), Ibadan, Nigeria; IARI, New Delhi in India; Chinese Agricultural University, Beijing in China; Beunos Aires in Argentina, etc.

Names in Different Languages

Maize is referred by different names depending on genotype, cob characteristics, language spoken, cultural predisposition, and geographic location [1,2,3,5]. Following are few examples:

Americas: *Mahiz* in Arawak, *Ixim* in Mayan dialects; *Corn, Indian Corn* in English;

Europe: *Maize, Corn, Indian Corn, Teosinte* in English; *Milho* in Portuguese; *Mais* in French; *Maiz* in German; *Maiz, Maio* in Spanish; *Kukuruza* in Russian; *Maiz* in Dutch; *Kukuruz* in Croatian; *Aravasito, Kallamboq* in Greek; *Kukurica* in Hungarian; *Mais in* Estonian; *Kukuruz*; *Popusoi, Porumb* in Romanian, *Majs* in Swedish; *Granturco* in Italian; *Kukurydza* in Polish; *Maissi* in Finnish; *Majs* in Danish; *Kukuruz* in Bosnian; *Kukurice* in Czech; *Kukuryz* in Serbian.

Asia: *Makka, Makki or Bhutta* in Hindi, Gujarati, Punjabi, Marathi, and Bengali; *Makka* in Urdu; *Muskina jola* in Kannada; *Makka jona* in Telugu; *Cholam* in Tamil; *Cholam* in Malayalam; *Bada Iringu* in Sinhalese; *Bap* in Vietnamese; *Sumi* in Chinese; *Tomorokoshi* in Japanese; *Oksusu* in Korean; *Kanga* in Maori; *Tiras* in Hebrew; *Misir* in Turkish; *Oora* in Arabic.

Africa: *Agwado* in Hausa, *Milho Zaburro* or *Saburro* in dialects of west coast of Africa.

Uses[1,3,4,5]

Maize is utilized to prepare a wide range of nutritious food items. In Mexico and North America, tortillas, bread, and biscuits are popular. Maize flour is used in salads and appetizing soups. Several types of maize flour based porridges are in vogue in South America, Africa, and Southeast Asia.

Maize is an excellent source of Industrial starch. Maize contains about 7–8% oil in its grains. Maize oil or "Corn oil" is an important cooking medium in North America. Maize steep is a viscous fluid rich in carbohydrates and protein. Corn step is an important base material for several kinds of industrial fermentation products like enzymes, vitamins, antibiotics, etc. Corn cobs are used in furfural production. Maize starch and flour serves excellently in alcohol production. In addition, maize steep is used to produce variety of industrial products like soaps, skin care items, absorbents, biodegradable plastics, and glues.

Maize stalks, leaves, and cobs serve as important feed items to farm animals. Maize fodder is utilized fresh or dried as hay. Maize silage is a richer and nutritious animal feed. Maize residues is also composted and recycled *in situ* to enhance soil quality and C sequestration. During recent years, it has become useful in biofuel production. Firstly, cobs and stalks are useful in generating heat upon burning. Kitchen stoves modified to fit maize cobs and stalks are available. Alcohol produced from maize is mixed with petrol to augment biofuel production.

Nutritional Aspects

Globally, maize is cultivated for its grains that serve as carbohydrate source. Average annual per capita consumption is 20 kg in developing countries (CGIAR, 2008). Its cultivation is predominant in North and South America. The per capita consumption of maize grains is 80 kg in Latin America and the Caribbean Islands. Maize contributes about 33% of calorie intake in Latin America. In Sub-Sahara and Southern Africa the per capita consumption is 60 kg. In Eastern Africa the per capita utilization of maize is 90–100 kg per year. In South Asia and China, per capita consumption is relatively low and only 5% of calorie intake is through maize or maize products.

Biochemical Aspects of Grains[3,4]

Sweet Corn: About 100 g of sweet corn grains provide 90 k cal energy. It contains Carbohydrates 19 g (sugar 3.2 g and dietary fiber 2.7 g); Fat 1.2 g; Protein 3.2 g; Vitamin A 10 µg; Thiamine (Vit B_1) 200 µg; Niacin (Vit B_3) 170 µg; Folate (Vit B_9) 46 µg; Vitamin C 25µg; Minerals 2.7%; Ca 10 mg; P 7 mg; Fe 500 µg; Mg 37 mg; K 270 mg.

Maize was accepted rather rapidly by Europeans, Africans, and Asians. Its consumption as major cereal without additives caused health problems. Initially, maize induced niacin deficiency in areas outside of Americas. Maize consumption resulted in a disease called Pellagra. Protein deficiency was also rampant among maize eating population. Amino acids lysine and tryptophan deficiency occurs if diets are exclusively made of maize. Actually, pellagra and protein deficiencies never occurred among Native Americans. They usually added ash or lime and consumed maize grains along with other items like amaranth and gourds. Alkali treatment of maize-based foods released vitamin-B (Niacin). Therefore, pellagra did not appear among Native Americans. This process of addition of alkali to release vitamin-B is called "Nixtamalization." It is a word derived from Native American language. In addition to maize, Native Americans consumed fish to overcome any sort of protein deficiency.

Sources: Hudson, 1994; Jenkins, 2002[3]; Turrent and Serratos, 2004; USDA, 2008; Krishna, 2010; 4www.fao.org/docrp/005/y3841e/y3841e04.htm. 5http://en.wikipedia.org/wiki/maize: 1,2www.ienica.net/crops/maize.pdf.

Origin and Botany

The origin, domestication, and spread of maize are topics interesting to crop specialists, agro-ecologists, geographers, and several others dealing with maize farming. It is generally accepted that maize was domesticated in Southern Mexico some 8000 years ago. This generalization is based on evidences derived from archeological sites and preserved specimens. In general, paleo-botanical studies suggest that cultivation of domesticated maize was in vogue some 7,000–8,000 years ago. Further evidences were derived through genetic studies involving chromosomal and molecular sequencing of DNA. There are several reports regarding relatedness and genetic distance between various races and genotypes of maize. More recently, that is in mid 2008, complete sequence of maize genome containing 50–60 thousand genes has been deciphered. This effort helps in identifying phylogenic relatedness, evolutionary pattern, and adaptability of various maize genotypes to environmental constraints. Knowledge about DNA sequence and genome organization is of immense value during classical breeding and molecular gene transfer that allows us to develop transgenics.

Southern Mexico is said to be the primary center of genetic diversity for maize. This region harbors several different derivatives of *Zea mays*. Both natural and man made factors might have induced such large genetic diversity of maize and its relatives. There are suggestions that in addition to natural evolutionary trends, human preferences for certain maize genotypes and their variants, ethnicity, migratory trends among natives, and exchange of maize grains between communities may have all induced large scale genetic diversity of *Zea mays* in Mexico (Perales et al., 2008). Botanically, modern maize evolved from Teosinte (God's Corn) (*Zea mays* ssp. *mexicana*). The wild type or ancestor of domesticated maize called "Teosinte" (*Euchlaena*) is found abundantly in Central America. Primitive types of cultivated maize were derived from teosinte through repeated selection. However, some botanists argue that maize evolved from a Mesoamerican maize variety called *Chapalote*. Obviously, a review of botanical characteristics of both maize and teosinte will provide a better understanding about how teosinte developed into modern maize. Actually, teosinte and modern maize differ profoundly with regard to vegetative characters and inflorescence. Distinctions with regard to male tassel (spikelet) are notable. Despite it, they are deemed variants. Mexican teosinte and modern maize form fully fertile hybrids and cross pollinate. The inherent differences in chromosomal structures and other genetic traits are comparable to variations observed among diverse races of maize.

There are indeed several theories regarding origin and domestication of maize in Mesoamerica. Following are few of them:

- Maize was domesticated through repeated selection of Mexican teosinte. This teosinte (*Zea mays* ssp. *parviglumis*) is annual in habit and is native to Balsas

River valley in Southern Mexico (Beadle, 1939). Support for teosinte hypothesis derives from classical genetics and cytology (Beadle, 1972; McClintok et al., 1981); systematics (DeWet and Harlan, 1972); morphology (Gallinat, 1983); molecular genetics (Doebley et al., 1987); quantitative genetics (Doebley et al., 1990); population genetics (Hilton and Gaut, 1998; White and Doebley, 1999); and phylogenetics (Benz, 1999).

- Maize is derived from hybridization (natural) of small domesticated maize and a teosinte belonging to section Luxuriantes, either *Z. luxurians* or *Z. diploperennis.*
- Maize was derived through domestication of wild type in several locations across Central America. However, there are suggestions that domesticated maize was derived from a single domestication event in Southern Mexico and it later spread rapidly across Mesoamerican locations.
- Maize could have been the result of hybridization of *Z. diploperennis* and *Tripsicum dactyloides.* This theory was put forth by Mangelsdorf (1939; 1974). However, there are refutations to the theory that *Tripsicum* sp. was involved in domestication of modern maize. There is total lack of support of any kind to the hypothesis that maize was derived through hybridization with *Tripsicum* (Bennetzen et al., 2001; Eubanks, 2001).

At least 25 primary races of maize are traceable in the tropical zones of Mesoamerica. Actually over 350 races could be identified in the Meso- and Central America. Several genotypes of maize were cultivated during ancient era. Among the various races of maize, spread of about 19 has been identified with Classic period spanning 300–900 A.D. Studies on population structure and genetic diversity have shown that maize races of Southern United States were actually intermediary between Mexican and North American races. There are indications that counter diffusion of maize races native to Northern United States and Canada toward Mexico and Andes occurred, especially during post Columbian period (Vigouroux et al., 2008).

Currently, over 17,000 germplasm lines are held in the nurseries at International Maize and Wheat Improvement Center (CIMMYT) in Mexico (CGIAR, 2008). During recent years, proliferation of hybrids, variants and genetically engineered genotypes have been rampant. It has lead to masking of many ancient strains that served the natives excellently. There are aspersions that in due course, spread of genetically engineered maize genotypes may excessively obliterate, displace, or replace existing primary strains of maize meant for grains or forage. As such transgenic maize varieties (or GM maize) could have evolved in the crop science laboratories, glass houses, or experimental fields any where in the world. Therefore, knowledge about their adaptation and proliferation in a given area too deserves importance.

PHYSIOGRAPHY, AGROCLIMATE, SOILS, WATER REQUIREMENTS, AND CROPPING SYSTEMS

Physiography and Agroclimate

Global maize growing regions could be demarcated based on several different criteria. In the present context discussions are confined to classifications based on physiography, soil fertility, and productivity. Based on physiography, "Maize Mega-Environments" identified by researchers at CIMMYT are as follows:

Mega-environment	Latitude	Elevation
		(m.a.s.l.)
Lowland Tropics	0–25°N and S	<1000
Tropical Highlands	0–25°N and S	>1800
Subtropics	26–36°N and S	1000–1500
Temperate Zones	>36°N and S	All elevations

Source: Gerpacio, 2002.

Maize is a highly versatile cereal crop. Maize cropping zones are found in all the four continents where agricultural cropping is wide spread. Globally, maize belts extend between 50°N latitude and 50°S latitude. Maize cropping zones extend from sea level (coastal plains) to over 3,000 m altitude in mountainous zones, for example in Andes. Maize is intensively cultivated as a mono-crop in the Northern Great Plains. This region forms the well known "Corn Belt of America." Here, maize dominates the agroecosystem in preference to other crops. In the "Corn Belt" maize cultivation occurs in riverine zones, plains, and undulated landscapes situated between 100 and 300 m.a.s.l. Maize cultivation also extends into other regions of North America. In the Appalachians, maize culture extends into mountainous terrain (500–1,000 m.a.s.l.). Temperate climatic conditions prevail in most parts of the corn growing regions of USA and Canada. Mean temperature ranges from –4 to 25°C. The mean maximum temperatures may reach 30°C for a short period. Evaporation loss is minimal because of cold climate.

The Canadian plains too support maize cultivation despite relatively colder climate. Here, the agro-environment is characterized by cold temperature and frosty soils. As such, extent of frost-free period dictates maize cropping. Usually, cold tolerant varieties perform better in such agroclimatic conditions. Annual precipitation ranges from 100 to 600 mm in cropping regions of Southern Canada. Relative humidity ranges from 30 to 60%. Atmospheric temperatures are relatively low at 5–15°C, but mean maximum temperature during crop season may reach 22°C. The temperate climate allows a single cropping season, spanning from June to October. Maize is grown as mono-crop in Canadian plains. Sometimes, it is intercropped with legumes like soybean.

In South American Andes and Mesoamerica (e.g., Mexico or Guatemala), maize is cultivated on terraced surface along mountains and valleys. Maize production also occurs on leached lateritic soils found in the high altitudes (1,500–2,000 m.a.s.l.) of mountainous regions. In the Andean highlands, maize is cultivated from 1,200 to 3,600 m.a.s.l. (Beck, 2000). Mostly, floury or morocho type maize is preferred. The white maize is cultivated in about 80% of the area because it suites their preferences. Cold temperature, frost, hail, and drought are major abiotic constraints encountered by maize grown in Andes.

The Brazilian Cerrado region supports large scale production of maize. These are mostly semi-arid or subtropical plains. The average precipitation in the savannas that are used for maize culture ranges from 800 to 2,000 mm annually. The Cerrados experience a strong dry spell during summer (April to September), when soils are mostly left to fallow or pastures are developed. Based on seasonality and precipitation pattern, Brazilian Cerrados could be classified as seasonal savanna, semi-seasonal savannas, and hyper-seasonal savannas. It is pattern of rains, drought, and fire that decides type of savannas and cropping systems (Amorim and Batalha, 2009). The average temperature ranges from 18 to 28°C. Soils are classified as Oxisols. They are mostly ferralitic with low pH and Al toxicity. Lime application is essential to correct soil acidity and Al toxicity (Lopes, 1996). The Brazilian Cerrados are dominated by large mono-cropping expanses of soybean, maize, and other cereals (Ratter et al., 1997; Oliviera and Marquis, 2002). The Cerrados contribute nearly 25–30% of maize production of Brazil. Maize production also occurs in other parts of Brazil like Matto Grasso and Amazonia. Incidentally, Brazil is among major maize producing nations contributing nearly 3–8% of global maize. The productivity of maize in the Cerrados is moderate at 2–4 t grains ha^{-1} plus 5 t forage ha^{-1}.

Maize production zone in Argentina thrives on plains with gentle slopes (Pampas) situated between 32–35°S and 58–62°N. Mostly, silty clay loams (Mollisols) support the maize cropping belt. In the Pampas of Argentina, maize thrives under subtropical climatic conditions. Actually, the gentle slopes of Rolling Pampas support wheat–soybean–maize sequence. The annual precipitation ranges from 900 to 1,100 mm. The annual mean temperature is 15–18°C. The relative humidity ranges from 35 to 40% during the growing season. Farmers are allowed at least 200–220 days of frost-free period. Cold tolerant genotypes could be traced relatively sparsely in the Southern cone region of South America. Diurnal variations in the receipt of radiation ranges from 14 to 10.2 h sunlight in Northern Pampas, and 15 to 9.3 h sunlight in the South. The influx of radiation is high during December (26 MJ m^{-2}) but recedes in June (8.9 MJ m^{-2}). Around Buenos Aires, early sowing by August/September helps in efficient utilization of solar radiation and soil moisture leading better gain yield (Otegui et al., 1995).

Maize is also grown in Chilean highlands. Maize culture extends till 50°S. It is grown on mountainous terrain using terraces and in valleys mostly as part of crop mixture. Precipitation is high at 1,500–2,000 mm annually and temperatures range from 5 to 18°C. Maximum temperature may reach 28°C. Major soil types, namely Oxisols and Entisols are prone to erosion. Maize productivity is moderate or low at 2–3 t grain ha^{-1} and 5 t forage.

The maize growing regions in Europe occupies agricultural zones that occur between 5°W and 140°E longitude and 38 and 55°N latitude. It experiences wide variations in agroclimatic conditions. The European maize belt is vast and occupies regions that are predominantly cold temperate, moderately cold temperate, and Mediterranean. In the cold temperate plains, mean temperature during crop season fluctuates between 3 and 15°C. However, maximum temperature can reach 20–25°C. Diurnal variations do affect maize-based cropping systems. Day length period varies from 8.5 to 15 h. Mean monthly precipitation ranges from 95 to 110 mm during crop season.

Total rainfall is around 550 mm in Western European countries like Spain and France. It is around 950 mm in German plains. Maize crop experiences cool Mediterranean climate in Southern European agricultural zones. Evapo-transpiration of moisture from the ecosystem is low. The temperature fluctuates between 7 and 18°C during cool season (January to March) and 24 and 28°C during June to September. Maize cropping in Russia and Ukraine depends much on frost-free period. The annual precipitation in Southern European plains (Spain, Portugal, Southern France, and Greece) ranges from 500 to 1,000 mm. Relative humidity ranges 30–60% during the cropping season. Maize belt in Europe thrives mostly on plains, where in altitude ranges from 300 to 500 m.a.s.l. Maize culture also occurs at higher altitudes in the fringes of Alps. In the Hungarian plains, maize is sown in April. Sowing date depends on soil fertility and temperature. A delay in sowing maize by one month leads to a loss of 20–25% grain yield. For optimum nutrient recovery and fertilizer use efficiency, maize should be sown by April 10 to May 5. The heat threshold for most maize hybrids sown is 10°C. However, there are genotypes that germinate at lower temperature of 6–8°C (Racz et al., 2003).

In West Asia and North Africa (WANA), maize culture proceeds on sandy soils found in plains, undulated landscapes, and coasts. The agroclimate of maize growing regions is highly variable. Arid lands with low levels of precipitation and relative humidity are common. As such, maize is not a major cereal like wheat or barley or millet in the region. Actually, sharp and pronounced variation in precipitation affects cropping systems. In the Mediterranean region, temperature during crop season ranges from 5 to 18°C. Cold winters with <5°C and slightly hotter summer with 25°C are a clear possibility during maize culture.

In West Africa, maize culture is pronounced in wetter savanna and guinea regions. Maize cultivation is sparse in Sahelian zone, because of relatively lower levels of precipitation (350–600 mm y^{-1}). Further, moisture holding capacity of sandy Oxisols is rather meager. Instead, Sahel is often occupied by drought tolerant pearl millet or sorghum or cowpeas. Maize cultivation zone extends between 3 and 10°N latitude and 10°W and 10°E longitude. Maize culture is often confined to rainy season, lasting from June/July to October. However, tropical zones of Southern Nigeria, Ghana, Cameroon, Gabon, and other countries closer to equator support larger scale maize cultivation. Maize culture proceeds in all the three cropping seasons. Maize is an important cereal in entire tropical Nigeria. Its planting time seems arbitrary and at times improper choice of dates results in yield depreciation. Most often, it is sown between April and June in the three major agroecological regions for maize production, namely Southern Savanna, Southern Guinea Savanna, and Sudano-Savanna (Ikena and Amusa, 2004). Early planting is generally aimed at harnessing precipitation and inherent soil fertility more efficiently (Kamara et al., 2009). Maize is mostly intercropped with legumes, cassava, or oilseeds to improve land use efficiency. It is cultivated from sea level to 500 m.a.s.l. Precipitation is congenial, but sometimes high ranging from 1400 to 2000 mm annually. Average temperature ranges from 22 to 30°C, but it could reach 38–40°C in summer. Relative humidity is high (60%) in Coastal plains, as well as in forested zones. Diurnal variations are small (12–14 h) owing to closeness of this region to equator. Plants reach maturity faster because thermal requirements are satisfied rather quickly.

Maize is an important intercrop in Central and East Africa. It is cultivated on plains, as well as on mountainous terrain, at altitudes ranging from 0 to 1,000 m.a.s.l. Major maize producing countries are Kenya, Tanzania, Congo, Central African Republic, Ethiopia, and Somalia. The tropical climate supports maize culture all through the year. Precipitation levels are high at 2,000–2,500 mm annually. Soils are moderately fertile and rich in organic matter content. Relative humidity during rainy season is high at >80%.

Maize is an important cereal crop in Southern African agricultural zones that extend from 15–35°S to 10–40°E. Southern African maize belt thrives on semi-arid plains, coasts, river valleys, undulated terrains, and mountains. It occupies regions that experience tropical, subtropical, and temperate agroclimate. It is preferred as mono-crop and equally so as an intercrop with vegetables/legumes. Small stretches of continuous maize are also found in southern Africa. Major countries that produce maize are South Africa, Angola, Botswana, Zimbabwe, Mozambique, and Zambia. Maize culture extends from sea levels to 1,000 m.a.s.l. The maize agroecosystem thrives on sandy soils and drier climate in regions closer to or on the fringes of Kalahari. Precipitation in the semi-arid region is low or moderate ranging from 400 to 800 mm annually. Growing season precipitation of 250–400 mm is distributed mostly during January to March. In South Africa, maize cropping extends into temperate climatic zones. Precipitation pattern allows only one crop during rainy season. Therefore, post-rainy and summer crops need supplemental irrigation. The dry regions with low relative humidity are predominant. Mean annual temperature ranges from 10 to 28°C, but mean maximum temperature reaches 36–40°C. In Namibia, white maize is an important cereal. It is grown both under rain fed and irrigated conditions. Maize is a preferred crop in the flood plains of Namibian rivers. Dry land maize is cultivated in a triangle between Tsumeb, Otavi, and Grootfontein, also in Summerdown, Omaheke, and Caprivi region. Maize planted in August/September is harvested during February next year. Maize is also planted during December/January and harvested by June/July (NAB, 2007).

In India, maize cropping zones stretch from 8 to 32°N latitude and 74 to 96°E extending into 23 states, but it is a major crop in only few states like Rajasthan, Bihar, Karnataka, Andhra Pradesh, and Gujarat. Karnataka and Andhra Pradesh are the main corn producing states in South India. Majority of maize belt thrives under dry land conditions that persist in Southern Indian plains. Moderately fertile Vertisol and Red Alfisol regions of South India support vast stretches of maize, mostly as mono-crop in rotation with legumes or cotton. Maize cultivation is sparse in Himalayan region, yet it is harvested regularly in Sivalik Mountains at altitudes around 2,500–3,000 m.a.s.l. Similarly, Western Ghats supports mixed farming that includes maize. Here, maize grows under evergreen tropical conditions at altitudes 1,500 m.a.s.l. The maize ecosystem spreads all through the humid coastal plains of South India. The maize improvement project in India identifies at least four different agro-ecoregions that support its production in significant quantities (see Krishna, 2010). They are:

Himalayan Zone: It is a temperate zone with elevations above 600 m.a.s.l. Crop duration is relatively longer.

Northwest Plains: This region is characterized by wet and arid tropics. Moisture stress at flowering and grain-fill is common. Soils are sandy or alluvial and need fertilizer replenishment.

Northeast Plains: This region is characterized with hot and humid weather. Soils are loamy and alluvial. Nutrient deficiencies need correction through fertilizer application.

Peninsula Region: It is characterized by tropical and subtropical climate in the plains and hill zones. The coastal area supports cultivation of maize throughout the year. Soils used for maize production in South India is sandy Alfisols, Vertisols, and Inceptisols. They experience deficiency of major nutrients. Micronutrient deficiencies are sporadic.

Source: Krishna, 2010.

The Nepalese maize belt spreads into at least four different eco-regions. Maize is an important cereal in "Tarai region". It is intensively cultivated in the flood plains. Maize cropping also extends into mountainous regions up to 1,000–3,000 m.a.s.l. Maize is grown on terraced fields. Maize is mostly sown during rainy season (*kharif*) spanning from June/July to November. In Tarai region, maize meant for fodder is grown in summer. There are many small rivers that support maize production on their banks. On river banks, maize is rotated after a rice crop.

The Indonesian maize production zones are spread in a Mosaic covering the entire length and breadth of the country. Much of the maize crop is cultivated in *Tegalan* (rain fed dry land) or raised lands within flooded wetland region. Maize is also grown under tidal swamp conditions called *Surjan*. Tidal maize is more frequent in Java. Actually, rice is grown in standing water in sunken beds, but maize is planted on raised beds close by.

The Philippines maize belt extends mostly into subtropical zones. Maize thrives predominantly as an intercrop with legumes or vegetable. It is grown in valleys and terraces. Rice is the major cereal in low lands. Maize is planted in uplands mostly prior to onset of monsoon. Maize is also grown in post-rainy and summer seasons. It seems, in Philippines, traditional distinction of maize production systems is not based on agroclimatic characteristics. It is easier to demarcate maize belt based on grain types. Mostly, maize with white colored grains is sown to regions with poor soil fertility and low productivity. Whereas, yellow colored grain types are confined to high soil fertility and high input zones. Here, maize is mostly used for forage and starch production.

In China, maize agroecosystem extends into almost all provinces. It spreads from 26 to 50°N latitude and 72 to 130°E longitude. It thrives on wide range of topography and agroclimatic conditions. Maize belt extends into plains, plateaus, and mountainous regions. Maize is grown on hills, slopes, and terraces at altitudes reaching 1500 m.a.s.l. Maize culture extends into low lands where generally rice predominates. The temperate region in the North and Northeast supports a large expanse of maize. It contributes nearly 66% of Chinese maize. Tropical and subtropical climate available in Southern China also supports maize production. (Dowswell et al., 1996; Pray et al., 1998). Mean temperature in maize cropping zones range from 11 to 20°C. In the northeast, exceptionally low temperatures occur reaching as low as –5°C. The number of frost-free days is important. It ranges from 130 to 330 days depending on region. Maize crop is cultivated both in rainy and post-rainy season. In the intensive agricultural zones of northeast China, maize mono-cropping occurs throughout the year. In the tropical regions of Southeast Asia, especially in Vietnam, Indonesia, Malaysia, and Burma, it is sown immediately after first rains during June. The temperature is warm at 18–22°C and relative humidity is high at 70–80%. The maize belt that thrives on plains, undulated landscapes, and mountains in the above countries may receive 1,200–1,700 mm annually.

The maize cropping zone in Australia is relatively small compared with those encountered in Americas. In Australia, maize is intercropped and/or rotated with wide range of crop species. Maize is often rotated with legumes or oilseeds. Maize is also sequenced with other cereals like sorghum or wheat. Maize–cotton rotation is gaining acceptance in the subtropical zones of Queensland. It has been reported that maize grown after cotton performs slightly better than continuous maize. Maize is relatively a shallow rooted crop compared with trees. It extracts soil moisture and nutrients better from upper horizons of soil (Devereux, 2008). Maize cultivation spreads into most parts of Queensland such as Wet Tropical Coast, Central Queensland, Darling Downs, South Burnett, and Moreton. Maize is planted in October/November and reaches harvest in 100–120 days. Mean daily temperatures range from 20 to 32°C. Maize is being introduce into dry land zones with <600 mm precipitation, especially in New South Wales in Australia. Maize is sown immediately after harvest of wheat. Maize seeds are dibbled in between wheat stubbles under zero tillage conditions. Maize cropping season commences in September and ends by mid-January. The net precipitation received during a season ranges from 190 to 240 mm. The long-term average for these dry lands is 270–340 mm per season.

Soils

Soil type and its fertility aspects have immensely affected maize production trends and nutrient dynamics in the ecosystem. Soil parent material, its mineralogy, nutrient availability, organic matter content, and water holding capacity are some of the basic traits relevant to maize cropping in any area. Soils congenial for maize production should actually meet minimum requirements. In simplest terms, most suitable soil type for maize production is effectively deep, arable, and possesses good morphological traits. Water holding capacity and internal drainage should be optimum. Balanced nutrients and favorable physico-chemical properties are of course most essential. Soil quality, especially soil organic matter and C:N ratio should also be optimum. Maize crop inherently adapts to wide range of soil textural classes. Maize crop growth and grain production is optimum on sandy soils with clay content <10%, as well as on loamy or even clayey soils with 30% clay. For example, maize production is low but commensurate with soil fertility on sandy Alfisols found in West Africa. Its productivity is moderate to above average depending on fertilizer inputs when cultivated on clayey soils (Vertisols) in South Indian Plains. On Mollisols in the "Corn belt of USA," its productivity is held high by impinging the ecosystem with high levels of nutrients and water. Maize is also produced on Spodosols and Histosols rich in organic fraction.

Soils types that support establishment and productivity of "Maize Agroecosystem" in different regions of the world are as follows:

North America

Corn Belt: Udolls, Udalfs;

Great Plains: Mollisols, Aqualls with Udalfs, Borolls, Entisols;

North east and New England zone: Udults;

South (Alabama and Mississippi): Uderts, Ustolls;

Southeast (*Georgia, Florida*): Ultisols, Quartzipsamments, Aquults.

South America

Brazil: *Cerrado region*—Ustalfs (Tropustults), Usterts, Aquults (Udults) and Psamments.

Eastern Brazil—Udic Alfisols, Entisols, Inceptisols; Dystrochrepts, Tropudults.

Peru and Columbia: Orthents, Ustochrepts, and Plinthaquults.

Argentina: Mollisols (Ustolls), Alfisols (Ustalfs), Aridisols (Argids), Ultisols (Udults), and Entisols.

Mesoamerica

Maize production occurs on Orthents and Ustochrepts in Mexico and surrounding regions.

Europe

North and Northwestern Europe: Eutric Cambisols (Britain); Drystic Cambisols (France); Orthic Luvisols (Germany); Fluvisols and Histosols (Netherlands);

Central European Plains: Loess soils (East Germany, Poland, Hungary); Sandy soils (Romania); Podzols (Poland, Ukraine);

Russia and Eastern Europe: Chernozems, Podzols, Chestnut soils, Solonetz, and Solonchacks;

Southern Europe: Cambisols, Alfisols, and Entisols.

Africa

West Africa: Alfisols with high sand fraction are common in Sahelian and Savanna zones.

Southern Africa: Psamments with loamy or sandy texture in South Africa; Argids with clay horizons in Zambia and adjoining areas; Argids with Torriorthents, and Aridisols with Psamment in Zimbabwe.

East Africa: Ustalfs with Usterts and Argids in Tanzania and Kenya, Ustalfs with Troporthents in the "Horn of Africa" (Ethiopia and Somalia).

North Africa and West Asia: Sandy Alfisols in North Africa; Xerasols, Calciferous Cambisols in Syria; Xeric Alfisols, Xeric Inceptisols, and Mollisols are frequent in Iran, Iraq and Fertile Crescent.

South Asia

Inceptisols and Alluvial soils support maize cropping zones in the Indus region. Mollisols, Inceptisols, and Entisols are frequent in Indo-Gangetic plains. Vertisols are frequent in the Central and South Indian plateau region. Alfisols abound in Southern Indian plains. Sandy alluvial soils occur in the riverine zone and coasts. Alluvial soils in the Eastern Gangetic plains and Delta region also support maize production.

Southeast Asia

Maize belt in China thrives on Aridisol with Psamments; Udalfs with hapludults in South China; Cryic Alfisols, Inceptisols, Entisols, and Ultisols are encountered in the Fareast; Udic Alfisols, Ultisols, Inceptisols, and Entisols are found in Korean region; Ustalfs with

Tropustults and Dystrochrepts are common in Burma, Vietnam, Kampuchea, and Indonesia. Maize is also grown in mountainous terrain with lateritic soils.

Australia

Maize is cultivated on Usterts with Ustalfs, Udic Alfisols, Udalfs with Haplaudults, Tropustults, Entisols, and Inceptisols in Australia.

Sources: Hall et al., 1992; Brady, 1995; Beyer, 2002; FAO-UNESCO soil maps.

Note: For greater details on soils see Chapter 2.

Atmospheric Conditions

Maize adapts itself to a range of temperature, relative humidity, precipitation, radiation, and diurnal patterns in different parts of the world. Yet, we should recognize that local atmospheric conditions in the maize fields and cropping zones may be specific. Generally, minimum soil temperature for seeds to germinate is 10°C, but emergence is hastened if soil temperature is about 16°C or above. Seeds emerge in 5–6 days at 20°C. Seedling growth is retarded if temperatures fall below 13°C in temperate regions. Maize is a warm weather crop and is usually not grown, if mean daily temperature is expected to fall below 19–23°C. Frost delays germination, damages seedlings, and retards their growth. In temperate areas, a minimum of 120–140 frost-free days are required for the crop to reach maturity. Maize is actually a C_4 crop well adapted to warm climatic zones. It is photosynthetically highly efficient in zones with warmer temperature and bright sunshine. Usually, short duration varieties need 80–110 days and medium duration varieties need 110–140 days to mature. The number of days required for maturity increases whenever mean daily temperature reaches below 20°C. When mean daily temperatures are 10–13°C, maize crop puts forth only forage. The seed set is severely affected and cobs mature late. In tropics, where mean daily temperatures reach above 40°C, crop needs higher levels of water, but it matures early. Thermal requirements of short duration varieties range from 1,800 to 2,200 degree days, for medium duration varieties it is 2,500–3,000 degree days and for late maturity genotypes it is much higher at 3,700 degree days (FAO-AGLW, 2002). Maize cropping zone extends into dry plains where relative humidity is low (Rh 30%). It also thrives well at high relative humidity (>60%) in the coastal plains and high tropical mountains.

Precipitation Patterns

Maize is a versatile crop with regard to adaptability to soil moisture and rainfall pattern. Its root system adapts well to variations in moisture in the soil profile. Maize adapts to regions with low precipitation. Maize is one of the best bets in dry land regions. Maize produces optimum levels of grains and forage even when precipitation levels are too low at 400–550 mm per season. It adapts to erratic rainfall patterns and intermittent droughts that are common to semi-arid regions. On the other hand, maize grows equally well in tropical high rainfall (2,000–2,500 mm) regions of South America, India, and China.

In the "Corn Belt of USA," growing season precipitation ranges from 95 to 290 mm depending on location. The annual precipitation received ranges from 150 to 600 mm. The Brazilian Cerrados receive 500–2,000 mm precipitation annually. During the season, maize crop may utilize 300–350 mm water out of 600–700 mm received. Precipitation use efficiency (PUE) depends much on planting dates and fallow system. Fallows with pasture enhance PUE. Maize is also cultivated in high rainfall zones of Amazonia (1,200–1,800 mm annually) and other regions like Planaltina. The precipitation levels are moderate in Pampas of Argentina where maize-based cropping systems thrive. Annual precipitation ranges from 900 to 1,100 mm. Maximum precipitation occurs between December and March. Precipitation is least during June–July. The annual precipitation is about 600 mm in South but reaches up to 1,000 mm in the North and Northeast Pampas. Water requirement of a maize crop during the season fluctuates between 550 and 600 mm. The annual evapo-transpiration is around 1,300 mm. Hence, it leaves a small deficit that needs to be derived via riverine irrigation or ground water. Caracova et al. (2000) suggests that maize hybrids occupy dry lands. Long-term maize cropping on such dry lands could induce deleterious effects on soil physico-chemical properties and moisture storage. However, modern hybrids with better nutrient and water use efficiency (WUE) could overcome agroclimatic constraint to a certain degree.

The annual precipitation in northern drier regions of West Africa could be low at 250–300 mm. Here, maize production is nil or rare. However, precipitation gradually increases and reaches 1,100 mm, as we traverse south toward Savanna and Guinea regions. In West Africa, especially in the semi-arid regions, precipitation pattern is mono-modal and skewed. It lasts for about 80–110 days. Much of rainfall occurs from June to end of September. On an average, maize growing regions of Senegal receive 500–700 mm annually. It ranges from 600 to 850 mm in Savanna and Guinea region. Usually, short duration maize genotypes are efficient in terms of PUE. Short duration genotypes avoid end-season drought which is quite frequently felt in West African cereal cropping zones. Typically, water balance in maize/peanut intercrops of Senegal reads as follows: Total precipitation in 90 days is 711.3 mm; loss as drainage water is 321.3 mm; crop use and evapo-transpiration is 385 mm. It amounts to PUE of 42–45% during a crop season (Sene and Niane-Badiane, 2001). Variability in precipitation is a prime factor affecting crop productivity in West Africa. Therefore, selecting versatile maize genotypes that tolerate intermittent drought stress and apt agronomic procedures are needed. In Ghana and Southern Nigerian maize cropping regions precipitation is again skewed. The rainy season may begin as early as mid-April, but peak precipitation months are June, July, and August when precipitation ranges from 150 to 200 mm per month. During September, peak rainfall reaches 250–300 mm per month (Abunyewa et al., 2004).

The annual rainfall ranges from 100 to 600 mm in North Africa and drier regions of West Asia. In West Asia precipitation is spread from November to May. Spring rains may range from 40 to 250 mm. Supplemental irrigations are necessary to achieve higher biomass and grain yield, despite measures that improve PUE. The rainfall pattern of WANA is actually not highly conducive to maize production. It is scanty in dry regions of Syria, Israel, Egypt, Libya, and Yemen. The annual precipitation ranges from 350 to 600 mm in much of the WANA region. Maize production is confined to rainy season. Post-rainy crop is dependent on irrigation. Maize cropping zones in

Egypt that are high yielding are almost entirely irrigated and localize in the Nile valley region. Maize is intercropped with peanuts or other legumes in order improve WUE.

The maize production zones in Southern African nations like Zimbabwe, Zambia, and Malawi receive 840–1,380 mm rain in a year (Nyamangara et al., 2003). Soil moisture loss due to percolation and seepage is higher on sandy soils. Long-term measurements on rainfall pattern and PUE achieved by maize crops during the past 80 years at various locations in South Africa, indicates that PUE varies with type of tillage, land forms, planting procedures, timing adopted, and mulches. For example, Zere et al. (2007) reported that PUE* for maize sown in November after conventional tillage ranged from 0.126 to 0.616 kg ha^{-1} mm (mean 0.260 kg ha^{-1} mm). Whereas, PUE ranged from 0.110 to 0.501 kg ha^{-1} mm (mean 0.354 kg ha^{-1}mm) in plots given conventional tillage and sown in January. The PUE is also affected by rain water harvesting and storage facility. The average PUE was 0.320 kg ha^{-1} mm in November sown plots with facility for rain water harvesting, but it increased to 0.440 kg ha^{-1} mm, if sown in January.

Maize cultivation in India is predominant during rainy season (*kharif*). The crop is sown with onset of monsoon rains during June and harvested by end of September, when rains fade. The annual rainfall in peninsular Indian is bi-modal and ranges from 700 to 1,100 mm. Rainfall during *kharif* ranges from 400 to 650 mm depending on location. The pattern of precipitation is skewed. About 200–300 mm is received during July, August, and September. The post-rainy or *rabi* crop of maize thrives on stored moisture and small amount of winter rains during November to January. It constitutes less than 20% of annual rainfall (100–180mm). Precipitation during summer is negligible (<50mm). Therefore, maize sown in summer is invariably irrigated. In the Western Ghats, maize thrives under high rainfall ranging from 2,000 to 3,000 mm annually. The *kharif* crop is sown

$$\text{Note:} \quad *PUE_y = \frac{Y}{P_g + P_f + (\theta_{h(h-1)} - \theta_{h(n)})} \, kg.ha^{-1}mm^{-1}$$

where, PUE$_y$ = Precipitation use efficiency for a particular year, including the fallow season, based on the grain yield kg ha^{-1} mm^{-1}; Y = Grain Yield kg ha^{-1}; Pg = Precipitation during the growing season (mm); Pf = Precipitation during fallow season; θ h(n) = water content of the root zone at harvest in year n (mm); θ $_{h(n-1)}$ = water content of the root zone at harvest in year n–1 (mm), with onset of monsoon in early June. The rainy season crop may receive up to 800 mm during June–November. Farmers practice variety of crop mixtures and sequences to maximize PUE.

In China and other Southeast Asian nations, maize culture depends on precipitation, temperature, and frost-free days. Precipitation is highly variable in northern dry areas of China. It ranges from 450 to 900 mm annually. In Yangtze and Yungui plateau, rainfall ranges from 800 to 1,800 mm y^{-1}. Annual precipitation is low in Loess and Northwest region. It ranges from 280 to 700 mm. Maize is a preferred cereal in this dry land area. Annual precipitation in Vietnam ranges from 1,200 to 1,700 mm. The rainy season lasts from April to October. The rainfall pattern is skewed. About 200–300 mm rain occurs during a short span from July to September. The evapo-transpiration is high at 700–800 mm due to tropical climate (Harsono and Karsono, 1997). In Srilanka, maize crop thrives in high rainfall regions that receive 1,600–2,300 mm annually. In Indonesia, maize is sown in all three seasons depending on precipitation pattern, soil moisture, and irrigation potential. Much of the maize belt thrives on dry lands, although in many parts of Indonesia annual precipitation ranges from 900 to 2,000 mm.

Water Requirements and Use Efficiency
Maize crop has to absorb water stored in soil or that provided via irrigation in order to satisfy the requirements. In areas with deficient rainfall and low stored moisture farmers usually adopt irrigation. It is generally observed that maize crop extracts about 70% of moisture requirements from top 0 to 60 cm depth, 20% from 60 to 90 cm depth, and rest 10% from deeper layers of soil. Incidentally, depletion of nutrients is linked to soil moisture uptake pattern. Maize crop produces 10–16 kg grain per mm of water used. A crop that yields 3 t grains requires at least 350–450 mm of rain per crop season. Under full irrigation, a commercial crop of maize yields 6–9 t grains ha^{-1} and 10–13 t forage ha^{-1} (FAO-AGLW, 2002). In the Chinese maize belt, WUE is said to fluctuate between 40 and 50% depending on cropping systems and location. In Indonesia, about 60% of maize is intercropped with peanuts. It helps farmers to raise WUE of intercrops to 1.44 kg m^3 compared to 0.62 kg m^3 with maize mono-crop.

Wherever be the maize cropping zone, the critical stages when optimum soil moisture is a necessity are seedling, knee-high stage, tasseling, silking, and grain filling. During the above growth stages, if precipitation is uncongenial then irrigation support is required. Generally, irrigation requirements of maize crop increases as crop duration stretches and yield goals get revised upwards. Timing of irrigations and quantity disbursed at different stages of the crop is of course crucial. Following is an example depicting timing of irrigations during a cropping season:

Irrigation	Stage of the Crop			
No.	Establishment	Vegetative	Flowering	Grain Formation
2	----------	----------		
3	----------	----------	----------	
4	----------	----------	----------	----------
5	----------	----------	----------	----------

Source: FAO-AGLW, 2002.
Note: Horizontal lines indicate the period when irrigation should be scheduled. First irrigation is done at sowing.

Reports from Southern Africa suggest that, on an average, maize crop needs 450–600 mm water per season. It is slightly higher than water requirements of drought tolerant cereals like millet or sorghum (350–500 mm) that occupy drier regions of Southern Africa. The maize crop grown in Southern Africa produces at least 15 kg grain for each mm of water consumed. Over all, maize crop may extract and utilize about 250 mm of water to reach physiological maturity (Jean du Plessis, 2008). Farmers in Southern Africa, South Asia, and elsewhere in dry land ecosystems of Argentina use variety of agronomic procedures that enhance WUE. Let us consider an example from Zambia in Southern Africa, where in, Sesbania fallows are known to affect both WUE and soil N dynamics.

Application of fertilizer and fallows with green manure improves WUE and maize grain yield:

	No Fertilizer	Fertilizer-N,P,K	Sesbania Fallow
Grain Yield (t ha⁻¹)	1.1	6.0	3.3
Water Use Efficiency (kg mm⁻¹ ha)	4.3	8.8	5.7

Source: Phiri et al., 2003.

Maize Genotypes

So far, each and every maize genotype sown either in small or large patches, any-where in the world must have caused proportionate alterations in the nutrient dynam-ics and productivity. During past 5 decades, genetic potential of maize cultivars/hy-brids released for production has improved. At the same time, it has induced high levels of nutrient supply into maize belts. The specific traits of a maize cultivar that affect different aspects of soil and/or above ground nutrient dynamics need due at-tention. For example, higher grain yield induces greater quantity of nutrient removal from the ecosystem. Large root system increases nutrient recovery from soil phase. It also enhances accumulation of SOC and C sequestration in the below ground por-tion of the ecosystem. There have been marked changes in harvest index (HI) that is forage:grain ratios of maize composites and hybrids cultivated during past century. Harvest index of a maize genotype has immense influence on the extent of nutrients recyclable, nutrients lost to farm animals as forage and nutrients removed as grains for human consumption. Generally, higher demand for grains has necessitated use of maize genotypes with higher harvest index. The need for greater land use efficiency has lead to use of genotypes with better response to chemical fertilizers and those with ability to produce higher grain yield. This has induced application of large quantities of nutrients into maize belts. Sometimes, it has resulted in undue accumulation of nutrients in the soil phase. It has made nutrients vulnerable to loss via leaching, run-off, and volatilization. High nutrient supply has also affected nutrient balance in soil. Greater details on maize genetic improvement and its influence on nutrient dynamics in the ecosystem are available in Chapter 7.

Cropping Systems in Maize Belts

Maize mono-crops are prevalent in the "Corn belt of USA." Continuous maize in-deed imparts its great influence on soils, vegetation, crop productivity, and even on several of atmospheric parameters like CO_2 evolution, C sequestration trends, nutri-ent dynamics, and most importantly grain/forage harvests. Maize productivity is held high at 7–10 t grains plus 10–15 t forage ha⁻¹ in most parts of Corn Belt. High input agriculture is practiced. Hence, nutrient turnover rate is relatively high within this part of maize agroecosystem. Maize is being increasingly accepted into cropping systems in Central and Southern Great Plains. An improvement in WUE seems to be the prime reason. The WUE determines productivity of various cropping systems. The traditional wheat–fallow system (143 lb ac in⁻¹) is being changed to wheat–maize–fallow (226 lb ac in^{-1}) because of better WUE. Over all, maize–wheat–fallow, maize–cotton, maize–soybean, maize–legumes, maize–tobacco, maize–legumes–fallow, maize–fallow are more frequent in Central and Southern Great Plains. Maize is also intercropped or strip cropped with several other crop species. Maize/soybean, maize/cotton, and maize/vegetable are useful crop mixtures. They enhance land use efficiency and add diversity to cropping systems. Many of the above crop mixtures are staggered at planting in order to suit the season, as well as to exploit soil moisture and nutrients more efficiently.

Cropping systems in Brazilian Cerrados are mostly soybean based. Rotations and intercrops of maize and soybean are common. Maize is also rotated with other legumes. In the Eastern plains of Columbia, maize is usually grown as a mono-crop or rotated with soybean. Common crop rotations encountered in this region are maize continuous; maize–soybean; maize–soybean-green manure; and maize-native savanna. During these sequences, farmers follow either no-till or conservation tillage. Maize productivity ranges from 2.3 to 4.2 t grains ha^{-1} (Basamba et al., 2006). In the Pampas of Argentina, cereal-based rotations dominate the landscape. Wheat/soybean intercrop is often rotated with a mono-crop of maize. Maize production in the Pampas is basically done under dry land conditions (Carcova et al., 2000). During past decades, this vast stretch of cropping zone has been intensified through higher supply of nutrients and water (Diaz et al., 1997; Plate 1.1). Productivity of maize has improved considerably. Major crop sequences followed in the Argentinean maize belt are wheat–maize–wheat/sunflower; mixed pasture–maize–maize–maize–wheat/sunflower; maize–soybean; maize/soybean strip cropping; etc. Intercrops with maize include legumes, maize/cotton; maize/wheat; etc. (Carcova et al., 2000).

Cropping sequences in Pampas of Argentina that include maize:

Subregion of Pampas	Dominant Sequence	Secondary Sequence	Maize Yield
			(t ha^{-1})
Rolling Pampas	Wheat–Soybean	Maize–Soybean	3.5
Southern Pampas	Wheat-Sunflower	Maize–Oats	2.8
Flat Pampas'	Wheat–Maize-Sunflower	Oats–Sorghum	2.8
Western Pampas	Wheat–Sorghum–Rye	Sunflower-Maize	2.5
Mesopotemic Pampas	Sorghum	Wheat–Maize	2.4

Sources: Hall et al., 1992; Krishna, 2003.

In the West African tropical zones, maize mono-crops are followed by a green manure or cover crop (guinea grass) or tropical kudzu (*Pueraria phaseoloides*). In terms of soil moisture storage and/or PUE, native fallows with mixed vegetation seem less efficient than fallows filled with green manure (Tijani et al., 2008). Maize/cowpea intercrops are most commonly preferred in Savanna zone. In order to preserve soil fertility, most farmers tend to practice maize–cowpea or legume rotation on sandy soils of Savanna zone. Maize–groundnut cropping systems are popular in Senegal and other nations of West Africa (see Krishna, 2008). Maize culture in Senegal is prominent in the so called "Compound Ring" or "Food Crop Ring." Whereas, peanut and cotton, the two main cash crops are confined to "Bush Ring." Rotations that include maize are common in Sudano-Sahelian region. Maize is progressively replaced by drought tolerant cereals like sorghum or pearl millet in Sahelian zone. Maize usually confines to relatively wetter regions. Here, precipitation pattern and quantity plays a vital role in deciding spread of maize cropping belt. Tropical climate in interior and Coastal West Africa allows maize/cassava intercrops. Maize is mostly a first season crop in West Africa. It is sown around June/July at the onset of rainy season and harvested by mid-October. Maize mono-crops and maize/cowpea mixtures are common in tropical Ghana and Nigeria (Horst and Hardter, 2006). In Nigeria, bulk of maize producing zone is situated in tropical southwest. About 50% of green maize is cultivated in the Southwest and other half emanates from cropping zones in North and East. During recent years, there has been a dramatic shift from green maize to grain maize in Northern Guinea Savanna regions of Nigeria (Ikena and Amusa, 2004). In this region, actually, maize is slowly but surely replacing erstwhile sorghum-based cropping systems. The effect is significant therefore maize belt extends even into northern Savanna of Nigeria. Maize/legume intercrops and maize/legume–fallow are important rotations in tropical Nigeria.

Plate 1.1 Maize crop being harvested in the Rolling Pampas of Argentina.
Source: Dr. Carina Alvarez, Universidad de Buenos Aires, Buenos Aires, Argentina.
Note: Expanses of maize mono-crop or maize–soybean rotations are common in these vast plains.

In Kenya, maize production is predominately accomplished during "Long Season." The "Long Season" actually corresponds to February–August. It includes the main rainy season. In 2008, nearly 80% of Kenya's annual maize production (2.3 m t) was derived during "Long Season." In the Rift Valley and Nyanza region, main maize season extends from May to August and it coincides with onset of rainfall. Maize/

legume intercrops followed by a green manure or short season vegetable is preferred in East African agricultural zones. Maize–legume–fallow rotations are also common in tropical zones. In the sub-humid highlands of Kenya, maize is sown as an alley crop along with agroforestry species like *Calliandra* or *Leucana* (Mugendi et al., 1999).

In the humid tropics of Central Africa (Congo, Uganda) and regions closer to Kenya, maize/cassava intercrops serve as carbohydrate sources. Maize is grown to maturity during first season and cassava continues to next season. Maize-green manure (*Stylosanthes guinensis* or *Macuna utilis* or *Crotalaria juncea*) are efficient in terms of water and nutrient use efficiency. Legumes like Stylo or Cajanus add to soil N. In forested areas, maize/cassava/perennial pigeonpea provides both carbohydrates and proteins to subsistence farmers. Nutrient recycling, especially that of soil-C and N is effective when green manures or pigeonpea are grown in a rotation.

Southern African farmers adopt a variety of maize-based cropping systems. Maize is a versatile crop and fits into several kinds of rotations and crop mixtures. Most common crop mixtures are maize/legume, maize/vegetables, maize/sorghum, etc. Maize mono-cropping, maize–legume/oilseed, or vegetable rotations are most frequent in Southern Africa. Maize–cereals like finger millet or sorghum is preferred in regions with unsure precipitation pattern. Maize–fallow or green manure is preferred rotation in Southern African dry lands. Maize after a *Sesbania* or *Gliricidia* fallow is supposed to attain better fertilizer and water use efficiency. Sesbania fallows conserve soil moisture better and enhance PUE during the fallow period. The WUE could be around 4–8 kg mm^{-1} ha^{-1} (Chirwa et al., 2007; Phiri et al., 2003). Green manures may also help sub surface recharge with water and nutrients.

In the WANA region maize cultivation is confined to wetter areas with relatively higher precipitation (500–800 mm). Here, maize is mostly rotated with legumes or vegetables. Maize farming is preferred because it offers higher amounts of both grains and forage. Productivity of maize and intercrops depends much on fertilizer inputs and irrigation. In the Nile Delta region, rotations such as maize–peanuts, wheat–maize–peanuts, or maize–berseem are popular. In North Africa, maize is again intercropped with legumes like cowpea or lentils. It is also sown in strip crops with agroforestry species (Faidherbia, Acacia, or Neem), since it allows better control over erosion and loss of nutrients from fields.

In South Asia, especially in India, maize belt is largest during rainy season (*kharif*). The *kharif* season begins with onset of monsoon. The *kharif* season extends from May 1st week to October in the Peninsular region. In India, maize is mostly a warm season crop. The *kharif* maize belt accounts for 85% of total annual acreage. The *kharif* crop is sown in April in Northern hilly region. Greatest fraction of Indian maize belt that occurs in Southern plains is sown during June/July coinciding with the onset of monsoon. Maize productivity during *kharif* is moderate at 2–3 t grains ha^{-1}. In the Gangetic plains, mixed crops of maize are sown during May/June. The post-rainy season crop is sown during October 15th to November 15th. The *rabi* sown maize belt constitutes only 6% of the Indian maize production zones. The summer crop of maize is usually sown during February/March in regions with assured irrigation. The crop is harvested before onset of monsoon in June. As stated earlier, maize agroecosystem is prominent in Southern Indian plains. Here, maize is intercropped with several other crop species. Maize-based rotations in Southern Indian plains include maize-pulse/oilseed; maize-vegetables, maize–cotton. Maize-*Rabi* sorghum is popular on Vertisols of North Karnataka and Andhra Pradesh. In Tamil Nadu, maize–rice–pulse (cowpea) is preferred in wet land zones. However, maize–finger millet is common to dry lands of Karnataka and Andhra Pradesh (Krishna, 2010). In the North Indian plains maize-torai/potato/pea-sunflower is frequently practiced. Similarly, in Tarai region, maize-torai-wheat and maize–rice-torai are common. Regarding intercrops, mixtures of maize/cotton, maize/pigeonpea, or any other pulse, maize/vegetable are most common in South India. Maize/oilseed brassicas or maize/pulse is preferred as intercrops in Gangetic plains. In the hilly zones of South India, maize/cassava or maize/tuber crop (e.g., potato) mixtures are common.

Southeast Asian farmers are known to adopt innumerable variations of cropping systems that include maize in sequence or as intercrop. Mixed culture with maize normally includes crops like legumes, oilseeds, and vegetable. Relay cropping with oilseeds is also common, wherein maize planting is staggered to help oilseed establish itself. Maize is also alley cropped with agroforestry trees. Alley cropping helps in sustaining soil fertility and ensuring proper SOC recycling in addition to improvement in soil N status. In North and Northeast China, maize is intercropped or rotated with peanuts, and cereals like wheat. Most commonly adopted two or 3 year cycle in Northern China is winter wheat-summer maize-spring wheat; or winter wheat-summer peanut-spring maize; winter wheat-sweet potato and spring maize (Wenguang et al., 1995). Maize is frequently intercropped with Cassava in Thailand. On low fertility Alfisols of Vietnam and Kampuchea, maize–peanut rotation is efficient both in terms of water and nutrient use efficiency. Productivity of maize is still low at 3.5 t grains ha^{-1}. In the Red river area of Vietnam, spring rice is usually followed by October/November sown maize/peanut intercrops. Rice–peanut maize rotations are frequent in Indonesia. In Malaysia, again maize is rotated and/or intercropped with peanuts and vegetables. It is alley cropped with agroforestry species such as *Leucana*, *Sesbania*, or *Gliricidia*. Rice is the main crop in much of Southeast Asian cropping zones. Maize is often rotated with rice during the fallow or dry season. Maize also serves as excellent short season summer crop and provides valuable forage to farm animals.

EXPANSE AND PRODUCTIVITY OF MAIZE AGROECOSYSTEM: CURRENT STATUS

Maize agroecosystem is large and it extends into vast stretches in each continent. World wide, annually, maize crop is sown in 140–150 m ha. Maize belt continues to expand (Bennetzen and Hake, 2009). Global maize acreage, average grain harvests, and productivity are shown in Figure 1.1 and 1.2. During 2009, 163 nations grew maize. During recent years, it has been possible to monitor maize belts in different continents/countries using remote sensing techniques. The spectral data are usually calibrated with annual area/production figures available from FAOSTAT. For each country, it is possible to monitor the maize cropping intensity using vegetative productivity indicators (VPIs) and prepare accurate forecasts about maize grain/forage yield by calibrating the remote sensing date with national area/productivity figures (FAO, 2006).

Figure 1.1 The maize agroecosystem of the world.
Note: Dark areas represent intensive maize cropping zones (6–8 t grain ha⁻¹); stripped areas indicate moderately intense production zones (2.5–6.0 t grain ha⁻¹). Dotted area indicates subsistence, low input maize cultivation zones (0.5–2.5 t grain ha⁻¹).

Globally, maize production strategies vary. Maize belts differ in terms of composition of genotypes, intensity of cultivation and productivity. In the USA and European agrarian zones, maize belt is highly intensive. In order to sustain high grain and forage productivity, appropriately high nutrient and irrigation inputs are practiced (see Table 1.2). Maize agroecosystem is only moderately intense with regard to nutrient supply and grain productivity in many Asian countries. Maize agroecosystem is held at subsistence levels of nutrient and water supply in Sahelian, Guinea Savanna, and other regions of Africa. Maize productivity under such semi-arid or dry land conditions is low. The ability to thrive in different environments is attributable to highly versatile nature of maize plant.

Table 1.2 Maize agroecosystem—A classification based on intensity of cropping and productivity.

Nature of Maize Agroecosystem	Productivity	Countries/Region
Intensive Mono-cropping Regions of Maize		
High Nutrient Inputs (300 kg ha⁻¹ NPK),	7–11 t grains ha⁻¹	USA (Corn Belt), Canada,
Irrigated, High Yielding Genotypes	12–20 t forage ha⁻¹	Israel, Egypt, New Zealand
		France, Italy, Austria, Spain,
		Belgium, Great Britain,
		Germany,

Modrately Intense Maize Belts		
Nutrient Supply Exceeds (200 kg ha^{-1} NPK)	2.5–6. t grains ha^{-1}	China, Thailand, Korea, Vietnam
Rain fed or Irrigation Restricted to Critical Stages	10–15 t forage ha^{-1}	Mexico, Brazil, Venezuela,
High Yielding Genotypes		Argentina, Hungary, Serbia, Russia Slovenia, Kazakhstan, Turkey, Syria, Uzbekistan, Morocco
Low Productivity Maize Farming Regions		
Nutrient Supply is Low 20–100 kg NPK ha^{-1}	0.5–0.2 t grain ha^{-1}	India, Nepal, Pakistan, Bangladesh
Low Seeding Rate, Rain Fed, Dry Lands	5–8 t forage ha^{-1}	Burkina Faso, Senegal, Botswana, Mauritania, Mali, Togo, Chad, Sierra.Leone, Congo, Ghana, Niger Saudi Arabia, Libya,

Sources: FAOSTAT, 2005; OKSTATE, 2009.

Note: Maize cropping zones follow a definite pattern based on natural resources and agronomic procedures adopted by farmers. Maize ecosystem is intense and highly productive in North America and European plains. Moderately intense cropping is confined to Asia, Mesoamerica, and Tropical Africa. Maize ecosystem is subsistent and least productive in Sahelian West Africa and other semi-arid regions.

During 2009, worldwide maize production was 796 million t (Figure 1.2). The USA supports largest maize cropping zone. Nearly one half of total global maize, about 332 million t grains equivalent to 47.2% was harvested annually in USA. Other major maize producing nations are Peoples Republic of China (151. 9 m t grains), Brazil (51.6 m t grains), Mexico (21.7 m t grains), Argentina (21.7 m t grains), India (16.8 m t grains), France (13.1 m t grains), Indonesia (12.4 m t grains), Canada (10.5 m t grains), Italy (9.9 m t grains) (FAO, 2008). Annually, on an average during the past decade, USA contributed 35–42% grains, China 15%, European countries together about 14%, Brazil 8.4%, and India 3% of the total global produce. About 60% global maize cultivation zones occur within six countries. Continent wise, during 2007, North America contributed 45%, South America 10%, Europe 12%, Asia, 26%, and Africa 7% of maize grain harvest. Total area (m ha) and productivity (t grains ha^{-1}) of maize farms in top 10 countries are as follows:

Country	USA	China	Brazil	Mexico	India	Nigeria	Indonesia	S. Africa	Rumania	Philippines
Area (m ha)	29.8	25.4	12.4	7.7	7.5	4.4	3.4	3.2 3.0	2.7	
Yield (t ha^{-1})	10.2	5.4	3.7	3.1	2.1	0.9	3.2	3.0	4.3	2.2

Source: IPNI, 2008.

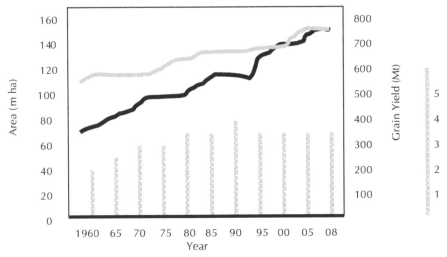

Figure 1.2 Maize agroecosystem: Expanse, grain yield, and productivity.

Note: Gray bars represent productivity (t grains per ha) of maize. Gray curve shows area and black refers to total grain yield.

Currently, the average productivity of maize crop in Africa is 1.5 t grains ha⁻¹ plus forage depending on harvest index. Maize farms in West Africa that are kept under low input and meant for subsistence produce a mere 0.7–1.5 t grain ha⁻¹ (Ofori et al., 2004). In South America, average productivity is 3 t ha⁻¹. In most parts of South Asia including India it is 1.7–2.3 t grains ha⁻¹. Maize productivity in China and Fareast is about 2.4 t grains ha⁻¹. Maize grain yields are relatively higher in the Corn Belt and other regions of North America, including southern Canadian plains. Maize is an excellent forage crop. Maize is also grown exclusively for forage. On an average, it yields 20 t fresh biomass (stalks and leaves), that is suitable for ensilage. The forage yield may fluctuate depending on the extent of irrigation and soil fertility conditions. In dry lands, maize forage harvest is about 18–29 t ha⁻¹, but under irrigated conditions it reaches 42 t ha⁻¹. Maize is predominantly a forage crop in some of the European countries. In European plains and loess region, maize forage yield fluctuates between 35 and 50 t ha⁻¹. For example in France it is 40 t forage ha⁻¹, in United Kingdom it is 25–35 t ha⁻¹ in Eastern European region, again it ranges from 25 to 35 t forage ha⁻¹. The grain yield in European nations ranges from 5.5 to 7 t ha⁻¹ (Ofori and Kyei-Baffour, 2008).

The acreage of maize in USA has fluctuated, perhaps depending on various reasons related to soil fertility, irrigation, crop genotype, and agroclimatic parameters. The demand/supply equations and governmental legislations too must have played a role in deciding size of maize cropping zone. Since 1900, the maize belt has reached a high of 45.5 m ha in 1917 to a low of 27.5 m ha in 1969. Factors like soil fertility, ensuing nutrient dynamics in farms, agronomic procedures, and crop genotype related factors have all influenced the productivity of maize during various years. Productivity of maize crop was 29 bu* ac⁻¹ in 1900 and it has since increased to 143 bu* ac⁻¹ in 2005

(FAOSTAT, 2005; Larson and Cardwell, 1999). The average gain in grain productivity has been around 1.0 bu* ac^{-1} year^{-1}. Since 1940s, that is after introduction of high yielding genotypes and chemical fertilizers, gain in grain yield has jumped to 1.8 bu* ac^{-1} year^{-1}.

The impressive improvement in maize grain harvest per unit area has been achieved through manipulation of factors like crop genotype, soil fertility, and irrigation. In USA, maize farmers started using chemical fertilizers, both nitrogenous and phosphorus fertilizers in early 1940. This indeed changed the soil nutrient dynamics. This step actually helped the farmers in matching nutrient demands of the crop with supply, based on the yield goals that was envisaged. Improved soil fertility status, it seems allowed maize farmers to plant closely. They achieved a better crop stand by increasing planting density from 1,2000 to 1,4000 plants ac^{-1} in 1,900 to over 27,000 plants ac^{-1} in 1940s. Also, traditional maize/small cereal grain intercrops got replaced by mono-culture of corn alternated with soybean (Allmaras, 1997). This lead to the development of an intensive maize growing region in Northern part of USA. This region is more commonly known as "American Corn Belt" (Figure 1.3). The states that support intensive maize production are aptly called "Corn States." They are Wisconsin, Minnesota, Missouri, Michigan, Illinois, Iowa, Ohio, Indiana, also parts of North Dakota, South Dakota, Nebraska, Kansas, Arkansas, Tennessee, and Kentucky. Obviously, maize dominates the agricultural horizon in the "Corn belt." Therefore, it has greatest impact on the nutrient dynamics, ecosystematic functions, and agricultural productivity of the region. The six states in "Corn belt," namely Iowa, Illinois, Nebraska, Minnesota, Indiana, and Ohio account for 82% of maize produced in USA annually. During 2007, national maize acreage was 35 m ha. Iowa is the top maize producing state in USA. Maize belt in Iowa extended into 5.2 m ha. It contributed 22% of annual grain harvest of USA. We ought to note that maize cropping zones in USA do extend beyond "Corn belt" into many other states. Over all, 60% of maize produced in USA is utilized to feed farm animals, about 20% is exported, remaining portion is used for human consumption and industrial uses, like starch production (OSU, 2009). The maize agroecosystem in Canada is large and spreads across southern region of the country. It occupies about 1.3 m ha and contributes 9.5 m t grains annually. The productivity is high at 7.82 t grains ha^{-1}, mainly due to chemical fertilizer supply and irrigation.

The maize agroecosystem of Central America took shape perhaps during ancient era or even earlier. Maize belt got enlarged through human migration. It seems development of citadels and need to feed larger population induced intensification of maize cropping zones. Currently, Mesoamerican maize belt thrives in regions endowed with rich vegetation and tropical or sub-tropical climate. Mexican maize farmers often clear forests and practice "slash and burn" or "shifting cultivation." Maize is cultivated along with beans, gourds, and other vegetables. Therefore, here maize thrives predominantly as an intercrop. Maize is an important crop for inhabitants of Andes and its slopes. As such, they specialize in raising maize belt on steep mountains by adopting terrace farming. For example, Guatemalan highland supports maize production through terrace formation. In Central Mexico, maize cropping zones evolved into *Chinampas* or floating gardens. It seems *Chinampas* allowed intensification of maize farming in this region. *Chinampas* is a unique maize farming technique. It was standardized much before Spaniards entered Mesoamerica. Currently, maize farming stretches into entire Mesoamerican agricultural zone. Maize is a dominant cereal among the crop mixtures grown in Central America. The maize belt extends into 7.7 m ha in Mexico and contributes 3% of global maize harvest (21 m t). The productivity of maize fields is about 3.1 t grains ha⁻¹ plus forage. Maize cultivation spreads into 600 thousand ha in Guatemala, 300 thousand ha in Honduras, and 130 thousand ha in Cuba.

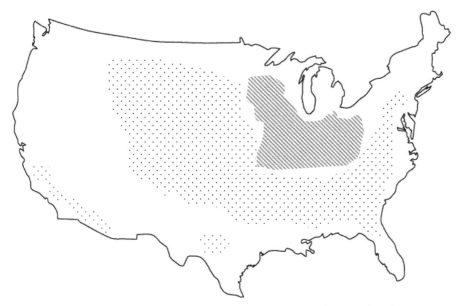

Figure 1.3. The intensive maize cropping zone of USA—"The Corn Belt" (striped area).
Note: Maize is grown in many other states of USA (dotted area), in addition to "Corn Belt States," wherever agroclimate is congenial.

The Argentine maize belt is concentrated in the Pampas (Figure 1.4; Plate 1.2). Maize belt of Argentina is large and spreads into 2.8 m ha. Annually, Argentine maize agroecosystem produces 19.5 m t grains, which is equivalent to 3% of global harvest. The productivity of Argentinean maize belt is relatively high at 6.5 t grain ha⁻¹ plus forage. Brazil supports a large expanse of maize cropping zone. It is mainly concentrated in the Cerrados. Maize belt also extends into other regions like Amazonia. The Brazilian maize belt spreads into 12.4 m ha. It accounts for 8.5% of global maize agroecosystem in terms of area. Productivity of Brazilian maize agroecosystem is relatively low compared to those in USA or China. It ranges from 3.2 to 3.7 t grains ha⁻¹ plus forage. Annually, about 51 m t maize grains are harvested from Brazilian maize growing regions (IPNI, 2008). In the Andes region of South America, white maize is most preferred. It is cultivated in high altitude regions ranging from 1,200 to 3,600 m.a.s.l. Maize encounters abiotic stresses like cold temperature and drought. Yet, the grain yield may range from 2 to 3 t ha⁻¹. Farmers tend to produce much of their maize under low input systems owing to environmental vagaries. On fertile soils, under high input systems these highland maize cultivars (white floury types) yield 9–12 t grain ha^{-1} (Beck, 2000).

Figure 1.4 Maize growing regions of South America.

Note: Brazilian Cerrados (striped area) and one below in Pampas of Argentina (dotted area) are vast expanses that support cultivation of maize and other crops.

Maize culture is prominent in France. It occupies 1.7 m ha and contributes 13.2 m t grains annually, which is equivalent to 2% of global maize harvests. The productivity is markedly high at 7.8 t grains ha. The German maize growing regions extend into 0.47 m ha. Maize is intensively cropped hence productivity ranges from 7.2 to 7.7 t grains ha^{-1}. Maize is an important cereal in Eastern European plains. In Romania, maize belt extends into 3.0 m ha and productivity is 4.3 t grains ha^{-1} plus forage. Ukraine supports maize belt that extends into 2 m ha on the plains and contributes 6.5 m t grains annually. Productivity is moderate at 3 t grains ha^{-1}. Hungary in Eastern Europe supports a large maize belt of 1.2 m ha. Productivity is moderate at 3.94 t grains ha^{-1}. The annual grain harvest is about 4.5 m t grains plus forage. The Italian maize belt spreads into entire country. It is predominant in agricultural zones of Northern Italy. Italy is actually a major maize producing nation in Southern Europe. Among European nations, Italy contributes 2% of global maize harvest (14.3 m t y^{-1}). The maize agroecosystem in Italy occupies 1.46 m ha. Again, productivity is rather high due to intensive farming techniques that employ high amounts of nutrient and water supply. Spain has maize belt of 0.47 m ha that thrives under intensive cropping system. The productivity is rather high at 9.5 t grains ha^{-1} plus forage.

Plate 1.2. Maize mono-crop in Pampas of Argentina.
Source: Carina Alvarez, Universidad Buenos Aires, Argentina.
Note: Vast stretches of maize mono-crop or maize/soybean intercrops are common in these Argentinean plains. The productivity of maize on Mollisols ranges from 3 to 6 t grain ha^{-1} plus forage.

Maize cultivation extends into several nations of North Africa and West Asia. The maize cropping regions are relatively small in most countries. It ranges from few thousand ha to one half of m ha in some nations. Maize agroecosystem in North Africa and

West Asia varies enormously with regard to intensity of cropping and nutrient supply. In Morocco, Libya, and Saudi Arabia, maize is cultivated with low input utilizing subsistence farming methods. It yields 0.8–2.4 t grains ha^{-1}. Whereas, in Algeria, Lebanon, and Syrian Arab Republic the maize belt is moderately intense. Farmers in these countries harvest 3–4.5 t grains ha^{-1}. The productivity of maize is highly dependent on soil fertility and nutrient dynamics. Maize crop yields 12 t grains ha^{-1} plus forage in countries like Israel, where the crop is invariably irrigated and nutrient supplies are high. In Egypt, there is a special drive to enhance maize area and productivity. About 20% of maize flour is being mixed with wheat flour regularly while making bread. This way it avoids excessive import of wheat from other countries (Bahr et al., 2006). Currently, maize hybrids released for cultivation in Egypt yield 2.2–2.9 t grain ha^{-1} and 8–10 forage ha^{-1} (Soliman, 2006).

Maize is an important cereal crop of West and Central Africa. The maize cropping belt in this part of the world expanded gradually yet perceptibly from 3.2 m ha in 1961 to 8.9 m ha. As a consequence, maize production increased form 2.4 m t grain in 1961 to 11.8 m t grains in 2008. However, we ought to realize that productivity of maize crop is quite low in Sub-Sahara. Productivity of maize fields in Senegal, Mali, Niger, and Chad is around 0.80–1.3 t grains plus proportionate stover. The average grain yield in tropical Ghana and Nigeria is much high.

Maize is a staple food for people residing in regions catered by Southern African Development Authority (SADC). The region includes countries such as Angola, Botswana, Lesotho, Malawi, Namibia, South Africa, Swaziland, Zambia, and Zimbabwe. Major constraints to maize production are drought and low soil nitrogen (Zambezi and Mwambula, 2006). In Southern African cropping zones, maize production is derived mostly from small and medium sized farms less than 10 ha. Based on input level and agronomic procedures adopted grain/forage harvests are commensurately low. Grain yield ranges from 1.2 to 1.3 t ha^{-1}. Productivity is much lower in regions where subsistence farmers abound. Low input and subsistence farming techniques results in only 700–800 kg grains ha^{-1}. Forage yield too is proportionately low. Yet, we ought to know that in Zimbabwe and other Southern African nations, larger farms adopting improved procedures and high inputs yield 8–10 t grain ha^{-1}. The potential yield of maize under high put conditions is much higher. It leaves a large yield gap for subsistence farms to cover.

Annually, African continent contributes 46–50 m t grains to the global maize harvest. Among the 50 African nations that produce maize, South Africa supports a large maize belt. Maize is an important dry land cereal of South Africa. South African maize belt represents 15% of African maize cropping zone. South African maize farming zones are spread into 3.2 m ha. It contributes 12 m t grains annually, which is equivalent to 2% of global maize harvests. The productivity of maize crop is moderate at 3.0 t grains ha^{-1}. Maize grain and starch serve a large portion of human population in South Africa. It also supplies ethanol and forage to a certain extent. Much of the maize belt in South Africa is planted to white maize (Jean du Plessis, 2008).

Kenya is among the major maize producing countries of East Africa. Currently, the "Maize Agroecosystem" in Kenya is highly expanded. It occupies large areas in the following provinces namely Central, Eastern, Western, Coastal, Rift Valley, and Nyanza region. Maize belt occurs in small patches of mono-cropping zones, as well as intercrop with legume or vegetable. Over all, maize cropping zone is relatively large in Kenya. It occupied over 1.2 m ha in 2008. Maize agroecosystem is concentrated in Rift Valley, Western Province, and Nyanza region. Together, these three provinces account for 0.83 m ha and contribute 1.6 m t grains equivalent to 60% of Kenya's total maize grain production. Maize belt in Eastern province is large at 0.23 m ha, yet total production is low because productivity of maize in this region is rather low at 0.6 t ha^{-1}. The average productivity of maize in Kenya is about 1.71–1.75 t grains ha^{-1}. There has been a perceptible decrease in productivity during the recent years. In some areas of Rift Valley, maize cropping is affected due to soil acidity that needs periodic correction (Mwangi et al., 2000).

Zambia received maize during 16th century, mainly from Latin America via Spain. Soon, its cultivation picked up and it replaced large patches of native cereals like sorghum and finger millet. In due course, Cuban Flint type maize became a staple carbohydrate source to human population in Zambia and surrounding regions. It also provided precious forage to farm animals. Maize belt extended into 0.77–0.82 m ha and produced 0.56–0.63 m t annually during 1960s. Introduction of chemical fertilizers in 1970 improved maize farming area to 0.90–1.10 m ha. Consequently, total production jumped to 1.64 m t annually. Diffusion of maize into fertile areas improved productivity to 1.8 t grains ha^{-1}. We may note that there was a slump in maize area and production during mid 1990s. Currently, maize ecosystem exists predominantly as a mixed crop with legumes. Its production hovers just around 1.0 m t annually. Whereas, annual demand for maize in Zambia is 1.2 m t grains. We may expect expansion, as well as improvement in productivity of maize in Zambia.

Malawi is yet another maize growing country from Southeastern African continent. It received the Cuban flint types around 15th century. Like, Zambia, soon maize cropping zones of Malawi expanded and in several places replaced the erstwhile staple cereals sorghum and finger millet. The maize belt in Malawi is mostly filled up with late maturing genotypes. It thrives during rainy season lasting from November to April. Maize belt has fluctuated considerably both in terms of area and productivity. Total maize grain production has ranged from 1.0 to 2.5 m t annually during past 16 years. Factors related to availability of seed and chemical fertilizers, precipitation pattern and farm subsidies may have all contributed to fluctuations. It is believed that lack of a stable soil fertility replenishment program has affected nutrient dynamics and productivity of the cropping ecosystem. During past 5 years, average area of maize agroecosystem in Malawi has ranged form 1.4 to 1.6 m ha. The productivity is about 1.8 t grain ha^{-1} compared with a global average of 4.2 t grain ha^{-1}.

In South Asia, India has the largest maize growing region. The Indian maize agroecosystem is filled with mostly four different types of maize, namely, dent corn, flint corn, flour or soft corn, and waxy corn. In addition, small areas of forage maize for animal feed, sweet corn for alcohol production, and popcorn for snacks are also encountered. It occurs both as mono-cropping area and as intercrop with legumes, oilseeds, and vegetables. Maize thrives mainly during rainy (*Kharif*) season. Nearly 90%

of the maize cropping is done during monsoon period between May/June and October. About 8–20% of maize belt thrives on stored moisture during post-rainy (*Rabi*) season. About 1–2% maize area occurs during summer as irrigated crop.

During past 5 decades, area and productivity of maize cropping zone has steadily increased in entire India. The annual growth rate has fluctuated from 1.4 to 4.9% over previous year. The percentage area irrigated has improved from 12 in 1950s to 24 in 2005. The average productivity of maize has jumped from 630 kg ha^{-1} in 1950 to 1770 kg ha^{-1} in 2005 (Maize Statistical Reports, 2006). It is partly attributable to fertilizer-based nutrient supply, irrigation, and constant allocation of maize crop to fertile zones in the farm. In particular, intensification and enhanced grain production of the maize cropping zones since mid 1960s is due to enhanced nutrient and water supply. Recent reports suggest that maize productivity may increase further and reach 2.4 t ha^{-1} by 2020 (Ikisan, 2007). Following are the size and productivity trends during past 5 decades within maize cropping belt of India:

Year	1950	1960	1970	1980	1990	2000	2005	2008
Area (m ha)	3.26	4.41	5.85	6.01	5.92	6.31	7.21	7.92
Production (m t)	2.05	4.31	7.49	6.96	9.65	10.85	12.72	15.4
Productivity (kg ha^{-1})	627	957	1279	1159	1632	1720	1852	1903
Irrigated Area (%)	12.8	9.9	15.9	20.1	20.8	22.7	22.8	24.5

Source: http://www.fadinap.org/india/productivity; Krishna, 2010.

During past 5 decades, maize agroecosystem in India has expanded considerably. Total grain harvested has increased from a mere 3 m t in 1950s to almost 15 m t in 2008. Rapid improvement in production technology, availability of high yielding genotypes and irrigation has contributed to enhanced productivity. Most importantly there has been a constant increase in demand for maize grains and products. We may note that, 75% of maize produced in India is utilized to prepare starch and starch-based products. About 20% is used as animal feed.

Maize is an important cereal in the Northwest Frontier Province of Pakistan. This region contributes 60% of national maize harvests. The cropping expanse is about 0.95 m ha and productivity fluctuates at 1,680–1,785 kg grain ha^{-1}. Total maize harvest of Pakistan is around 1.7 m t annually (Asif and Anwar, 2007). About 65% of maize cropping zone is irrigated but rest 35% is grown under rain fed condition.

Maize belt of China extends into many of its major agricultural cropping zones. It is predominantly grown in Northeast and Southern China. The Chinese maize agroecosystem spreads into 25.5 m ha. It constitutes 17.5% of the global maize agroecosystem in terms of expanse. The productivity of Chinese maize belt is relatively high at 5.4 t grains ha^{-1}. Actually, maize cropping is intense in Northeast where productivity is much high at 8–10 t grain ha^{-1}. However, in other regions it is around 3–4 t grain ha^{-1}.plus forage. Total maize harvest is about 151 m t grains ha^{-1} which is equivalent to 15% of global harvest. Maize harvests have actually increased due to improved varieties/hybrids, fertilizers, and improved agronomic practices (Qiao et al., 1996).

It is interesting to note that in Jilin province, average grain yield increased from a mere 2835 kg ha^{-1} in 1974 to 6329 kg ha^{-1} due to hybrids, fertilizers, and appropriate cultural practices.

Maize is grown in several regions of Australia, including Queensland and New South Wales (Birch et al., 2007). The maize cropping zones extend into 60 thousand ha, mostly confined to small areas in Queensland, New South Wales and South Australia. The productivity is moderate at 5.4 t grains ha^{-1} plus forage. Maize sown solely for forage may yield much high at 15–25 t fresh forage ha^{-1}.

Introduction of Genetically Modified (GM) Maize into Agroecosystem

Maize cultivars utilized by farmers have generally been selected for higher grain/forage productivity, duration, morphological traits, palatability, nutritional aspects, and tolerance to disease, insects, and drought. During the past century, much of the crop improvement has been achieved using classical genetic principles. Composites and Hybrids with ability for higher grain/forage production were introduced in many parts of the world during mid 1900s. It induced use of greater amounts of nutrients and this lead to rapid turnover of nutrients in the ecosystem. Since past decade, mainly from mid 1990s, maize geneticists have been employing molecular (gene splicing and cloning) techniques to introduce specific traits like tolerance to herbicides, insects (*Bt* gene), diseases, etc. So far, such GM modifications have not been aimed at or have drastically affected the mineral absorption or accumulation/partitioning traits. However, if GM techniques are aimed at improving nutrient absorption, nutrient translocation index, nutrient accumulation pattern in grains, or changing mineral composition of forage, then introduction of such GM maize genotype(s) will affect nutrient dynamics. The nutrient dynamics that ensues due to introduction of a GM genotype, then, needs to be carefully monitored. Since 1997, GM genotypes have been spreading rapidly into the maize agroecosystem. Worldwide about 12% of the maize agroecosystem has been constituted by GM maize (GMO Compass, 2009). In countries like USA (80%) and South Africa (57%), GM maize has perceptibly replaced erstwhile genotypes, rather, dominant enough to affect ecosystematic functions perceptibly. Sooner or later much of global maize agroecosystem could be filled up with GM maize. Yet, in many countries, GM maize occupies only 1–3 % of total acreage (GMO Compass, 2009). We ought to carefully monitor its effects on nutrient dynamics, if any, and on the ecosystematic functions.

USES

Firstly, maize cultivation, rather its expanse and intensity are highly depended on human preference, its utility in various forms and total demand for maize products. Human food habits are among most important factors that dictate extent of maize belt. Similarly, demand for forage, various industrial uses and immediate need for various by products also affect extent of maize cropping in a given region. Economic aspects, like subsidies, pricing, export of maize products may also induce expansion/reduction in maize belt. Let us consider various uses of maize that are more common in different continents.

Maize flour is most commonly cooked into *tortillas* in Mexico and other countries of Americas. Maize flour is mixed with other cereal powders to prepare a variety of breads and baked products. Maize is a major ingredient in biscuits and snacks. Maize flour is used in preparations of salads and appetizing soups. It adds to consistency and nutritional value of food preparations. Maize flour is also used in preparation of ice creams. Maize grains and cobs serve as snacks. Roasted cobs and cooked grains are salted and consumed as snacks. Sweet corn cobs are consumed fresh when still small in size (unripened) or cooked as vegetable [3, 4, 5].

In India, maize flour is used to prepare *Roti* or *Chapathi*, enleavened bread and porridges called *Ambli*. It is also used to brew local liquor either solely or in combination with millet. "Sattu" or "champa" prepared by grinding maize to powder form is consumed alone or with wheat and/or barley.

Africans prepare a range of porridges using maize flour. For example, it is called *Bidia* in Zaire, *Sadza* in Zimbabwe, *Putu* in Zululand, *Meali* in South Africa, *Posho or Ugali* in East Africa, *Kpekpele* in Ghana. Porridges made from maize flour are common in other continents too. *Polenta* in Italy, *Mamaliga* in Romania, *Angu* in Brazil and *Mush* in USA are few examples. Maize flakes are staple breakfast items in the USA and many other parts of the world. Maize grits (*kus kus*) are common in Southeastern USA and several African nations. Maize is also prepared as hominy where kernels are soaked with lye. The Brazilian dish *Canjica* is made by boiling maize kernels in sweetened milk. Popcorn is an important and popular snack in most parts of the world.

In West Africa, maize is an important cereal contributing nearly 50% of carbohydrate requirements of the local populace. It is said that almost every part of maize has some use for the villagers. Maize is mainly consumed as starchy base for a variety of porridges and pastes. Maize grits are consumed as break fast. Maize flour is used in preparing biscuits, and other baked products. Maize cobs are also consumed fresh. Maize beer is an important fermented beverage. Maize porridges cooked into thick fluid are popular in Southeast African Nations like Zambia, Malawi, and Botswana. In Zambia, maize flour is boiled in hot water and thick dough is prepared. In Malawi, staple dish prepared using maize flour is called *nsima* in Chichewa a Malawian language. Both white and yellow flint types are used to prepare *nsima*. Two types of *nsima* are known depending on milling procedure. *Mgaiwa* is good in taste, while *Ufa* appears and tastes well. *Chibulu* and *Kachaso* are fermented alcoholic beverages and *Chindoungowa* is fermented soft drink prepared by Malawians. *Mbuluuli* is pop corn like food item.

Maize has been excellent source material to produce industrial starch. Maize grains and flour are rich in starch. Starch production is most efficient from maize grains since it almost leaves no byproducts and waste material. In fact, 75% of world starch production is derived from maize grains. Maize starch is used in adhesive paper, surfactant polymers, textiles, pharmaceuticals and building industry. Maize starch is used as native starch, esterified starch, thinned starch, oxidized starch, dextrins, maltodextrins, glucoses, sorbitol, mannitol, cyclodextrins, etc.

Maize oil is obtained from corn germs. Maize grains contain 3–7% oil. The fatty acid content of corn oil is as follows: 8–12% of C16:0, 1–3% C18:0, C18:1, and

C18:2. Maize oil is used as a cooking medium and for salad dressing. Margarine is obtained through hydrogenation of corn oil. Maize oil is also used in pharmaceuticals.

Maize steep is a viscous fluid obtained through low temperature concentration of steeping water. The concentrated liquid contains 35–50% dry matter, 44–48% crude protein, 16% ash, and 25% lactic acid. The crude protein fraction contains free amino acids, polypeptides of various chain length, vitamins, and enzymes. Corn steep is an important ingredient utilized during industrial fermentation, especially in enzyme production, antibiotics, nutritional yeast, and amino acids.

Maize Gluten is obtained as a byproduct after extracting starch. It contains fiber, gluten, starch, and a small amount of oil. It is used as a protein base in animal and poultry feed. It does not offer rich protein, yet it is an important source of Xanthophylls and Provitamins.

Several types of alcoholic beverages are prepared using maize grains and flour. Traditionally, maize is the main source for preparing bourbon whisky. In Central and South America, native Indians prepare a fermented beverage called *Chicha*. Maize starch is an excellent base carbohydrate source during production of beer.

Maize is a useful ingredient in several industries. For example, in ethyl alcohol production, biofuels, beverages, food additives, soaps, skin care products, absorbent material, biodegradable plastics, etc.

Maize cobs available as residue after removal of kernels (grains) are an excellent source material for furfural production. The fibrous residue of maize cobs is a suitable raw material to produce pentosans like xylan and arabinan. The pentose content reaches 25–40%. Furfurals and pentoses are used as base material in several industries. Alkaline treatment yields natural glues.

Maize is an important starter material in bio-fuel production. Maize residue (stacks and cobs) are used to generate heat. Specialized stoves that improve efficiency of heat production are available. Pelleted maize cob residue is used in home heating in many European villages. Maize stalks, leaves, and cobs after separation of grains are used as base material during bio-gas production. Maize stalks and cobs, with or without grains are used to produce alcohol. Alcohol is then mixed in low concentration, at 8–10%, with gasoline. It enhances fuel efficiency and reduces carbon mission. It is known to increase octane rating. During recent years, maize is being increasingly used in bio-fuel production. It replaces gasoline to a certain extent.

Maize produces good quality forage for farm animals. It is consumed fresh or as hay in dried condition. Ensilage or fermentation of maize tops results in fodder of superior digestibility and palatability. The above ground portion of maize (stalks, leaves, and dry cobs after removal of seeds) could be removed from field and composted. The compost so prepared serves as organic manure. Upon incorporation in fields, it adds to soil carbon, minerals, and growth factors.

SUMMARY

Like most other crops, maize too was initially confined to area of its origin and domestication—that is Mesoamerica. It stayed in Mesoamerica and surrounding regions for considerably long period until human migrations to farther locations induced its diffusion. It is said that maize diffused into Great Plains rapidly and caused definite alterations in the food habit of the populace. During early medieval period, maize cultivation got intensified in the plains and it lead to formation of a large mono-cropping stretch in Northern plains—known better currently as "Corn Belt of USA." During 1500s, Spanish sea farers shipped maize to many locations of the world and caused drastic expansion of maize belt. During 19th and 20th centuries, maize cultivation progressively spread to as many locations. The intensity of maize cropping increased rather markedly with the improvement in water resources (irrigation), availability of fertilizers, and hybrids.

It is reasonable to believe that domestication, expansion, and later intensification of maize agroecosystem through the ages, must have commensurately influenced the nutrient dynamics of the ecosystem. We have no idea regarding nutrient dynamics during early Neolithic age, when maize cultivation was confined to relatively fertile and wet or marshy locations surrounding the human settlements. It is not easy to assess soil fertility or nutrient turnover rate that occurred during pre-historical, ancient, or even medieval times. Yet we can guess that, maize as a crop must have induced changes in soil fertility and nutrient dynamics within the surroundings. *In situ* turnover of nutrients at low subsistence levels must have determined crop growth and grain formation. However, spread of maize into Great Plains and New England zone during 1st millennium, must have induced clear and perceptible changes on not just landscape and food habit, but in addition affected soil nutrient dynamics, nutrient turnover, and maize productivity. Gradual improvement in maize cropping area and productivity during medieval period could have induced proportionate changes in soil fertility. In many locations depletion of soil fertility could have induced migrations and adoption of "shifting agriculture". During early to mid-20th century, mechanized farming, introduction of chemical fertilizers, hybrids and supplemental irrigations induced marked changes in soil and above ground nutrient dynamics. It improved forage and grain productivity. Nutrient recycling via crop residue got enhanced and net nutrient turnover improved enormously. Overall, a comparative study of nutrient dynamics prevalent during various stages of development of maize agroecosystem should be useful. Currently, it is relevant in the context of global climate changes and green house effects.

There are several reports that have assessed and forecasted global maize grain/forage requirements. They have often based it on trends in maize consumption, preferences, and possible shift in food habits of population in different continents. Computer-based models have added accuracy to such forecasts. A few of such forecasts made perhaps a decade ago or prior to it may need careful re-evaluation. They may not have considered the rapid upsurge in demand for maize to support bio-fuel production. At present annual maize production stands at 796 m t. Reports by IFPRI (2005) indicate that by the year 2020, global demand for maize would reach 837 m t annually. The developed regions need 501 m t, East and Southeast Asia–280 m t, South Asia 23 m t, Sub-Saharan Africa 52 m t, Latin America, 123 m t, and WANA 25 m t. So far, steady increase in both, maize area and productivity has satisfied the enhanced demand for maize products. This trend seems to continue in future. The average productivity of maize could be enhanced marginally from the present 4.2 t ha^{-1} to 5 t ha^{-1}. We should note that each extra ton of maize grain requires 18–21 kg N, 4–6 kg P, and 18–20 K and other nutrients. Nutrients impinged into fields should proportionately increase. It results in certain degree of alteration in nutrient dynamics in the ecosystem. Expansion of maize agroecosystem could be attained in Brazilian Cerrados, Pampas of Argentina, Sub-Sahara, India, and South Asian countries. Marginal increase in maize cropping zone could also be achieved in Eastern European plains. Obviously, in some areas, maize may proportionately displace a previously cultivated cereal. Intensification of mono-cropping zones leading to higher forage/grain harvest is yet another suggestion.

The main theme of the following seven chapters is to enlist our knowledge about consequences of maize cropping and its intensification on nutrients in the agro-environment.

KEYWORDS

- **Corn belt**
- **Maize agroecosystem**
- **Maize cultivation**
- **Maize culture**
- **Mono-crop**
- **Precipitation patterns**
- **Teosinte**

FOOT NOTE

*One bushel = 27 kg grains

Chapter 2

Soils of Maize Agroecosystem: Tillage and Nutrient Dynamics

Soil is an important ingredient of any agroecosystem. It provides anchorage and acts as repository of nutrients essential to sustenance of an agroecosystem. Soil regulates most of the important functions of maize agroecosystems and plays a vital role with regard to nutrient dynamics and productivity. The depth of soil influences extent of below-ground portion of an agroecosystem. It also affects below-ground nutrient storage and physico-chemical transformations. Soils harbor microbes and other biota that mediate innumerable biochemical transformations. Many of these transformations have direct bearing on nutrient dynamics in the ecosystem. The soil types, their textural class, soil quality, organic matter content, and mineral nutrients often referred collectively as soil fertility, is a crucial factor that decides perpetuation and productivity of an agroecosystem. Soil also holds the all important moisture. Soil moisture and its interactions with nutrients are key aspects that influence nutrient dynamics in the ecosystem. Aspects like nutrient availability to roots, physico-chemical transformation, recycling, accumulation, and losses are highly dependent on soil type and moisture regime. Soil moisture is required to support physiological activity of the crops both in the below- and above ground phase of the ecosystem. Maize agroecosystem is no exception to the above soil related phenomena.

As defined earlier, global maize agroecosystem is actually a conglomerate of several large or small stretches of maize cropping zones. Such maize cropping zones could be either mono-crops or crop mixtures. As such, it should be clear by now that maize occupies and thrives on several different types of soils (see Chapter 1). Soil types encountered in different continents vary enormously with regard to various physical, chemical and biological characteristics. The soil fertility and moisture related aspects too differ. The soil fertility related constraints to maize production may fluctuate in a given area and season. Over all, knowledge regarding soils, their characteristics and productivity in terms of crop yield is of utmost importance. Within the context of this book, it is difficult to discuss each and every soil type and its variations that are encountered all across the globe. Therefore, discussions are confined to major soil types found in large cropping zones like Corn Belt of USA, Cerrados of Brazil, Pampas in Argentina, European plains, tropical West Africa, Southern Indian plains, Temperate regions of Northeast China etc. Soil management practices adopted in a maize field need due attention, since they have direct impact on nutrient dynamics and productivity of a crop belt. Soil management practices like tillage, mulching, and amendments applied to correct soil maladies are discussed in greater detail.

Discussions within this chapter are intended to provide an introduction to various soil types that support maize culture. Basic facts about physico-chemical transformations that occur in soils are made available. Further, recent knowledge regarding influence of soil management procedures on nutrient dynamics that ensue in the maize belts has also been delineated.

SOILS OF MAIZE AGROECOSYSTEM

Maize belts across different continents thrive on variety of soil types that differ with regard to parent material, texture, structure, bulk density, pH, mineral nutrient distribution, organic matter, moisture holding capacity etc. Soil fertility *per se* and ability to support cultivation of maize and productivity levels attained also vary. Let us discuss the major soil types that support maize production in different continents.

Soils in Maize Belts of America's

The "American Corn Belt" thrives mostly on fertile soils classified as Udolls and Udalfs. Soil types, such as Mollisols, Aqualls, Ustolls, Fluvents, Aqualfs, and Haplaquepts also support maize farming (Table 2.1; Plate 2.1). Soils are inherently rich in organic fraction. Despite it, organic matter is often maintained at optimum levels through crop residue recycling. We should note that intensive cropping in the "US Corn Belt" necessitates removal of massive quantities of nutrients into above ground portion of ecosystem, mainly into grains/forage. Soil nutrient loss is much higher in "US Corn Belt" than in other maize growing regions of the world. Incessant cropping depletes soil fertility and diminishes SOM, hence, irrespective of soil type; much of corn growing region is provided with inorganic and organic fertilizer replenishments. In New England zone, maize belt thrives on Udults. Soil productivity is held high through incessant use of fertilizers and farm yard manure. Maize grain yield reaches 5–8 t ha^{-1}. Maize grown exclusively for forage may produce 25–30 t fresh forage ha^{-1}. In Southern Plains, maize thrives on Uderts and Ustolls. In the Southeast of USA (Georgia, Florida) maize is grown on Psamments, Quartzipsammets, Ultisols, and Aquults. Piedmonts of Georgia, Carolinas, and Virginia also support large scale maize farming.

Major soil types encountered in the Maize Agroecosystem of South America are Mollisols (Ustolls), Aridisols (Argids,), Ultisols (Udults), Alfisols (Ustalfs), and Entisols (Psamments) (Brady, 1995; Plate 2.2). Soils in Argentina comprise 8 orders of soil taxonomy, but Mollisols are most common in Rolling Pampas (Diaz et al., 1997). Entisols and Aridisols are other soil types used for maize production (Table 2.1). Cordoba region in Argentinean Chaco is the prime maize growing zone in Argentina. Maize is cultivated at several locations in "Gran Chaco" region of South America. Haplustolls and Hapludolls are more frequent in Southwest of Argentina. Soil erosion and runoff could be rampant in the agroecosystem. This may reduce soil organic carbon (SOC) in the top soil. Annually, 4–5% soil-N and 20% SOC could be lost from the soil profile due to erosion. Soils in the Cordoba region are classified as Hapludolls. They are light textured, loamy, and fertile. Soil fertility status is deemed moderately good. Inorganic nutrient supplements to maize fields are minimal. Fertilizer inputs to crops cultivated after soybean, such as maize or sorghum is relatively low. Nitrogen supplements to cereals grown after peanuts are usually small. Soils in the Cordoba region are sandy loams with organic matter content ranging from 1 to 3%. They are low in available-P, but high in K. Mollisols are also used to cultivate peanuts. They are loamy with 1–3% organic matter and once again low in available-P but high in K content. Intensive cultivation practice has lead to soil deterioration. Wind and soil erosion, loss of organic matter, and soil crusting are major soil related constraints to nutrient recovery (Garbulsky et. al., 2005; Pedelini, 2002; Riveros, 2005). Incessant cultivation of soil without proper nutrient replenishment has caused severe imbalances in availability of nutrients to plant roots.

Plate 2.1. Left: A view of soil profile (Mollisol) found in maize growing regions of Iowa (US Corn Belt). Mollisols support intensive cropping of maize. They are impinged with large amounts of chemical fertilizers and FYM in order to achieve high yield goals envisaged by farmers. Right: A soil profile from Northern Illinois. These soils classified as Endoaquolls are black silty clay loams and extensively used to produce maize.
Source: Dr. Jerry Hatfield, National laboratory on Agriculture and Environment, Iowa, USA; National Resource Conservation Service, USDA, Washington DC.

Plate 2.2 Maize is predominantly grown on Hapludolls (Mollisols) (Left) and Haplustols with Calche (Right) in the Pampas of Argentina.
Source: Dr. Alberto Quiroga and Dr. Alfred Bono, National Institute of Agricultural Technology, Anguil, Buenos Aires, Argentina.

In the Western Chaco region, soils are loamy but slightly acidic or neutral in upper horizons. Soils are suitable for maize cropping as well as inter-cropping it with leguminous trees like *Prosopis*. Soils are generally fertile, but water is limiting. Soil types and their characteristics encountered in the Paraguayan Chaco are as follows:

Soil Type	Texture			pH	EC	SOM	P	Ca	Mg	K	Na
	Sand	Silt	Clay		dSm	% mg kg⁻¹ soil	-------	-------	%	-------	
Luvisol	31	28	41	6.8	0.3	2.6	78	7.1	3.1	1.4	0.2
Cambisols	35	37	28	6.6	0.2	2.2	59	2.0	1.6	0.7	0.3
Solonetz	15	29	56	7.3	2.7	2.9	73	8.9	3.2	1.3	2.1

Source: Glatzle et al., 1999.

Note: Other soil types encountered are Arenosols, Regosols, Gleysols.

The Cerrados of Southeastern Brazil is a vast expanse of savanna vegetation and agricultural enterprise. It has predominantly a drier sandy or loamy stretch of soil. Highly weathered Oxisols dominate covering up to 47% of Cerrado region (see Table 2.1). Other major soil types encountered are Ultisols (15%) and Entisols (15%). These soil types exhibit serious limitations with regard to agricultural cropping and are classified as low in soil fertility and mineral contents. The Oxisols are acidic and low in available N, P, and K. Secondary nutrients such as Ca, Mg, and S are also deficient. Micronutrients like Zn, B, Cu, and Mo are found in concentrations less than threshold for maize/soybean production (Sanchez, 1978; Lopes, 1996). Major constraint to maize production on these soil types is the high Al content. It leads to P-fixation and reduction in fertilizer efficiency. High Al content actually affects rooting and nutrient acquisition by maize. Regarding soil acidity, since most soils are highly acidic with pH 4.5–5.5, correction using lime is almost mandatory in any region of Cerrado. Liming essentially increases soil pH and at the same time corrects Al toxicity. It is useful to apply Dolomitic lime because it contains both Ca and Mg. Firstly, pH gets corrected plus it adds Ca and Mg that are usually deficient in soils of Cerrados. Actually Al/(Ca + Mg) ratios in soils are carefully assessed before planting maize or other crops. On virgin soils, a large doze of lime is applied to correct pH and Al toxicity. One way of improving soil P status is to apply fertilizer-P in quantities slightly more than that required for crop growth and yield formation. It helps in satisfying P fixation and then allows some to accumulate in soil, so that P deficiency does not get expressed repeatedly. Similar principles could be adopted in building N and K reserves in maize cropping zones.

Table 2.1. Descriptions of major soil types encountered in various maize cropping zones of the world.

Corn Belt of USA: Soil types—Ustolls and Udolls

Kastanozems (Kastanozems—WRB[*]; Ustolls—US[#]; Kastanozem—FAO[$]):

Kastanozems are found in steppes or prairies. They have thick dark brown top soil. Kastanozems are chemically rich with saturated bases. Soil reaction (pH) may be slightly alkaline. Soil aggregates in a kastanozem is susceptible to erosion through wind and water. Kastonezems are commonly found in the North American plains and in Pampas of Argentina (Beyer, 2002). Potentially, Kastenozems are fertile soils. Yet, crop production may get constrained due to paucity of water. Irrigation needs to be regulated since it could induce salinization. Adequate tillage and repeated fallows are common methods to avoid deterioration of soil fertility. Soils in the "Corn Belt of USA" have been consistently cropped. Nutrient inputs have been steadily increasing in order to replenish massive removals by high yielding corn and rotation crops (e.g., cotton or soybean). Nutrient turnover rates are naturally high. Maize productivity is high at 6–8 t grains ha^{-1} plus 10 t forage ha^{-1} in the Corn States of USA, but moderate at 3–4 t grain ha^{-1} plus 6–8 t forage ha^{-1} in Pampas of Argentina.

Chernozems are also encountered in Northern plains of USA. These soils are rich in organic matter, nutrients, and biotic component. Several of the soil physico-chemical properties are congenial for crop production. Soil aggregates are stable and relatively resistant to wind and water erosion. They possess fertile surface horizon, hence support maize production excellently.

Cerrados of Brazil: Soil types—Luvisols, Alisols, Oxisols, and Laterites

Cerrados of Brazil is a large stretch of semi-arid plains in the eastern half of the country. It supports large expanses of dry land cereals, legumes, and pastures. Maize is an important cereal grown mostly on Alfisols, Ultisols, and Oxisols,—see descriptions under South Indian plains for details on Alfisols.

Udalfs (Luvisols—WRB; Udalfs—US; Luvisol—FAO):

Luvisols are characterized by surface horizons depleted of clay fraction and nutrients. They exhibit subsurface accumulation of high clay activity. Most traits of Luvisols favor crop production particularly if well drained. Luvisols contain minerals and organic fraction sufficient to support cropping. However, Drystic Luviosls may need amendments with lime to improve fertility. They are sometimes associated with Argids, especially in dry land belts. When situated closer to water sources, these soils are associated with gleyic or stagnic luvisols.

Alisols (Alisols—WRB; Ultisols—US; Alisols—FAO):

Alisols have brown colored surface horizon. Alisols possess high clay activity and Al exchange capacity. Alisols are less weathered compared to Lixisols or Nitisols. Usually, they are well drained but deficient in plant nutrients. Al toxicity in surface horizons is very common. Soil erosion is rampant. Crop productivity is generally low on Alisols. Nutrient buffering capacity is low hence fertilizer and lime supply are necessary.

Ferralsols or Oxisols (Ferralsols—WRB; Oxisols—US; Ferralsol—FAO):

Oxisols are highly weathered mineral soils rich in sesquioxides and well suited for cropping. They have a ferralic horizon extending from 30 to 200 cm depth. Such feralic horizon results from intensive weathering (ferralitization). The clay fraction is dominated by low activity clays. The silt and sand are constituted by highly resistant goethite, gibbsite, and hematite. The physical aspects are quite congenial for maize production. Yet, they have low nutrient reserves and high exchangeable Al and Mn. Major constraints to maize cropping are acidity and Al toxicity. These factors actually result in improper rooting and low P uptake. Chemical fixation of P gets accentuated due to high Al^{++} and Mn^{++} activity. Liming and higher dosages of P fertilizers are usually recommended. Split application of nutrients improves crop growth and yield. Deficiencies of Mg, Zn, B, and Mo are encountered. Maize productivity depends much on fertilizer and organic matter supplements.

Pampas of Argentina: Soil types—Mollisols (Ustolls), Ultisols, Entisols, and Aridisols (Argids)

Mollisols: see Ustolls under Corn Belt of USA

Ultisols: see Alisols under Brazilian Cerrados

European Plains: Soil types: Udalfs (Luvisols); Cambisols; Fluvisols; and Chernozems

Cereal production in the European plains occurs on soil types classified as Cambisols, Luviols, and Fluvisols in Western Europe; on Podzols and Black Earths in Central Europe and Russia; and on Calcixeralfs in Mediterranean region of Europe.

Chernozems: (Chernozems—WRB; Udolls—US; Chernozem—FAO):

Chernozems are most commonly encountered in the maize cropping zones of Europe including Russia and parts of Eastern European plains. Soil physico-chemical properties like bulk density, porosity, water holding capacity, cation exchange capacity (CEC), high SOM and neutral pH are congenial for maize cultivation. Chernozems are blackish soils rich in organic fraction and prominent calcarious subsurface. A large part of temperate European maize belt is supported by such Chernozems. Chernozems possess deep, very dark grey, humus, and nutrient rich surface horizon. They are supposedly few of the best soil types for agricultural cropping (Blume, 1998; Beyer, 2002). The biotic fraction of soil is well stabilized and this results in surface horizon as thick as 2 m. Therefore, it supports good rooting of maize crop leading to high amounts of nutrient absorption. The soil aggregates are relatively more stable than those encountered in Kastanozems. Soil aggregates easily withstand wind and water erosion.

Udalfs (Luvisols—WRB; Udalfs—US; Luvisol—FAO): Descriptions for Luvisols or Udalfs are available under Cerrados of Brazil.

Podzols (Podzols—WRB; Spodosols—US; Podzol—FAO):

Podzols are quite frequently used for cultivation of cereal crops in Eastern European and Russian plains. They are acid soils with blackish-brownish-reddish subsoil with Illuvial Fe-Al-organic compounds. The name podzol means soil is with subsurface that has appearance of ash due to strong acid aided bleaching. It has a spodic Illuvial horizon. Podzols could be sandy or loamy. Those with coarse texture obviously have low water retention capacity. The CEC is moderate or low depending on SOM content. The C:N ratio ranges from 20 to 25. Podzols of Eastern European plains could be infertile and support moderate maize yield despite fertilizer application. This is attributed to sandy texture, low water retention, and low soil fertility status (Beyer, 2002).

Solonetz (Solonetz—WRB; Several soil orders—US Solonetz—FAO):

Maize agroecosystem in Ukraine, Russia, and parts of Eastern Europe thrives on Solonetz or saline humic soils. Solonetz possess subsurface horizon with clays rich in sodium content. Actually, inherent salt contents, nature of parent material and ground water together induce formation of solonetz or saline/alkaline soils. High Na content and pH are major constraints to maize culture. Usually, salt tolerant cultivars are preferred. Maize productivity depends on nutrient and water supply and measures that correct Na effects. Generally, maize is grown for forage on these soils.

West African Savannas and Guinean Zones: Soil types—Sandy Oxisols, Ultisols, and Alfisols

For descriptions on Oxisols see under Brazilian Cerrados

West Asia and North Africa: Soil types—Aridisols and Cambisols

The cropping regions of West Asia are endowed with Entisols, Mollisols, Vertisols, Inceptisols, and Aridisols. North Africa is dominated by Inceptisols, Lithosols or Shallow soils, Entisols and Aridisols.

Aridosol (Arenosols—WRB; Psamments—US; Arenosol—FAO):

Aridosols are common in the dry land cropping belts of West Asia and India. Aridosols are rich in salts. Such salt accumulation can impair nutrient recovery leading to depressed growth. Major plant nutrients are very deficient. Boron toxicity is quite frequently reported in West Asia and Semi-arid India. Maize crop needs both fertilizer supplements and irrigation in order to reach moderate level of productivity (1.5–2.5 t grain ha^{-1}).

Cambisols: (Cambisols—WRB; Inceptisol—US; Cambisol—FAO):

Cambisols possess horizons that indicate moderate weathering. They are difficult to characterize because of large variations in physical and chemical traits. They could be sandy, silty, or clayey depending on parent material. In general, Cambisols are good for cropping, particularly when they are rich in bases. The acid and coarse textured Cambisols could be improved by fertilizer and organic manure addition.

Southern Indian Plains: Vertisols and Alfisols, other soil types are Inceptisol and Entisol

Vertisol (Vertisol—WRB; Vertisol—US; Vertisol—FAO):

Maize belt in South India occupies large tracts of Vertisols, especially in North Karnataka and Andhra Pradesh. They are commonly called Black cotton soil or *Regur*. Vertisols are deep, black colored, clayey (30%) soils. Clay minerals like montomorillonite, smectite, and illite predominate. These Vertisol stretches in India are fertile and productivity is relatively high. They show remarkable swell and shrink characteristics. Vertisols show tendency to crack during drought or paucity of moisture. Nutrient and moisture buffering capacity is relatively high. Soil becomes very sticky during rainy season. Therefore, agronomic procedures like tillage, planting, and interculture could become difficult or impossible during early *kharif* season. Post rainy crops of maize are possible because of better moisture storage capacity. Maize productivity ranges from 3 to 5 t grains ha^{-1} and proportionate forage depending on cultivar.

Alfisol (Acrisol—WRB; Alfisol—US; Alfisol—FAO):

The major portion of maize belt in South India thrives on Alfisols or commonly called Red soils. Alfisols are deep, either loamy or gravely at times with relatively higher Fe content. Soils are rich in mica, quartz, feldspar, and hematite. Moisture holding capacity is moderate or low. Tillage induces hard pan and loss of SOC. As such, SOM is low owing to high soil temperature, oxidized state, and semi-arid environment. Soil (top layer) and nutrient loss through erosion, seepage, and percolation could be high. Moisture infiltration rates are high due to sandy/gravely nature of Alfisols stretches of South India. Maize productivity ranges from 800 kg to 4000 kg grain ha^{-1} plus 3–10 t forage ha^{-1} depending on fertilizer and irrigation supplements.

Southeast Asia and China: Soil types—Alfisols, Entisols, Inceptisols, and Ultisols

Alfisol (Acrisol—WRB; Alfisol—US; Alfisol—FAO): Descriptions are provided under South Indian plains.

Alisols (Alisols—WRB; Ultisols—US; Alisols—FAO): see under Brazilian Cerrados.

Cambisols: (Cambisols—WRB; Inceptisol—US; Cambisol—FAO): see under West Asia and North Africa.

Soils of Australia: Soil types—Red and Yellow Ferralsols

The major soil types of Australia where maize is cultivated are classifiable as Quartz-ipsamments, Calciorthids, Palexeralfs, Haplaustalfs, Paleusterts, Chromousterts, Torrerts, and duplex soils (Rhodoexralfs, Natustalfs, and Paleustalfs). Black Earths of Queensland, Red and Yellow Earths of Western Australia, and New South Wales are also used for maize cultivation.

Ferralsols (Ferralsols—WRB; Oxisols—US, Ferralsols—FAO)

Feralsols possess a ferralic horizon at some depth—say 20–30 cm from surface. The ferralic horizon is derived from long and intensive weathering commonly termed as "Ferralitization". High ambient temperatures and rainfall induces formation of ferralic horizon. The Ferralsols have a weakly expressed soil structure. The clay fraction is constituted by low active clays. The Ferralsols are amenable for maize production. They are less prone to erosion due to good depth and stable structure. Chemically, Ferralsols have variable CEC that is dependent on pH. Nutrient supply capacity decreases with cultivation. Liming becomes essential under acidic conditions. The Red Ferralsols of Queens land has been incessantly tilled and cropped. Tillage has induced loss of soil structure and fertility. Above it, organic matter recycling tends to be low causing loss of SOC. Periodic droughts have accentuated loss of soil fertility and quality.

Sources: Sanchez, 1978; Brady, 1995; Beyer, 2002;. Gerpacio, 2002; FAO-AGLW, 2002; Krishna, 2003.
Note: WRB*= World Reference Base; US# = United States Soil Taxonomy; FAO$ = Food and Agricultural Organization's soil taxonomy.

Soils of European Plains that Support Maize Culture

Maize is cultivated in different locations with in the European plains. It serves as a grain and forage crop. Major soil types encountered are Cambisols in Britain and France, Luvisols in Germany, and Fluvisols in Netherlands (see Table 2.1). These soils are arable and suited well for maize production. In the Central Europe, major soil types encountered are Loess soils, Podzols, Light Brown Steppe soils, Black Earths, and Sandy soils. Chernozems are predominant in Russian and Eastern European plains. In Russia, nearly 48% of soil is classifiable as Chernozems, 14% as Podzols, 10% as Solonetz, and rest as Grey soils. Chernozems are frequent in Ukraine, Bylorussia, Lithuania, and Estonia. In the Czeck Republic and adjoining maize growing regions Chernozems, Fluvisols, and Luvisols are encountered (Maly et al., 2002). In the Mediterranean region of southern Europe, calcareous soils with sediments are utilized to produce maize. Argillic Cambisols, Alfisols, and Entisols are most common in the Mediterranean Europe.

Soils of West Asia and North Africa

Maize is cultivated as an intercrop or rotated with other major crops like wheat or barley, lentils, berseem etc. It is mostly grown under low input subsistence farming conditions. In general, regions that support maize culture in North Africa (Algeria, Sudan, Egypt, and Libya) possess Alfisols (Ustalfs, Psamments). Agricultural soils in Egypt, where in wheat, maize, and peanuts get rotated are moderate in fertility and invariably deficient in organic matter. Further, in newly reclaimed zones, organic matter content is <1%. Such sandy soils are calcareous with high $CaCO_3$ content. In other parts of North Africa, maize grows on Aridisols, Entisols, Inceptisols, and Vertic Inceptisols (e.g., in Morocco). The productivity of soils in North Africa depends much on fertilizer replenishments, crop rotations and most importantly irrigation. In the Middle-East, especially in Syria, Iran, Iraq, and Israel, soils encountered are Xerolls (Xerorthents) and Cambisols. They are calciferous soils. Actually, such gypsiferrous soils are common in the semi-arid and arid regions of West Asia. These soils are low in fertility, especially major nutrients and organic matter. Deficiency of micronutrients like Zn and B could also be discerned. In WANA region, maize is mostly confined to relatively wetter regions and productivity ranges from 2.5 to 3 t grains ha^{-1} plus forage 5 t ha^{-1}.

SOILS IN MAIZE GROWING REGIONS OF WEST AFRICA

Soils that support maize cultivation in West Africa are mostly sandy and classifiable as Paleudalfs, Tropequents, Eutropepts, and Dystrpepts (see Table 2.1; Plate 2.3).

Plate 2.3 Sandy Oxisols of West African Cereal Production zones.

Above: Sandy soil sown with cereal in transitional zone of Southern Niger. Below: Cereal roots are easy to harvest.
Note: Profuse rooting and tight adherence of soil particles. It is known to conserve soil moisture. Drier areas support pearl millet while wetter zones are planted with sorghum or maize. Sandy soils are prone to wind erosion and loss of nutrients via percolation, seepage and emissions. Cereals root profusely in order to explore and enhance soil moisture extraction. Soil moisture retention is very low owing to sandy nature and low organic matter content. Loss of C as CO_2 is accentuated due to high soil temperatures and microbial activity.
Source: Dr. K.R. Krishna, ICRISAT, Hyderabad, India.

The percentage sand fraction is high reaching 35–40%. The sandy soils are prone to erosion through wind and water. Erosion is high immediately after a precipitation event. Percolation of nutrients and water to subsurface could also be severe. Soil nutrient status is highly variable, especially in area where crop production trends are still under subsistence level. Soil organic component, so important for cropping is in fact scanty. The SOC ranges from 0.3 to 0.6% much below the minimum requirement of 3–5%. Further, clay fraction of sandy soil that supports maize culture is predominantly kaolinite. Therefore, cation exchange capacity (CEC) depends more on soil organic fraction. Soils are acidic and pH ranges from 4.2 to 6.5. Base saturation ranges from 55 to 86%. Residue recycling, mulching and farm yard manure (FYM) addition are key techniques that sustain West African maize farming. Deficiency of nutrients, in general, and more specifically major nutrients like N, P, and K are rampant. Total-N in soil may range from 75 to 180 mg N kg^{-1}, but available pool is low at 2–18 mg N kg^{-1}. The availability of major nutrients decreases with soil depth. Phosphorus deficiency is easily corrected by addition of fertilizer P, but response to its addition is immensely controlled by other soil traits such as SOC content, mulching and residue recycling. In the Savanna and Guinea region, soil exhaustion could be rampant, hence nutrient replenishment schedules and fallows are important (see Buerkert et al., 2002; Krishna 2008; Yost et al., 2002).

Soils of South African Maize belts

In South Africa, maize is cultivated on a wide range of soil types. Sandy soils with <10% clay and 30% sand provide arable conditions that allow moderate levels of maize harvest. Clayey loams with 10–30% clay are also amenable to raise good crop of maize. The productivity of such clayey soils is generally high because of better water and nutrient buffering capacity. In South Africa, usually light or medium textured soils are preferred by maize farmers. Soil erosion and nutrient loss could be rampant in regions where subsistence farming is practiced. As such, heavy precipitation can cause considerable loss of nutrients from the maize fields. Carbon sequestration methods are important, since inherent SOC is low (0.5–1%) in most of these sandy soils. Soil pH ranges from 5.5 to 7.8. Soil-N content is to be replenished using chemical fertilizers and organic manures. Total and available P pools too are meager to support a high yielding maize crop. Phosphorus replenishment should satisfy chemical fixation as well crop's need. Soil K might be optimum, yet incessant cropping could deplete this element swiftly.

Soils of South Indian Plains and Hills used for Maize Production

Maize is cultivated on a wide range of soil types in the Indian subcontinent. In the Indo-Gangetic plains, maize agroecosystem occupies Inceptisols, Entisols, and Alluvial soils. Maize is rotated with wheat or rice in the North Indian plains. It is frequently intercropped with legumes and vegetables. In the Coasts, sandy soils support maize production. Maize agroecosystem of South India is accentuated in the dry land region and it is supported predominantly by Alfisols and Vertisols (Plate 2.4). Maize is a major cereal intercropped with cotton or legumes in the Black soil region of North Karnataka and Andhra Pradesh. Incidentally, Karnataka and Andhra Pradesh are major

"Corn States in India". Together they contribute nearly 30% of nation's annual maize produce. The Vertisols (or Black Cotton soils) are deep, clayey (montomorillonite) and rich in plant nutrients. They are slightly alkaline in reaction. The CEC is high. Therefore, nutrients are buffered better and moisture holding capacity is again high. They exhibit swell-shrink characteristic, hence drought spells may affect soil structure and rooting. The productivity of maize is relatively high in the Vertisol belt. It ranges from 3 to 4 t grain plus 6–8 forage ha^{-1}. Maize cultivation proceeds in all three seasons namely—rainy, post rainy, and summer depending on precipitation and/or irrigation.

The Alfisols, more commonly known as Red soils or Red Sandy loams, are wide spread in South Indian plains and well suited to maize culture. Maize is intercropped with variety of legumes, oilseeds, and vegetables. Maize culture in South Indian plains is accomplished predominantly during rainy season (*kharif*). About 80% of maize production in the Indian subcontinent is obtained from *kharif* harvests. On Alfisols, maize productivity is dependent on inherent fertility status, fertilizer replenishments and most importantly precipitation pattern during *kharif* season (June to October). Productivity of Alfisols ranges from 2 to 3.5 t grains plus 5–8 t forage ha^{-1}. Alfisols are deep, mostly sandy or loamy, low in organic carbon, rich in Fe and Al salts. Soil pH ranges from 5.5 to 7.5.

Plate 2.4. *Top:* A Typical Red Alfisol field at Hebbal Agricultural Experimental Station, University of Agricultural Sciences, Bangalore in Karnataka,—a Corn State of South India. Maize is produced on such Red Soil strips during *kharif* season. Maize is often intercropped or rotated with legumes, mainly Pigeonpea, Cowpea, Field bean and Horse gram. Bottom: A Vertisol field prepared for sowing maize near Thungabadhra dam Project Area in the South Indian Plains.
Source: Dr. Krishna, Bangalore.

Alfisols are deficient in N and P but rich in K. In addition, chemical fixation of nutrients like P could reduce fertilizer-P efficiency. Nutrient percolation to lower horizons is rapid. Nutrient loss through seepage, surface flow and erosion can be rampant, especially after a heavy precipitation event. Alfisols found in the dry land regions of South India are prone to SOC loss via oxidative processes. Hence, repeated application of FYM is essential. During past decade, maize agroecosystem that thrives on Alfisols and Vertisols has expanded consistently and the spree continues. In many places it has replaced erstwhile crop species like sorghum or pearl millet. It is attributed to better grain harvests, lots of forage that serves to feed farm animals and good remunerative price because of Industrial uses. Obviously, chemical fertilizers and farm yard manures are being impinged into soils at a greater intensity. Its impact on long term upkeep of soil characteristics needs investigation.

Lateritic soils are predominant in maize growing fields of Western Ghats of South India. Lateritic soils are highly leached due to incessant rainfall. They are deficient in many of the essential nutrients. They are rich in Al and Fe compounds. Often, maize is grown on slopes (terraces) or valleys as a summer crop for forage or as an intercrop with legumes.

In the West Coast of South India, maize farming occurs on laterites. The laterites are low in fertility and prone to heavy leaching due to incessant rains. They are rich Al and Fe salts that resist leaching. Maize is mostly intercropped with vegetables under the prevailing tropical conditions.

Soils in Maize Cropping Zones of Southeast Asia

Soil types encountered in Chinese maize cultivation zones are Alfisols, Entisols, Inceptisols, and Ultisols. In the Northwest, Aridosols and Orthents are utilized to grow maize. In the Northeast, maize cultivation occurs on isohumisols with long history of cropping. In Shandong and nearby regions, maize is grown on gravely soils derived from weathered granites, shale, and gneiss. In Jiangxi, Hennan, and Hubei provinces maize is produced on red and yellow acidic soils. In Myanmar, major soil types encountered in the maize cropping zones are Fluvisols, Gleysols, Lithosols, Cambisols, Andisols, Vertisols, and Luvisol. Rice is the major cereal, yet maize is found intercropped with legumes like black gram or peanuts. Ferralsols acidic in reaction are most frequent in Vietnam. Around Mekong delta soils are heavy.

Soils used for Maize Culture in Australia

Soil types utilized to produce cereals like wheat and maize are generally coarse textured Quartzi-Psamments or Torri-Psamments. Calciorthids, Palexeralfs or Haplustalfs termed as earths are also good for maize production. Cereals are also cultivated on Duplex soils (Rhodoxeralfs, Natustalfs, and Paleustalfs) (Perry, 1992). In Western Australia, cereal crops thrive on duplex and sandy soils. These are generally low in organic-C and have restricted soil fertility status. Application of fertilizers is important on such soil types. The Brown soils of New South Wales, Victoria and South Australia are used mainly to produce wheat, but maize is also cultivated for grain and forage, although in relatively small areas. The Black Earths of Queensland supports large tracts of cereal production that includes maize. In addition, Australian cereal belt contains Yellow and Red Earths, especially in Western Australia and New South Wales. In Northwest Australia, Vertisols rich in exchangeable-K have supported maize production. However, recent reports suggest that incessant cropping has lead to depletion of K. Potassium deficiencies have been observed (Carter and Singh, 2004).

PHYSICO-CHEMICAL TRANSFORMATIONS OF NUTRIENTS IN THE SOIL: A SUMMARY

Physico-chemical transformations that occur in maize fields are of great importance, since they influence extent of nutrients available to maize root; extent fixed into soil crystal structure or organic fraction; that lost from fields through percolation, erosion, and seepage; also that lost to atmosphere via volatilization. Maize is an arable crop. Soils are held in oxidized state during maize cultivation except immediately after an irrigation or heavy precipitation event that causes inundation for a while. Nutrient transformations that occur in arable soils literally have a major say on nutrient dynamics in the ecosystem. In this section, it is intended to provide salient facts and outlines about physicochemical processes involving major nutrients essential for maize cultivation.

Major Nutrients: Nitrogen, Phosphorus, and Potassium

The nitrogen status of soils used for maize production varies enormously. Accumulation of soil-N is possible in intensive cropping zones where fertilizer is replenished repeatedly, for example in Corn Belt of USA or European plains or Northeast China. On the contrary, incessant cropping with meager nutrient replenishment can reduce soil-N status drastically. Depletion of soil-N is a feature in many areas that support maize production. Deficiency of soil-N could be rampant in Oxisols of Latin America or Alfisols of South India. Whichever be the soil type, there is a strong need to understand about various forms of soil-N, physicochemical transformations involving soil-N and N availability to roots during various stages of crop. Major forms of N encountered in soils from maize cropping zones can be grouped into inorganic and organic forms. They are as follows:

- NH_4-N is found in exchangeable form adsorbed to clay surface and colloids or it is traced as fixed NH_4-N in the crystal lattice;
- NO_3-N is found in soil solution, thin layers on root and in rhizosphere. It is highly mobile and prone to loss via leaching and percolation;
- N_2O-N is feebly traced and is confined to pockets of reduced layer in soil. Where as, N_2 and N_2O are found in the soil air space and solution.
- Organic form of N is traceable in amino acids, amino sugars, nucleic acids, and humins.

Nitrogen transport in soil is mediated through diffusion and/or mass flow. It affects availability of soil-N near the root zone. Movement of soil NO_3 is governed by mass flow and NH_4-N by diffusion. In the arable soils used for maize production NO_3-N translocates faster than other forms of N. Incidentally, maize extracts N mostly in NO_3 form. On the contrary, in wetlands with anaerobic conditions, a crop like rice absorbs most of its N in NH_4-N form. Therefore, diffusion of NH_4-N becomes important during lowland rice cultivation. Coming back to maize fields, percolation of soil-N occurs rapidly and it can remove sizeable quantities of N away from root zone. Runoff could also be severe in maize fields. Runoff removes soil and soil-N with it. Factors such as percolation, seepage, and runoff can together reduce fertilizer-N use efficiency in maize fields. Sandy soils are prone more to loss of N through percolation. Where as, clay loams with better CEC and those with larger organic fraction may retain soil-N better. Sorption and desorption are also important physico-chemical processes that affect NO_3-N availability. In most arable soils NO_3-N is weakly held hence it is easily available to maize roots. Ammonium-N is held to soil lattice structure. Mulching, spot or band placement of fertilizer-N reduces loss of N from root zone. Volatilization of soil-N or fertilizer-N as NH_3-N, N_2, or NO_2 is an important process that results in loss of N from maize cropping belts. Ammonia volatilization is dependent on soil-N status, humidity, temperature, and wind. Ammonia volatilization gets accentuated under warm tropical conditions. It is said loss of soil-N via volatilization increases by 0.25% for every rise of temperature by 1.0°C. Maize belts may loose 20–40% of fertilizer-N input through volatilization, depending on environmental factors (Hood, 2002). As an example, volatilization of N could be relatively more in tropical/semi-arid regions of India, West Africa, in Cerrados of Brazil, and in dry regions of Pampas of Argentina.

Mineralization is a soil chemical process that affects N dynamics in the maize agroecosystem. Mineralization involves conversion of organic forms of N to mineral-N. Mineralization occurs both in aerobic and anaerobic microsites of the soil. Mineralization reactions in soil culminate at formation of NH_4-N in wetlands. Hence, it is termed as "ammonification." Where as, in arable soils that support maize cultivation, mineralization continues further leading to formation of NO_3-N. In maize fields, oxidative conversion of soil organic N forms to mineral-N predominates. The mineralization mediated by aerobic soil microbes involves conversion of organic-N to NH_4-N and then to NO_3-N.

Mineralization is affected by various factors related to soil (clay, colloids, aggregation, CEC, C:N ratio, pH, salinity, soil moisture, temperature, and microbial flora), cropping procedures (tillage, liming, rotations) and environment (precipitation, temperature, etc.). Mineralization and release of inorganic-N is rather rapid in tropical and semi-arid zones because of warm temperature, oxygenated soil, and enhanced microbial activity (Krishna and Rosen, 2002). Tillage may further accentuate microbial activity and mineralization rates. Maize roots explore substantial quantities of N released through mineralization. Usually, this fraction is called N_{min}. In fact, we should regularly consider the extent of N_{min} derived while deciding on fertilizer N supply to the crop.

Soil microbes mediate immobilization of mineral N (NO_3-N or NH_4-N) applied to soils into organic forms. Fertilizer-N applied to soil is firstly immobilized into soil microbes. Mineralization and immobilization reactions are usually coupled. Immobilization/mineralization reactions are dependent on C:N ratio, SOC, soil-N status, soil microbial load, moisture, temperature, pH etc.

Nitrification is an important soil chemical transformation that has far reaching effects on N dynamics in the root zone. Nitrification involves conversion of NH_4-N to NO_3-N. It is mediated by soil microbes that are collectively termed as Nitrifiers. Maize fields are generally held in oxidized state. It aids rapid conversion of NH_4-N to NO_3-N. *Nitrosomonas* sp. mediates conversion of $NH_4^+ + 1\frac{1}{2} O_2 \rightarrow NO_2 + H_2O$ and *Nitrobacter* spp. mediates $NO_2 + \frac{1}{2} O_2 \rightarrow NO_3$. Nitrification supports build up of NO_3 in soil that could be easily absorbed by maize roots. However, in flooded soils or under inundated conditions that ensue immediately after an intense precipitation event, NH_4-N is not converted to NO_3-N. Accumulation of NO_3-N makes it vulnerable to loss via percolation, seepage, or volatilization. Hence, nitrification inhibitors (e.g., Ethyle pyridine, Etridiazole, Nitrapyrin) are added. Nitrification inhibitor that persists in fields restricts or delays conversion of NH_4-N to NO_3-N. It avoids accumulation of NO_3-N.

Soil P is an important ingredient that affects growth and productivity of maize belts. The soil P status varies widely in different soil types encountered in the maize cropping zones. Soil P is deficient in sandy soils of South America and West Africa. Alfisols found in South Indian plains too suffer from P deficiency. Soil P could be optimum or above threshold requirement of crops in areas like European plains or US Corn Belt.

Soil P exists in different forms. They can be grouped into inorganic-P and organic-P forms. Inorganic-P exists as salts of orthophosphoric acid. Inorganic P forms encountered are Ca-P, Al-P, and Fe-P. Incessant cropping depletes soil P drastically. Among various forms of P in soil, plant available-P is important with regard to maize cultivation. The above inorganic forms of P may constitute between 55 and 60% of total-P, depending on soil type. Insoluble-P is mostly attached with Al and Fe. The Al-P and Fe-P are abundant in sandy soils of South America and Alfisols of South Indian Plains. However, in soils with pH 6.5–7.5 or above, Ca-P predominates. The organic-P forms occur mostly as inositol-PO_4, phospholipids, nucleic acids, and their derivatives. The total-P, organic-P and available-P are usually higher in the surface layer (0–15 cm). The total organic-P in soils varies enormously. It is dependent on soil organic fraction, cropping history, residue recycling procedures, and P inputs to the maize fields.

The behavior of P in most soils that support maize cultivation could be explained as follows: Solution P $\leftrightarrow$ Labile P $\leftrightarrow$ Non labile P. Maize root absorb soluble P found in labile fraction. Non-labile P occurs in chemically bound organic and inorganic form. The threshold available-P in soil may vary widely depending on crop species. In most soils, available-P <5–10 ppm P is considered as low, between 10 and 20 ppm P as medium and >25 ppm P as high.

Chemical fixation of P in soil is an important phenomenon that reduces its availability to roots. Chemical fixation also affects efficiency of fertilizers applied to soil. The fixation involves reactions that incorporate P into crystal lattice structure of soil. The extent of P fixation is dependent on soil type, especially clay contents, Al-P and Fe-P fractions. In the Alfisols, chemical fixation of fertilizer-P due to Al and Fe fraction ranges from 20 to 40%. In the maize belts of South America, Al and Mn toxicity may affect absorption of P by maize roots. Hence, fertilizer-P recommendations are calibrated to satisfy both Al and Mn mediated fixation plus amount required for yield goals envisaged.

Soil P undergoes variety of physico-chemical transformations. Such transformations affect available-P. Conversion of inorganic-P or fertilizer-P into organic forms constitutes immobilization of P. Soil microbes assimilate inorganic-P and convert them into organic-P. Usually, immobilization occurs if Org-C/Org-P ratio is 300:1 or more. Where as, if Org-C/Org-P ratio is less than 200:1, then mineralization of P sets in releasing P into soil solution. Mineralization of organic forms of P either prior to or during maize cultivation is helpful. It improves available-P pool in soils.

Rock phosphates are P ores. They could be used directly in powdered or granulated forms. They are often partially treated with sulfuric or phosphoric acid. Acidulating rock phosphates releases P into soil solution. It increases P availability to maize roots. The solubilization of rock phosphates in soil depends on factors like SOM, pH temperature, etc. Soil microbes too mediate dissolution of rock phosphates by releasing various organic acids. There are alternate mechanisms to explain solubilization of P in soils that support maize production. One mechanism suggests that microbial exudates and organic acids get adsorbed into clay particles. Therefore, it decreases sites for inorganic-P to adsorb. Another mechanism suggests that adsorption of Ca^{+2} causes shift in mass equilibrium of sparingly soluble Ca-P.

Dynamics of soil K has immediate effects on maize crop growth and productivity. Soil K exists in different forms, such as fixed-K or non-exchangeable-K, exchangeable-K, and soluble K. Fixed-K is chemically bound to soil crystal lattice, hence it is not easily available to plant roots. In most soils, about 60–70% K is held in non-exchangeable form. The amount of fixed-K is dependent on soil parent material. The K available to plant roots is found in exchangeable and soluble fraction. The availability of K is affected by soil characteristics like texture, nature of clay, sorption/desorption mechanisms, residue recycling, and SOC content. Incidentally, solution-K which maize roots most easily exploit is the smallest of K fractions found in soils. Maize roots may absorb soluble-K fraction rather rapidly. It leads to depletion of soluble-K and soon deficiency of K may set in. Hence, in most soils used for maize, although total-K is optimum, maize crop may still suffer due to K deficiency. Further, although soils in maize production regions of different continent are classified as medium or high in soil K content, incessant cropping and tendency not to replenish K periodically creates its dearth. Soil K absorption is also affected by interactions that ensue with other essential elements in soil. For example, Ca + Mg concentration in soil solution affects soil K absorption. Hence, it is often reported that excessive use of lime to correct pH may affect soil K availability. In organically rich soils, fertilizer-K applied could get immobilized into organic fraction, depending on C:K ratios.

Soil K found in exchangeable form and that in solution is vulnerable for loss due to erosion. As such, soluble-K is highly mobile in soil profile. Such loss of K due to erosion, seepage, or percolation is accentuated after a heavy rainfall event. The K buffering capacity of soil affects maize productivity. Soils with larger clay fraction tend to buffer K better. Type of clay found may also be important. For example, montomorillonite clays buffer soil K better compared with other types. On Vertisols of South India where maize intercrops are predominant, soil K replenishment is not a major problem. These Vertisols are endowed with high buffering capacity due to montomorillonite clays. There are instances where soil-K depletion has not affected crop yield for almost 13 crop seasons.

SOIL TILLAGE AFFECTS NUTRIENT DYNAMICS IN THE MAIZE BELT

Maize farming methods adopted differ based on factors related to geography, topography, soil types encountered, season, soil moisture status, maize genotype, fertilizers, and yield goals. Maize farmers have in fact improvised their tillage methods periodically. During recent period, a major trend is to reduce tillage or even adopt no-till system in order to thwart loss of SOC. Aspects such as preservation of soil structure; conservation of SOC, and better rates of C sequestration seems to be the prime aim (Walters, 2001). Maize farming involves use of different types of tillage equipments. Each of these implements is designed to suit a specific purpose. The effect of tillage operation on soil characteristics and nutrient dynamics invariably depends on the kind of implement and intensity of its use. Commonly used implements are:

Primary Tillage Implements:

Moldboard Ploughs: They are used to turn soils up to 30 cm depth especially slightly heavier soils. Turning the soil helps in uprooting weeds. It also helps in incorporation of crop residues into deeper layers of soil. Moldboard ploughs are not recommended on sandy light textured soils. Soil aggregates that are poorly held could be easily destroyed. It may also induce wind erosion of soil and along with it certain amount of nutrients from the field. Plowing with Moldboard ploughs can alter nutrient distribution in the surface layers of soil.

Disc Ploughs and Discs: The disc plough has a slicing action on soil. It allows better penetration of hard and dry soils. It could be effectively used on soils where soil structure is not a primary concern. It is not recommended on sandy soils.

Chisel Ploughs: Chisel ploughs are used mainly to dig soils up to 25 cm depth. It loosens soil. Chisel ploughs are suited on dry soils, especially to break soil crusts and clods. Chisel ploughs do affect soil structure and aggregates. Chiseling of soil influences SOC accumulation, soil respiration, and loss of SOC. It also affects mineral nutrient availability to maize roots.

Rippers: Rippers are used for deep tillage. Repeated use of rippers to the same depth can induce formation of soil hard pans. Such hard pans usually impede root penetration and spread. Hard pans have to be destroyed periodically in order to achieve better usage of soil moisture and induce optimum infiltration/drainage of water. On wet or clayey soils, rippers may induce compaction of soil and limit lateral spread of maize roots.

Secondary Tillage Implements:

Rotary tillers are useful on heavy and slightly clayey soils. On dry soils it may destroy structure rather quickly. Use of rotary tillers induces soil aeration and loss of SOC.

Spiked or Tined cultivators are useful in breaking crusts and clods. They are used during preparation of seed beds. It also helps in controlling weeds. Tined hoes are apt only on moist soils.

Harrows: Harrows are used to prepare seed bed once the soil has been pulverized. The primary objective of disc harrows is to break crusts and clods and obtain fine tilth for seed bed formation (Jean du Plessis, 2008).

Major types of tillage systems adopted by maize farmers can be grouped into Conventional Tillage systems, Conservation or Reduced Tillage systems and no-till systems.

Tillage Systems used during Maize Production:

No- or Zero-tillage (NT): Soils are left undisturbed from planting till harvest (Plate 2.5). One of the pre-requisites is that soil should be covered with crop residues or mulches in order to avoid loss of soil moisture and suppress undue erosion. No-till system improves C sequestration. Weed infestation could be higher; hence timely weeding early during the crop is necessary. If not, weeds may affect fertilizer/soil nutrient recovery by maize.

Stubble-mulch Tillage: Soils are disturbed before planting maize seeds. Usually, chisel ploughs and spring-toothed implements are used to loosen soil. Weeds may not get effectively reduced hence hand or chemical weeding is necessary. Mulches are needed to reduce loss of soil and nutrients due to surface erosion.

Reduced Tillage (RT): Reduced or Conservation tillage involves moderate disturbance of soils using regular ploughs. About 15–30% stubbles or residues are left on surface. It helps in recycling SOC and reduces soil erosion. Weeds are reduced to a certain extent. Chemical or hand weeding is recommended in order improve fertilizer use efficiency.

Conventional Tillage (CT): Conventional tillage involves repeated discing and plowing, may be 3–5 times before the onset of rainy season. Soils are disturbed and turned up to 15–30 cm depth. It results in soil mixing and provides greater uniformity with regard to soil nutrient and pH. Crop residues need to be applied to improve SOC and reduce erosion. It also helps in conserving soil moisture. Repeated use of traction and heavy implements often results in formation of hard pans. Hard pans can impede rooting resulting in improper exploitation of soil nutrient in the subsurface layers. It affects percolation and drainage of water. Nutrient stratification is also common.

Source: Jean du Plessis, 2008.

As stated above, maize farming involves a wide range of soil management procedures. Soil tillage is perhaps the first step in the series that has direct impact on soil nutrient dynamics, crop growth and grain productivity. Maize farmers world wide usually adopt at least three different types of tillage systems. They are Conventional tillage—deep discing/plowing several times (CT), Reduced or Conservation tillage (RT)—mild plowing on surface with small ploughs and no-tillage (NT)—seeds are dibbled after the harvest of the previous crop or fallow, without disturbance to soil (Plate 2.5). Farmers tend to select tillage systems that suit the soil type and environmental parameters expected to be encountered during the season. Tillage practices and their combinations adopted indeed differs enormously based on geographic location, topography, soil characteristics, fertilizer schedules, irrigation, cropping systems, crop genotype, and yield forecasts.

Soil tillage may specifically affect a series of soil traits relevant to nutrient availability and accumulation in surface layer of soil. Indirectly, subsoil nutrient distribution and availability may also be influenced by tillage operations. There are indeed several reports on long term effects of tillage on mineral nutrient distribution and soil quality. Firstly, tillage improves distribution of nutrients applied to soil via fertilizers or FYM. For example, Halvorson et al. (2006) have reported that in the Great Plains of USA, maize crop is provided with 240 kg N ha^{-1}. The fertilizer N applied got distributed rather uniformly through out 0–180 cm deep soil profile under CT system, but it was uneven in case of fields under NT systems. The intensity of tillage related effects on soil traits may vary in time and space. The number and degree to which soil traits are assessed may have differed based on purpose of study. A recent report by Moebius-Clune et al. (2008) states that out of 25 soil traits evaluated, about 15 of them were adversely affected by repeated tillage. At a given stage, all soil traits may not be equally important in terms of nutrient dynamics. However, tillage and its effects on soil structure, organic fraction, moisture and nutrient availability to roots are considered more often and on priority.

Plate 2.5. Maize sowing machine meant for NT fields in the Rolling Pampas of Argentina.
Source: Dr. Carina R. Alvarez, Universidad de Buenos Aires, Argentina.

Among various soil traits evaluated, tillage immensely affects soil aggregation process. Traits like soil aggregate size, its strength, distribution, and persistence are important. Soil aggregation is defined as "arrangement of individual soil particles into compound particles of specific shape and size under the influence of natural, chemical, physical, and biological factors". Aggregation is actually a combined effect of soil organic matter, biological activity, surface charges on clay particles, formation of Al and Fe oxides, etc. (Wolkowski, 2006). Factors that affect soil aggregate formation based on their importance are as follows: microbial gums > iron oxide > organic carbon > clay. Nakamoto and Suzuki (2001) have reported that on Andosols, maize roots have significant influence on aggregates and its stability. Penetration of growing maize roots into aggregates and soil drying seem to cause disruption of soil aggregates.

Tillage and Soil Organic Carbon

There are several reports suggesting that tillage affects soil nutrient dynamics rather immensely. It has direct effect on soil aeration, oxidative state, rates of nutrient transformations, especially mineralization of SOC and mineral nutrients held in organic fraction of the soil. These soil nutrient transformations are aided by both chemical reactions and microbial activity. Such transformations in turn affect rooting, nutrient recovery, crop growth, and productivity. Foremost, soil tillage seems to have great impact on SOC dynamics. The rates of mineralization/immobilization reactions in soil are also affected. It is generally accepted that excessive tillage induces microbial respiration and loss of SOC as CO_2. Often, under arable conditions prevalent during maize production, SOC gets depleted through mineralization reactions. Hence, C sequestration procedures are preferred. Most commonly, farmers adopt restricted or conservation tillage practices. Sometimes they adopt NT systems in order to reduce loss of SOC. Such reduced tillage systems tend to thwart loss of SOC and soil nutrients compared to CT systems. Conventional tillage systems involve deep discing, plowing up to 30 cm depth, and turning of soil in order to achieve uniform soil pH and nutrient distribution in the upper layers. These soil maneuverings induce rapid loss of SOC from upper horizons. It may also result in loss of soil structure (aggregates), bulk density, redox potential, soil microbial transformations, rates of respiration, and SOC loss etc. We should expect significant variations in the above listed phenomena based on field, its location, soil type, cropping systems, fertilizer and organic manure inputs, residue recycling, etc.

Long term studies in the corn belt has shown that under repeated chisel plow (CP), surface residues (organic matter) tend to decrease rather drastically. As a consequence, soil is prone to erosion and nutrient loss from surface horizons. Where as, Strip tillage and no-till plots accumulated greater amount of surface residues that eventually added to SOC and improvement in soil quality (Wolkowski, 2006). Following is an example, wherein tillage influences surface residue, soil loss, and grain yield of maize:

Tillage	Surface Residue	Soil Loss	Grain Yield
	(%)	(t ha⁻¹)	(t ha⁻¹)
Chisel Plow	28.2	4.67	9.96
Strip Tillage	62.8	0.28	9.70
No-tillage	77.3	0.05	9.03
CD at 5%	9.1	NA	0.11

Source: Wolkowski, 2006.

Note: Values are means derived from a 9 year long term evaluation of Corn-Soybean rotation at various locations in Wisconsin, USA.

Various other long term trials, for example those in Missouri conducted in 1980s and 1990s have shown that NT plots contain higher levels of SOM in the top 2.0 cm layer compared with disked, chiseled, or plowed maize fields. The grain yield average for 13 years was higher in no-till plots compared with plowed fields. Following is an example:

Tillage System	Soil Organic Matter	Grain Yield
	(%)	(t ha⁻¹)
Plow	2.5	6.59
Chisel	3.3	6.64
Disk	3.5	6.68
No-till	4.1	6.70

Source: Quarles, 1994.

Again, long term trials in the Corn Belt have shown that during practical maize farming, it is the interaction between tillage systems and nutrient supply schedules that affects SOC accumulation pattern and maize productivity. According to Poirer et al. (2009), influence of NT on SOC gets accentuated in the upper 0–20 depth of soil. Where as, if moldboard plow (MP) is used, SOC accumulation is greater in the lower layers of soil. Therefore, if entire soil profile from 0 to 60 cm is considered, difference between NT and mold board plow may get nullified. Regarding, nutrient inputs and tillage interaction, it was found that plowed soil had low amount of SOC compared to NT systems. The highest SOC stocks were recorded at 0–20 cm depth in NT systems. It shows that in NT systems there is a tendency for crop residues and SOC to localize in the surface layers.

Let us consider a few studies about tillage and its effects on carbon dynamics in soil during maize cultivation. Within the Corn Belt of USA, precisely in Illinois, studies on seasonal dynamics of SOC fractions and soil aggregates have shown that tillage system affects net accrual of SOC. Evaluation of NT and CT system on a silty clay loam at Dekalb in Illinois has shown that, variations in SOC accumulation within top 30 cm layer could be ascribed to aggregate dynamics (Yoo and Wander, 2008). Basically, soil aggregate turn-over affected SOC accumulation pattern. Slow turn over of aggregates induced accumulation of humified fractions. Whenever, aggregate turn-over is greater it induced better amalgamation of aggregates into SOM. Further investigations on various fractions of aggregates and their interaction with soil organic component indicated that, NT systems improved SOC sequestration by increasing micro-aggregate "occluded particulate organic matter (OPOM)". Over all, tillage systems seem to affect SOC accruals, firstly by affecting soil structure, especially aggregate dynamics.

Brye et al. (2001) have pointed that a great deal of effects of tillage on leaching of SOC and N is really linked to measures that affect drainage from fields. A 4 year analysis on Typic Arguidoll in Wisconsin, USA has shown that tillage firstly induces higher levels of drainage loss of water received through precipitation. Consequently, it causes greater loss of SOC and N from soil. Following is an example:

	Drainage Loss of Water		Leaching Loss of Nutrients	
	mm	% Precipitation	Nitrogen	Soil Organic Carbon
			(kg N ha⁻¹)	(kg C ha⁻¹)
Prairie	461	16	0.6	19
No-tillage Maize	116	33	201	435
Chisel Plow Maize	1575	47	179	502

Source: Brye et al., 2001.

If entire maize belt of Northern USA is considered, then effect of tillage on drainage loss of water (soil moisture) and nutrients in dissolved state, especially N and C is enormous. Tillage could easily become an important factor affecting both nutrient and water dynamics.

According to Bono et al. (2008), tillage systems adopted could affect C sequestration and SOC balance in the maize farming zones. In the Pampas of Argentina, maize is actually rotated with other cereals like wheat or barley and legumes (e.g., *Vicia villosa*). Farmers adopt both CT and NT systems (Plates 2.6–2.9). Long term evaluations (3–6 years) suggest that maize grain yield is higher under NT than CT. Higher grain productivity under NT was attributable to better SOC sequestration and improved soil moisture retention. No-till induced higher SOC accruals. The SOC accumulation averaged 4 Mg C ha^{-1} yr^{-1} under NT but only 3 Mg C ha^{-1} yr^{-1} under CT. At the end of experimental period SOC balance was nearly equilibrated under NT, but greater losses were perceived under CT. It was concluded that in the semi-arid region of Pampas, we should adopt NT to obtain better SOC balance and induce higher C sequestration. This may in turn improve C dynamics in the maize ecosystem.

On-Farm trials in Ohio, USA, indicate that impact of NT systems on SOC and N could be soil specific. The SOC and N concentrations were higher under NT in only five out of 11 soil types tested. The NT effect on C sequestration is dependent on soil depth. In many cases, below 10 cm depth, soil under NT possessed lesser SOC than under CT. Overall, NT seems to improve SOC concentrations in upper layers of some soils, but may not store SOC more than CT, if entire profile is judged (Blanco-Canqui and Lal, 2008).

Tillage and its interaction with other agronomic practices might be important in terms of SOC dynamics and its turnover. For example, Higgins et al. (2007) suggest that crop sequences followed in the Corn Belt may interact with tillage method and intensity, and bring about changes in SOC content. Tillage systems like MP, CP, and NT and their interaction with established crop sequences common to Northern plains were evaluated on long term basis for 14 years at a stretch. Tillage by crop sequence interaction effect was significant with regard to SOC accumulation. No-till plots with continuous corn accumulated higher amounts of SOC at 165 Mg SOC ha^{-1}. Continuous soybean plots accumulated greater quantities of SOC compared to continuous corn. It was attributed to better decomposition of SOC under soybean than under maize. Regarding tillage type, both NT and CP maize fields possessed higher levels of SOC compared to MP. Actually, substantial losses occurred under high intensity plowing. The main suggestion was to enhance C supply via residues and reduce tillage, in order to preserve SOC reserves in the soil. Based on a 6 year trial in Nebraska, Follet et al. (2009) have suggested that NT systems could reduce loss of SOC plus aid C sequestration in areas where grasslands have been reclaimed into maize farming.

According to Walters (2001), tillage affects SOC in a number of ways. Adoption of conservation or reduced tillage in particular decreases SOC oxidation. Therefore soil-C loss is curtailed. Conservation tillage may also enhance C sequestration.

However, it depends on amount of residue generated and recycled. There are instances where conservation tillage has resulted in slightly lower residue yield compared to CT systems.

Plate 2.6. Maize crop in Pampas of Argentina grown under conventional tillage.
Source: Dr. Alberto Quiroga and Dr. Alfred Bono, National Institute for Agricultural Technology, Anguil, Buenos Aires, Argentina.
Note: Pre-planting plowing removes stubbles and weeds if any from the field. Mulches are absent hence soil erosion could be severe. Repeated tillage induces soil compaction.

In fields with continuous maize, methods of crop residue (carbon) management may have to be altered to suit the type of tillage system adopted. In NT system, large amount of crop residue is left in field. It protects soil from erosion, loss of soil moisture, and nutrients. Actually, under no-till system, accumulated crop residue needs to be utilized efficiently. The NT, mono-cropping, and lots of crop residue may have large impact on soil water, nutrient dynamics, and crop productivity. Al Kaisi (2007) suggests that an integrated approach to seeding, plant density, nutrient application, quantity of nutrients, and cropping sequences may help in improving SOC dynamics in fields under continuous maize. One example is to leave the crop residue standing at 12–24 inches height after harvest. It allows free access to machinery during planting a no-till plot. Upright residue can provide better protection to soil from wind and water erosion. It is said fall application of N enhances crop residue decomposition.

Plate 2.7 Maize crop grown under no-till system in the Rolling Pampas of Argentina.
Source: Dr. Alberto Quiroga and Dr. Alfred Bono, National Institute for Agricultural Technology, Anguil, Buenos Aires, Argentina.
Note: Maize field is *without* a cover crop or mulch.

Plate 2.8 A no-till maize crop sown after a triticale crop in Pampas of Argentina.
Source: Dr. Alberto Quiroga and Dr. Alfred Bono, National Institute of Agricultural Technology, Anguil, Buenos Aires, Argentina.
Note: Rows of maize seedlings in between triticale stubbles.

Plate 2.9 Maize crop grown under no-till system with a rye cover crop in the Rolling Pampas Argentina.
Source: Dr. Alberto Quiroga and Dr. Alfred Bono, National Institute for Agricultural Technology, Anguil, Buenos Aires, Argentina.

Tillage intensity has direct impact on SOC, soil microbial activity, nutrient transformation and availability levels. Minimum or zero tillage is best suited when time gap after the harvest of previous crop is small, also when soil disturbance intended is least. Minimum tillage enhances C sequestration in soil. In South India, shift from conservation tillage to NT had an effect on maize productivity. However, it has been observed that interaction of tillage intensity versus nutrient additions, especially N input affects maize production. Generally, NT systems result in marginal decrease of maize grain productivity. Despite it, NT is preferred because of advantages with soil fertility and environmental concerns, such as CO_2 evolution and loss of SOC. In the US Corn Belt, long-term evaluations have clearly shown that shift from conservation to no-till system reduces yield by 16%. Further, high N inputs beyond 85 kg N ha^{-1} may mask effects of tillage intensity if any (Halvorson and Reule, 2007). A detailed understanding of interactive effect of tillage intensity, N and P inputs to maize fields seems crucial (Al Kaisi and Kwaw Mensah, 2007). Basically, NT systems reduce soil erosion and loss of SOC which otherwise occurs due to rapid oxidation. Carbon sequestration due to residue recycling is maximum, if NT systems are adopted.

Maize adopts well to minimum or conservation tillage practices. Minimum tillage involves one deep plowing/discing after the harvest of the previous crop. On light soils, tillage intensity is kept low. At best one round of plowing is done. Where as, on heavy soils, 2 or 3 rounds of plowing are essential. Farmers in South India tend to adopt single or zero tillage before sowing maize (Ikisan, 2007). Minimum tillage suits if maize is intercropped and planting is staggered. Whatever be the intensity, plowing essentially aims at reducing weeds, plus improving soil aeration, soil moisture holding capacity, microbial activity, and nutrient transformation. Excessive plowing may induce loss of SOC due to oxidization of soil chemical environment.

Soil tillage, especially its type, depth, and intensity affects oxygen tension, moisture content, SOC, and microbial activity immensely. In turn, it may have consequences on several other soil nutrient transformations relevant to maize growth and production. During recent years, farmers in South India tend to adopt zero tillage that has beneficial effect on SOC. Zero tillage enhances C sequestration in the maize fields. It also enhances microbial load and their activity. Following is an example:

Tillage	SOC	Microbial Biomass C	Microbial C/SOC
	g kg^{-1}	mg kg^{-1}	%
Ploughed	19.6	120	0. 61
Zero Tillage	21.5	237	1.10

Source: Subba Rao and Sami Reddy, 2005.

Computer-based models have been utilized to study effects of tillage on dynamics of C in soil that supports maize production. Turnover of SOC is actually a slow process and requires measurement for extended periods to quantify the effects of fertilization, tillage and cropping pattern on SOC. Ganesh et al. (2006) have used RothC-26.3 model to understand C sequestration potential and organic- C turnover during finger millet-maize rotation. The measured SOC pool agrees with predicted values derived by using soil fertility and productivity sub routines. Long-term effects of soil tillage on C sequestration could also be modeled and simulated.

During recent decades, either conservation tillage or NT system that reduces loss of SOC from soil profile has been practiced more commonly in the maize belt of China (Zengjia and Tangyuan, 2009). Reports from Chinese maize growing regions suggest that soil erosion and nutrient loss is greater under CT. About 140–190 kg C and 5–90 kg N ha^{-1} yr^{-1} could be lost from fields maintained under CT (Wang, 2006). Soil erosion and nutrient loss is reduced significantly if NT systems and mulching are practiced. Reduced tillage decreased loss of major nutrients N, P, and K by 50–90% compared to repeated tillage.

Tillage, Soil Compaction, and Nutrient Distribution

Repeated tillage of soil has its impact on nutrient distribution in the soil profile. Often, it is said that NT systems are so popular with maize farmer all over world, that it may actually induce nutrient stratification. Further, tillage equipments, for example tractors, as they ply could induce soil compaction. Hard pans created by repeated traction could impede root growth and penetration below a certain depth in the soil profile. This aspect has direct influence on extent to which nutrients and water that could be garnered by roots. Soil nutrient depletion and distribution pattern may alter. Garcia et al. (2007) examined if one time tillage, in other words, just soil disturbance could avoid undue stratification of fertilizer based nutrients. They examined effects of tillage systems usually adopted by maize farmers in Northern plains, mainly in Nebraska. One time chisel plough or disk tillage did not redistribute nutrients effectively within 0–20 cm layer. Greater redistribution of nutrients like P was possible with MP. Actually, it seems better to adopt tillage in order to obtain a better redistribution of nutrients, if the fields were continuously under no-till systems. Whatever be the procedure adopted, nutrient stratification has to be avoided to achieve better recovery by maize crop. Tillage intensity too might affect nutrient availability and recovery by maize roots (Mozafar et al., 2000).

Soil compaction occurs due to repeated tillage or if soil is trafficked when moist or immediately after rains. Fine textured soils with high clay and silt content are prone to soil compaction. Maize crop that requires field operations at different stages in the early in spring and late fall season may induce soil compaction (Mclaughlin et al., 2004). Soil compaction could be surface or subsurface. Both types of compactions do significantly affect nutrient dynamics. It affects nutrient distribution, stratification, and percolation to vadose zone. Surface compaction is easily visible and it reduces drainage or seepage of water and nutrients to lower layers of soil. Tillage can break up surface compaction. Soil compaction can also be reduced by using lighter machines and decreasing usage of heavy tractors. Subsurface compaction is also called "plough pan". It is difficult to identify. It induces nutrient stratification, flooding, and stagnation of water and reduces drainage. Subsurface compaction obstructs root growth beyond the pan. Poor crop performance due to subsurface compaction can be over come by deep plowing.

A report on profile structure of Mollic Arguidolls found in Pampas of Argentina suggests that agricultural traffic (traction) can induce compaction of both surface and subsurface layers (Taboada and Alvarez, 2008). They compared effects of conventional-till (CT) and no-till or zero-till (ZT) maize fields (Plates 2.6 and 2.7). Plough pans were easily induced in plots given CT. The CT also affected bulk density and root penetration. Clay pans developed in the BT horizon impeded root penetration and spread. Maize root abundance reduced by 40–80% in field given CT compared to those under ZT. Maize roots in fact got stratified at 0.2 m depth. Conventional tillage therefore resulted in reduced nutrient recovery. There is a need to avoid excessive tillage and its ill effects on soil profile. In a different study on clay loams of Iowa, maize rooting was again severely reduced due to CT procedure. The traffic pattern affected bulk density and hydraulic conductivity adversely. Here, root length density got reduced by 33% compared to un-trafficked plots (Kaspar et al., 1995). In European corn growing regions too farmers tend to adopt either reduced or no-till systems. Comparative studies suggest that overall root length density may still be higher in CT. However, in the upper 5–10 cm, roots garner high amounts of nutrients. Root length density was greater with NT systems compared to CT. Mean root diameter (root thickness) was also greater in NT than in CT (Qin et al., 2005). Tillage effects on roots may still be more complex considering that in maize farms, in addition to tillage, factors like cropping sequences, fertilizer inputs, organic manures, and irrigation may all act simultaneously.

According to reports from Missouri Experimental Station, NT induces stratification of nutrients. As a consequence, rooting pattern may get influenced based on nutrient and moisture distribution under NT systems. Following is an example:

Soil Depth	Root Weight		
	(mg CC^{-1} Soil)		
	Plow	Chisel	No-till
0–7.5 cm	0.75	0.83	1.88
7.5–15 cm	0.98	0.98	0.75
15–23 cm	0.51	0.50	0.50
23–30 cm	0.23	0.22	0.23

Source: Hoette, 2009.

There is a clear localization of roots in upper layers of soil. Further, in fields under NT systems the density of roots is greater in the top 7.5 cm. Rooting density under NT is almost double that found with plow or chisel treated plot. This could help the plants in garnering as much nutrients stratified in upper layers.

Tillage, Soil Nutrients, and Nutrient Recovery by Maize

Tillage has its impact on soil-N fractions, their transformation rates and availability to maize roots. Sharifi (2008) reports that soil under no-till plots contained higher levels of potentially mineralizable N (N_0). The labile and intermediate mineralizable N pools were larger under no-till plots than those under CT. Actually, adoption of NT affected quality of active organic-N fraction and it was mediated by higher levels of labile N pools. Regarding N availability indices, KCl-extractable-NH_4-N, NaOH extractable-N, Illinois Soil-N test were all sensitive to tillage induced changes in soil organic fraction. Over all, it is believed that tillage induced changes in soil particulates, quality (organic fraction) and active organic-N fraction may all affect N supply to maize roots.

Tillage systems influence extent of nutrients garnered by maize grown in semi-arid climates. Among the four types of tillage (zero, minimum, conventional, deep) examined, uptake of N, P, and K was higher in plots given tillage compared to that held under no-till system. Soil-N status decreased if deep tillage was adopted. Soil P was higher if minimum tillage or reduced tillage systems were adopted. Over all, Iqbal et al. (2006) have suggested that in semi-arid regions it is preferable to adopt reduced tillage systems in order to conserve SOC and improve nutrient recovery by maize. According to Boem (2008), no-till reduces P buffer capacity of surface soil. The capacity of surface soil for P sorption and P retention was related to its clay content rather than organic matter.

In Japan, adoption of zero-tillage plus cover crops improved nutrient recovery by maize. Cover crops reduced weed infestation and loss of nutrients. Zero-tillage, a cover crop and application of only 50% of recommended fertilizers seemed to offer best results in terms of nutrient (N, P, K) availability, nutrient recovery, crop growth, and grain productivity (Zougmore et al., 2006).

In the maize cropping zones of USA, both NT and CT systems are being adopted by farmers. The tillage systems may have influence on the net N recovered and grain yield. Long term (5 years) evaluations have shown that maize crop responds to N inputs irrespective of tillage systems adopted. The maximum grain yield was attained at 276 kg N ha^{-1} under CT and at 268 kg N ha^{-1} supply in case of NT. On an average 43% of fertilizer-N supplied was recovered by the maize crop. Total N recovered by maize was 19 kg N under CT and 20 kg N under NT in order to produce 1.0 t grain (Halvorson et al., 2006). Clearly, tillage has an impact on N uptake and maize productivity.

We should note that tillage effects on nutrient uptake, use efficiency and its consequences on productivity are perceivable only within a certain range of fertilizer supply. For example, if fertilizer-N inputs are high then N recovery and forage production were unaffected due to tillage systems adopted. The N-use efficiencies observed for NT and CT fluctuated between 49 and 55% and N recovery rates were almost similar. It is said that fertilizer-N input masks the tillage effects if any (Staley and Perry, 1995).

Maize is often part of cropping systems that are maintained under Integrated Nutrient Management (INM). According to Singh (2006), 3-way interaction involving tillage x water regime x nutrient supply occurs during maize-wheat rotation maintained under INM. Such, 3 way interactions significantly affected several physico-chemical aspects of soils. Tillage effect on loss of SOC was pronounced if organic manure inputs were higher. The NT plots accumulated greater amount of organic-C, mostly in macro-aggregate fraction. Application of organic manures reduced downward translocation of NH_4 and NO_3. It prevented loss of soil-N from root zone. Tillage intensity and water regimes seemed to affect P absorption capacity of maize roots.

Computer aided simulations have also been utilized to study effect of tillage on rooting, moisture, and nutrient recovery by maize. Oorts et al. (2007) utilized data from long term trials (33 years) on maize in Northern France to simulate tillage effects on soil nutrient dynamics. They adopted "PASTIS model" that effectively simulates soil moisture, N and C dynamics. Simulations using PASTIS model showed that presence of mulch under NT systems played vital role in reducing cumulative water evaporation. Drainage below 25 cm depth was improved. Further, larger cumulative CO_2 flux under NT was attributable to CO_2 emissions from decomposition of residues. Such CO_2 was not derived from inherent SOM.

Filho et al. (2004) have reported that NT supports better rooting. Maize roots accounted for 1324 kg C ha^{-1} and 58 kg N ha^{-1}. Maize roots were mostly localized in the upper 0–10 cm layer. Maize roots exude a range of C compounds that serve to improve soil microbial population and activity. Microbial C and N were concentrated around roots and it coincided more with distribution of thin roots. It seems, thin roots were active in nutrient recovery. At the same time, thin roots were also prone to rapid decomposition. The nutrients released actually induced better microbial activity.

Land preparation methods such as flat beds (FB), ridges and furrows (RF), or broad-bed and furrows (BBF) may also affect nutrient availability and their recovery by maize roots. Nutrient recovery by maize planted on BBF gets enhanced by 20–30% compared to FB. Hence, nutrient balance too is affected adversely. Actually, due to higher depletion of N, balance shifts to –57 kg N ha^{-1} in plots with BBF. Where as, under FB, N balance was only –48 kg N ha^{-1}. The mechanism involved in higher nutrient recovery under BBF needs greater attention (Sridevi et al., 2004).

Tillage systems adopted may directly affect various physiological stages of the maize crop. For example, a comparative study at Chipinge in Zimbabwe involved 4 different types of tillage systems namely, hand hoe (*badza*), NT, inter row furrow, and tied ridge. Firstly, tillage systems seem to affect days taken to emergence, physiological maturity, above ground biomass accumulation (includes nutrients), and grain harvest (Gwenzi et al., 2008). Interestingly, there were significant interaction effects between tillage systems adopted and maize genotype selected. Maize hybrids had better precipitation use-efficiency and harvest index. Generally, any genotype performed better with regard to nutrient recovery and growth if it were planted on ridge/furrow systems compared with flat surface. We should also note that tillage x rainfall pattern affected growth and nutrient recovery patterns observed in a maize crop.

Clearly, maize genotype selected for cultivation may influence farmer's decision regarding tillage method and its intensity (Duiker et al., 2006). Tillage, primarily affects root development, soil exploration efficiencies, and nutrient recovery by maize. Further, effect of a particular tillage system may vary based on maize cultivar that grows in the field. Aspects like seed emergence, root penetration, and spread may be affected by soil characteristics like bulk density, compaction, temperature, SOC, etc. Therefore, tillage by genotype interaction may have relevance regarding nutrient dynamics that ensues in maize fields. On a wider scale, tillage system and genotypes that dominate a maize belt zone may together partially influence nutrient dynamics within the agroecosystem. Further investigations are required to confirm tillage by genotype interactions.

SUMMARY

Maize is a highly versatile cereal crop. It adapts and thrives well on a wide variety of soil types and fertility ranges encountered in different continents. Maize crop grows luxuriantly and produces high grain/forage harvest (10–11 t grain plus 15–20 t forage ha^{-1}) on Mollisols found in the Corn Belt of USA and Chernozems of European plains. Maize adapts well to Alfisols and Vertisols traced in the tropics of South India. Maize is a preferred crop on Mollisols of Pampas and Acidic Oxisols found in Cerrados of Brazil. Soil types of these regions support moderate levels of productivity at 3–5 t grain plus 10–15 t forage. Maize is also a most useful crop grown on sandy soils of West Africa. The subsistence farming methods allow low or moderate productivity at 0.8–2 t grain ha^{-1} plus forage. Maize negotiates several types of soil related constraints to its productivity. Soil acidity and Al toxicity are problems encountered in Brazilian Cerrados. Intermittent drought and dearth for nutrients is a concern in Pampas of Argentina. Nutrient deficiency, especially N and P is major problem most parts of West Africa, Asia, and Fareast. We should note that many of these maladies occur at different intensities and for different durations during crop growth and yield formation. Of course, there are also innumerable techniques devised to overcome many of these soil maladies. Genetic variation available within maize also helps to overcome a few of the soil constraints.

Foremost of the soil management techniques is tillage and its intensity. Soils used for maize production are mended using a wide range of implements and equipments. Each of these tillage implements has its specific effect on soil structure, nutrient dynamics, and crop growth. Repeated tillage is known to expose soil and induce oxidative reactions in soil. It induces microbiological processes that mediate rapid transformations of nutrients. Mainly, repeated tillage causes rapid loss of soil-C. It is deleterious and reduces soil quality. Yet, deep tillage using discs may be necessary to turn over soils and achieve uniformity with regard to soil pH and fertility. During recent years, maize farmers are enthusiastic in adopting conservative/reduced tillage systems or totally NT systems in preference to conventional deep tillage. The NT has its advantage regarding conservation of soil nutrients and SOC. The NT reduces soil erosion and nutrient loss. Generally, it is believed that NT for a few seasons followed by a deep tillage once in 3–5 years provides better soil conditions during maize production. Specific attention is also required to understand interaction of tillage intensity versus nutrient supply during intensive cropping of maize. Overall, we should note that soil and its tillage influences nutrient dynamics that ensues in the maize agroecosystem.

KEYWORDS

- **Ammonification**
- **Calciferous soils**
- **Fertilizer replenishments**
- **Mulching**

Chapter 3

Nitrogen, Phosphorus, and Potassium in Maize Agroecosystem

MAJOR NUTRIENTS—NITROGEN, PHOSPHORUS, AND POTASSIUM IN MAIZE BELTS

The three major essential nutrients, namely nitrogen (N), phosphorus (P) and potassium (K) have played a key role in expansion, sustenance and productivity of maize agroecosystem at large, in all the continents. On occasions, major nutrients have literally decided the nutrient dynamics and productivity levels of maize cropping belts. Major nutrients inherently contained in soil and that supplied through chemical and organic fertilizers, have of course influenced the maize-based cropping systems adopted by farmers. Maize cropping in dry lands and arable irrigated conditions, both are highly dependent on major nutrient supply. During recent years, research trends as well as farmers' preferences have been to intensify maize production systems wherever feasible. This has necessitated adoption of improved soil fertility and crop management techniques. Fertilizer formulations and supply methods have consistently aimed at enhancing productivity levels and maximizing efficiency. Most suggestions aim at manipulating major nutrient dynamics in order to match nutrient needs of the maize crop at different growth stages. It is believed that such exact matching will improve fertilizer efficiency, grain/forage productivity and at the same time sustain optimum dynamics in soil. Currently we have maize production techniques that suit subsistence farming with low nutrient input. For example, consider dry lands of India, Pampas, Brazil, Southern, and Eastern Africa. Low levels of nutrient supply results in grain yield ranging from 2 to 4 t ha^{-1}, at the most. In comparison, adoption of agronomic procedures like rapid soil tests, high nutrient inputs, split application, and proper placement of fertilizers have resulted in high productivity (8–10 t grain ha^{-1}) within Corn Belt of USA, Northeast China or European plains produces. It is generally accepted that intensification of maize farming was achieved by modifying nutrient dynamics in soil and farm *per se*. Supplementing crop with irrigation and high density planting were other key procedures that rapidly converted moderately yielding farms into highly intensive agricultural zones. At this stage we have to appreciate that nutrient dynamics that ensues in maize cropping zones of the world varies enormously. Such variations are attributable to geographic location, its physiographic characteristics, soil types, inherent fertility status, cropping trends, maize genotype that dominates the agricultural landscape, human preferences and at times economics and legislations regarding cropping, irrigation, and fertilizer supply may all affect nutrient dynamics and productivity of maize agroecosystem.

Incidentally, this chapter presents recent knowledge accrued on dynamics of major nutrients and their relevance to productivity of maize agroecosystem. Initial sections are devoted to explain research results from studies that pertain to all three major nutrients—N, P, and K. It is followed by detailed discussions on each major nutrient exclusively regarding various aspects of nutrient dynamics like distribution, availability to maize roots, transformation, etc. Aspects like nutrient supply, recovery rate, accumulation, loss, turnover rates, and grain productivity form the core of the chapter.

Nutrient Supply into Maize Agroecosystem

Nutrient supply into maize agroecosystem, either through natural sources or extraneous inorganic and organic manures is an important aspect that sustains productivity. Nutrient supply firstly primes the maize belt, leading to improved nutrient turnover and higher productivity. In addition to crop productivity, nutrient input induces and stabilizes a variety of physico-chemical and biological process that are important for perpetuation of a cropping ecosystem. The extent of nutrients impinged into maize belts across the world have varied enormously through time. It has been based mostly on location, season, soil type, crop genotype, yield goals, and economic viability. Let us consider a few examples.

The "Corn Belt of USA" is an intensely cropped zone. Here, nutrients are generally supplied at a higher rate based on soil type, cropping system, and yield goals. Most importantly, it is revised periodically as suggested by soil test results. The nutrient supply on an average fluctuates at 180–240 kg N, 40–70 kg P, and 180–200 kg K ha^{-1}. Nutrient inputs into maize crops grown in the Northeast of USA have been commensurate with moderate grain/forage harvests per unit area. Reports indicate that maize crop is supplied with 180 kg N, 70 kg P and 148 kg K ha^{-1}. Liming at 4 kg lime kg^{-1} fertilizer neutralizes the acid forming effect of fertilizer-N. Soil pH returns to 6.5–7.0 from 5.5 to 6.0. According to Dobermann (2001), fertilizer based nutrient supply in major corn producing states of USA has fluctuated at 95–185 kg N ha^{-1}, 10–34 kg P ha^{-1}, and 0–95 kg K ha^{-1}.

In the Rolling Pampas, maize mono-crops or that grown in rotation with wheat receives 50 kg N ha^{-1} and 16 kg P ha^{-1} at the time of sowing. Top dress N of 20 kg N ha^{-1} each is applied at 2 or 3 stages depending on soil fertility and yield goals. Soil K is usually sufficient. Yet, K availability levels may reach below threshold because of incessant cropping. Therefore, 50–60 kg K ha^{-1} is also applied at sowing based on soil tests.

Nutrient input to grain crops in Europe is high owing to intensive cropping practices. Average fertilizer supply in western European plains is 280–320 kg N, 80 kg P and 220 kg K ha^{-1}. However, maize is mostly a forage or silage crop in the European agricultural zones. It is grown for its good quality forage. A forage crop in European Plains receives N, P, K and micronutrients based on yield goals. Nitrogen is the key element that determines productivity. In Netherlands, nutrient supply to maize crop meant for silage is about 60–240 kg N, 20–40 kg P and 80–120 kg K ha^{-1}. Micronutrient supply depends on particular field and cropping system adopted. Usually Zn and Fe are supplied. On Dernopodzolic soils of Belarus, maize farmers are advised to supply 180 kg N ha^{-1} in 2 splits plus 60 kg P ha^{-1} and 120 kg K ha^{-1}. Organic manure (FYM) is incorporated before sowing at 20–30 t ha^{-1} (Sholtanyuk and Nadtochaev, 2004).

Maize is an important cereal in the semi-arid regions of Southern Africa. Most of the small farms adopt subsistence cropping systems. At best, these fields are provided with small amounts of chemical fertilizers and organic manures. Organic manures in fact remain the main source of nutrient replenishment. Cattle Manure and FYM are the chief sources of nutrients. Hence, farmers tend to supply them in larger quantities. At times high organic–N loading through FYM (40 t ha^{-1}) equivalent to 240–440 kg N ha^{-1} plus other nutrients necessitates rapid mineralization so that inorganic N is released. However, large commercial farms do adopt high input technology. According to Nyamangara et al. (2003), small holders apply about 53 kg fertilizers ha^{-1} that includes N, P, K, and S. Whereas, large farms apply about 150–205 kg fertilizers ha^{-1} in a cropping year. In some maize farms, a relatively smaller dose of organic manure (12 t ha^{-1}) and 60 kg inorganic fertilizers is supplied before sowing.

Maize cultivation in South Indian plains is predominant during rainy season. Nutrient supply is based mostly on State Agricultural Agency recommendations, Maximum yield goals or Site-Specific Nutrient Management Methods. The relatively high input systems that yield 4–5 t grains ha^{-1} are supplied with 170 kg Urea, 130 kg Di-ammonium Phosphate, 70 kg Muriate of Potash, and 10 kg Zinc sulfate ha^{-1}. Whereas, rain fed maize that yields about 1.5–2.0 t grains ha^{-1} is supplied 120 kg Urea, 70 kg Di-ammonium Phosphate, 45 kg Muriate of Potash and 10 kg Zinc sulfate ha^{-1}. Nitrogen fertilizers are usually split into starter and at least 2 splits to improve efficiency (ICRISAT 2004; Krishna, 2010). In the dry tracts of Northwest India, fertilizer inputs recommended by State Agricultural Agencies are not high. It is commensurate with grain/forage yield possible under harsh agroclimate. Generally, 80–112 kg N ha^{-1}, 40 kg P ha^{-1}, and micronutrient mix is supplied. In India, forage maize (e.g., African tall, Varun, Vijay) with a planting density of 1,25,000 seedlings ha^{-1} is supplied with 20–25 t FYM and 60–80 kg N ha^{-1} in two splits, first at sowing and second at tasseling stage. An irrigated forage crop may receive higher amounts of N at 120 kg N ha^{-1} in order to support rapid biomass accumulation (Ikisan, 2007). In general, matching nutrient recovery rates with biomass accumulation is important during forage production. Forage maize sown in May just prior to onset of monsoon puts forth peak growth during July and August. During the entire season lasting until October it may accumulate 35–40 t forage ha^{-1}. The peak nutrient recovery therefore coincides with dry matter production.

In China, nutrient input to maize is again dependent on range of factors related to geographic location, soil type, its fertility status, genotype, and environment. In North-Central Plains (Shanxi Province) of China, maize crop receives 195 kg N, 70 kg P, and 120 kg K ha^{-1} (Wang et al., 2008).

In Australia, especially in Queens land, maize is cultivated both under dry land and irrigated conditions. On an average, about 40–180 kg N, 0–35 kg P, and 0–50 kg K ha^{-1} is supplied to maize fields. Inputs are altered depending on water resources and yield goals.

Initially, that is before the advent of chemical fertilizers, much of nutrient input into maize belts anywhere, occurred predominantly through crop residue recycling and extraneous supply of organic manures. It equated to about 10–20 kg N, 1–2 kg P, and 10–15 kg K ha^{-1} depending on source. Sometimes, it was higher if large amounts of organic manures/residue were dumped into fields. At present, farmers in low input subsistence farming zones of West Africa, India and elsewhere in Southeast Asia still tend to supply entire or most of the nutrients using FYM. Nutrient supply in such subsistence farms is estimated at 20 kg N, 2–3 kg P and 20–30 kg K ha^{-1}. At present, medium and high input maize belts all over the world practice Integrated Nutrient Management (INM) techniques. Maize fields actually receive a combination of chemical (inorganic) and organic manures. The extent of nutrients impinged into maize farming zones via FYM or organic manures depends on nature of organic manure, amount added, timing, decomposition rates in soil and environmental conditions. Following is a list organic nutrient sources and amount of nutrients on dry matter basis that they may supply into maize fields:

Organic Source	Nitrogen	Phosphorus	Potassium
	(%)	(%)	(%)
Cereal Straw	0.30–0.65	0.08–0.75	0.12–2.20
Tree Leaves	0.65–1.67	0.12–0.45	0.40–2.30
Animal Manures	0.30–0.50	0.10–0.40	0.10–1.20
FYM	0.40–1.50	0.30–0.90	0.30–0.23
Vermi-Compost	1.20–6.35	5.15–6.25	1.20–7.30

Source: Krishiworld, 2002.

The amount of FYM or Animal manures added has varied enormously with geographic location, topography, soil type, season, inherent soil fertility, agronomic procedures, crop genotype, and yield goals. Subsistence farms may add 5–10 t FYM ha^{-1}. Farmers aiming medium level of harvest (2–5 t grains ha^{-1}), for example in Meso-America, Eastern and Central Africa or South India add 10 t FYM ha^{-1}. Farmers in intensive farming zones supply 15 t FYM ha^{-1}. Almost, 50–150 kg N ha^{-1} plus proportionately high amounts of other nutrients could be channeled via organic manures. In some of the commercial farms of Zimbabwe, there is a tendency to supply as much as 200 kg N ha^{-1} plus other nutrients through organic sources (Nyamangara, 2007). The nutrient ratio in the organic source decides the extent to which each nutrient is supplied.

Nutrient Input Ratios

We may try to ascertain and re-ascertain the ratios at which maize crop needs N:P:K in soil, also ratio at which it absorbs and accumulates them into various tissues. However, we ought to realize that nutrient absorption rates by plants vary with growth stage. Soil factors that induce nutrient loss may also vary in intensity and timing. Therefore, rate of depletion of each nutrient in soil differs and it definitely alters the original ratios attained through fertilizer or that planned by farmers. In other words, natural processes such as leaching, percolation, seepage, and volatilization, along with nutrient depletion brought about by maize crop alters the nutrient dynamics immensely and affects the ratios at which nutrients are available during mid and late stages of maize crop.

In the Corn Belt of USA, nutrient supply ratios have varied through time based on yield goals. Inherent soil nutrients, soil moisture status, and crop genotype too have influenced nutrient ratios. Currently, in most states of Corn Belt, N:P:K is supplied at ratio of 1.8–2.4:0.6:1.8. In Europe, high input forage crops may receive nutrients at 3:0.8:2.0. The forage yield goal in European plain is high at 35 t ha^{-1}. However, in the Hungarian plains maize crop may be provided major nutrients N, P, and K at 2:1:1 (Szolokine and Szaloki, 2002).

The nutrient ratios adopted by maize farmers vary enormously based on geographic location, soil type, maize genotype and environmental factors. Let us consider an example. The ratio of major nutrients that occur in various fertilizer formulations used has been dynamic throughout the past decades. The N: P: K ratio utilized on maize has ranged from 5.9:2.4:1 to 9.7:2.9:1 in the maize belt of South India. The ratio of N:P has been relatively more stringent. It is because soils are generally richer in K. The N:P ratio has fluctuated between 2.4:1 and 3.4:1 during the past decade. The ideal ratios are said to be 4:2:1 for N:P:K and 2:1 for N: P (Dass et al., 2008; Tiwari, 2001).

Australian maize farmers have modified the nutrient ratios to suit the inherent soil fertility. A maize crop grown with supplemental irrigation receives major nutrients at a ratio of 2.0:0.1:0.25. A rain fed crop in Queensland is provided with much low levels of nutrients. The ratios of major nutrients fluctuates around 1:0.1:0.5. Over all, we should note that nutrient ratios adopted have fluctuated but within limits depending on several factors related to crop genotype, soil fertility status and yield goal.

Nutrient Recovery by Maize Crop

Nutrient absorption by maize crop is actually a process that moves sizeable fraction of nutrients held in soil phase into above-ground crop phase of the agroecosystem. A portion of nutrients recovered from soil profile is held in roots. Nutrient recovery rates literally determine productivity of maize belts. Subsistence cropping involves relatively low nutrient recovery. For example, in Sub Sahara, average productivity of maize is low at 0.8 to 1.2 t grain ha^{-1}. It means about 20 kg N, 3 kg P and 18–20 kg K is absorbed ha^{-1} from the sandy soils. In regions with moderate productivity like Mexican highlands or South Indian plains where in about 3 t grain ha^{-1} are harvested, proportionately higher quantities of nutrients (60 kg N, 7–8 kg P, and 55–60 kg K ha^{-1}) are recovered per season.

Let us consider nutrient recovery in an intensively cropped zone. On a wider scale, maize productivity in US Corn Belt has increased from a mere 1.25 t ha^{-1} in 1900 to 11.5 t ha^{-1} in 2007 (Abendroth and Elmore, 2007; Sawyer, 2007). This maize belt has graduated from being a subsistence farming zone to one of the highly intensely cropped areas of the world. We know that maize absorbs 18–21 kg N, 2.5–3.0 kg P and 18–20 kg K ha^{-1} to produce a ton grain. In 100 years, "US Corn Belt" has been intensified enormously, almost 10–11-folds to be accurate. The nutrient recovery by a corn crop in US Mid west has increased from 25 kg N, 4 kg P, and 24 kg K ha^{-1} in 1900 to 235 kg N, 50–60 kg P, and 240 kg K ha^{-1} due to higher productivity (Abendroth and Elmore, 2007). We have no accurate idea regarding limits to intensification or extent to which nutrient recovery could be increased further. Irrespective of morphogenetics or planting geometry maize crop has to absorb N, P, and K at higher rates per unit time, if it has to produce more foliage/grain. In the Northeast of USA, maize production is moderately intense with grain yield ranging from 5 to 8 t grains ha^{-1}. Reports indicate that maize crop removes 16.6 kg N, 3.3 kg P, 4.2 kg K, 2.8 kg S, and 2.8 kg Mg t^{-1} grains produced. The forage types of maize are said to recover 10 kg N, 3.3 kg P, 6.0 kg K, 0.6 kg S, and 1.6 kg Mg per t^{-1} dry matter produced. Reports by Potash and Phosphate Institute of Canada suggest that a maize crop that produces 18.7 t ha^{-1} dry matter absorbs and accumulates 240 kg N; 44 kg P; 200 kg K; 34 kg S; 45 kg Ca; 56 kg Mg; 3 kg Fe; 0.6 kg Mn; 0.2 kg Cu; 0.1 kg B; and 0.1 kg Mo ha^{-1} (Johnston and Dowbenko, 2009). According to Smith (1995), a maize crop that produces 7.5 t grains recovers 190 kg N, 39 kg P, and 196 kg K ha^{-1} from soil.

In South Africa, for each t of grains produced by a maize crop it recovers 15–18 kg N, 2.5–3.5 kg P, and 3.0–4.0 kg K from soil (Jean du Plessis, 2008). Rapid absorption and assimilation of major nutrients coincides with flowering. The total nutrient recovered by a single plant of maize averages 8.7 g N, 5.1 g P, and 4.0 g of K.

Regarding maize in India, a crop that produces 1.0 t grains recovers at least 18.0 kg N, 4.5 kg P_2O_5, and 18 kg K_2O from soil. It demands relatively higher quantities of N and K. Yet another report based on several studies in the South Indian plains suggests that a crop that produces 6.27 t grain removes 168 kg N, 57 kg P_2O_5, and 130 kg K_2O. It amounts to 20.7 kg N, 9 kg P_2O_5, and 20.3 kg K_2O to produce 1.0 t grains. In addition to factors like soil fertility, moisture status, crop genotype, and environment, often yield goals set by farmers decides amount of nutrient removed from soil into aboveground portion of the crop. For example, a report by Potash Research Institute of India suggests that maize crop that yields 6 t grain ha^{-1} absorbs 120 kg N, 50 kg P_2O_5, and 120 kg K_2O (IPI, 2000). A maize crop may yield 2.1 to 2.3 t grain ha^{-1} under rain fed conditions. It removes about 61 kg N, 10 kg P, 67 kg K, and 7 kg S ha^{-1}.

Minor variations in quantum of nutrients required per ton of grains and critical levels of nutrients in soil are to be expected. Nutrient requirements are known to fluctuate within limits based on soil fertility status, genotype, and agro-environment, in addition to fertilizer supply. Kaore (2006) suggests that knowledge about nutrient absorption, pattern of absorption rates as well as ratio of nutrients recovered at various stages is important. Major nutrients removed by maize grown under rain fed conditions in semi-arid tropics is 26.3 kg N, 13.9 kg P_2O_5, and 39 kg K_2O per ton grains. The ratio at which nutrients get absorbed is 1:0.53:1.36- N:P:K. Rajendra Prasad et al. (2004) have summarized that in the Gangetic plains, maize crop removes 24.3 kg N, 6.4 kg P, and 18.3 kg K to produce a ton grain. Actually the crop partitions about 63% N, 63.9% P, and 12.5% K into grains and the rest is held in stalks and leaf. About 35–40% of nutrient recovered into shoots can be recycled. The ratio of major nutrients (N:P:K) drawn into maize shoot is 1.2:0.66:1. The variations in nutrient recovery, both in terms of quantity and ratio could be ascribed to soil types, their fertility status, crop genotype, agronomic practices, environment, and yield goals. Maize grown on fertile soil with high input technology that yields between 9 and 10 t grains and 11 t forage ha^{-1} absorbs 270 kg N, 50 kg P, 182 kg K, 31 kg S, 124 g Cu, and 520 g Zn ha^{-1}.

The average N uptake by forage maize ranges from 180 to 240 kg N ha^{-1}. Typically, maize removes 1.4 kg P t^{-1} fresh forage or about 50–55 kg P_2O_5 for 40 t ha^{-1} forage. The demand for K_2O is much higher. Maize crop that produces 40 t forage removes about 320 kg K_2O ha^{-1}. It is equivalent to 8 kg K_2O day^{-1} during the crop season. However, maximum uptake rates occur during July and August and it coincides with rapid dry matter build up. Forage maize grown under rain fed conditions that produces 30 t forage ha^{-1} removes 140–160 kg K_2O.

In North Central China, maize crop provided with optimum fertilizers produces about 7–7.5 t grain plus forage based on harvest index. On an average it recovers about 191–210 kg N, 23–27 kg P, and 148–189 kg K ha^{-1} (Wang et al., 2008).

Nutrient Loss from Maize Belts

In general, nutrient loss from a maize field or cropping zone occurs due to a range of natural factors related to soil and environment. In addition, cropping also causes loss of soil fertility. A significant amount of nutrients are removed from field when ever grain/forage is harvested and transported away from the location. Soil fertility loss perpetuates and gets accentuated due to incessant cropping without appropriate fertilizer-based nutrient replenishment schedules.

Nutrient mining may not be a problem in the "Corn Belt of USA", because replenishment schedules are commensurate or at times more than required by the crop. Yet, sizeable nutrients are removed and transported out of the ecosystem. Macronutrient removal due to maize grain and stover harvest is as follows:

Macronutrient	N	P	K	Ca	Mg	S
Grain kg ha^{-1}:	147	26	35	3	14	12
Stover kg ha^{-1}:	78	15	179	39	38	18

Source: Patzek, 2008.

However, in the "US Corn Belt" nutrient loss via natural processes is also severe. For example, average soil erosion during past 2–3 decades has fluctuated between 12.0 and 16.9 t ha^{-1} yr^{-1}. The surface soil loss depreciates soil fertility perceptibly depending on nutrient carrying capacity of soil particles (Patzek, 2008).

"Soil Nutrient Mining" or "Soil Mining" is a term used to denote loss of soil minerals and fertility in general. It is a situation commonly experienced in maize belt when nutrient removals exceed beyond that supplied. In many areas, fertilizer and/ or organic manure replenishment is either nil or meager. Soil mining has been felt as a generalized phenomenon in most parts of Asia, Sub-Saharan Africa, Brazilian Cerrados, and Pampas, etc. For example, Van der Pol and Traore (1993) reported that in Mali, semiarid cereal belt experiences negative nutrient balance. The nutrient extraction by maize crop is consistently higher, added to it loss via erosion, leaching and volatilization surpass that replenished through crop residue recycling, atmospheric dusts and organic matter application. Nutrient depletion in maize fields was about 25 kg N, 20 kg P, 20 kg K, and 5 kg Mg ha^{-1} per crop. As time lapses, cumulative loss results in soil mining and loss of soil fertility, rather drastically. In general, within Sub Sahara, annual loss of soil nutrients during 1990s was 22 kg N, 2.5 kg P, and 1.5 kg K ha^{-1} (Weight and Kelly, 2009). It could be attributed to nutrient removals through crops and losses via soil erosion, leaching, volatilization, etc. In addition, there was a perceptible long term depletion of soil organic matter (SOM). In some of the savanna zones where maize is important, SOM loss amounted to 30% of original levels. No doubt, soil mining gradually reduces crop productivity.

Nutrient mining is rampant in small farms of eastern African. The subsistence farming measures may not suffice to replenish nutrient removal through cropping and natural factors. Survey by International Maize and Wheat Improvement Center (CIMMYT) has clearly shown that subsistence and low input techniques have lead to depreciation of soil fertility because of nutrient mining (Kumwenda et al., 1996). In Uganda, soil nutrient depletion, in other words soil mining is supposedly the main factor affecting optimum nutrient dynamics in the maize farming zones (Nkonya et al., 2004). Further, soil nutrient mining is more severe in the small farms that practice subsistence, no or low fertilizer input technology. Nutrient balance in almost 95% of small holding maize farms was negative showing nutrient depletion. Next, in the Indian maize belt, initially around 1970s when maize cropping was getting accentuated, only N deficiency was perceivable. As time lapsed, P and K deficiency appeared due to mining. Obviously, replenishment was not commensurate to loss via natural factors and cropping. During 1970s and 1980s, major nutrient deficiencies were overcome by chemical fertilizer application. Then, deficiencies of secondary nutrients especially S and micronutrients like Zn and Fe became conspicuous by 1980s. This phenomenon of depletion of soil nutrients was mostly caused by incessant cropping and lack of appropriate replenishments. Information on extent of nutrient mining and balances in the maize growing states of India is available (see Krishna, 2010). Almost a similar trend in nutrient deficiencies were noticed in maize cropping zones of other South Asian countries, such as Bangladesh and Pakistan. A recent study dealing with nutrient budgets suggests that in the maize farms, usually N and K inputs are less than outputs, hence in a long run N and K mining seems inevitable. Whereas, Nutmon—tool box based estimations point out that a positive balance exists for soil-P status. Following is an example from maize farms of Tamil Nadu, in India:

Nutrient	Input	Out Put	Balance
	-----------	kg ha^{-1}	------------
Nitrogen	32.6	41.5	− 8.9
Phosphorus	9.2	8.9	+ 0.3
Potassium	18.0	9.2	-1.2

Source: Surendran et al. 2005.

In the Southeast Asian maize belts, in addition to soil mining via regular cropping, soil erosion is a major cause of deterioration and loss of fertility. Nutrient removals via runoff, erosion, and leaching are significant. In addition to deterioration of physico-chemical traits of soils, appreciable quantity of SOM is lost due to erosion. It results in loss of soil quality (Hermiyanto et al., 2007). In the Chinese maize growing regions, soil mining is mostly felt in subsistence farms of Northwest and South. However, in Northeast, high input technology envisages sufficient levels of nutrient replenishment. In the low input zones, incessant farming has lead to depletion of soil nutrients and fertility in general. Decline in soil fertility and nutrient mining is easily attributable to both cropping and natural factors. Field trials were conducted at Shandong Agricultural Center to assess nutrient loss from maize fields. Nutrient loss was easily attributable to high rainfall events and irrigation. The extent of N, P, and K loss was greater during early stages of maize crop. Nitrate (12.9–46.5 kg N ha^{-1}) was the major form of N loss followed by NH_4^+ (1.66–10.10 kg N ha^{-1}). During the same period, available-P lost was 0.15–0.24 kg P ha^{-1} and available-K lost was 7.1–13.0 kg K ha^{-1}. The leaching loss got reduced if inorganic nutrients were supplied along with FYM (Li et al., 2008). In Nepal, maize cultivation proceeds both in Plains of Tarai and Hills. Soil and nutrient loss from maize fields is indeed a major problem in the hills. Annual loss of soil from agricultural fields ranges from nil to 105 t ha^{-1}. Atreya et al. (2005) state that, annually about 300 kg organic matter, 15 kg N, 20 kg P, and 40 kg K are lost ha^{-1} from the maize growing regions. Soil loss was greater during May. Strip cropping, application of mulches, incorporation of legume residues and adopting suitable cropping sequences are some of the suggestions to thwart soil erosion and nutrient loss (Acharya, 1999).

The examples stated above make it clear that during recent decades, incessant cropping has removed large quantities of nutrients out of maize belts via grains/forage. A sizeable loss of nutrients via natural processes like soil erosion, leaching, seepage, volatilization, and emissions has accentuated the problem. The nutrient mining that ensues has definitely impaired nutrient dynamics in many locations. This has lead to low nutrient turnover rates and reduced crop productivity. Appropriate measures to replenish nutrient in amounts that match the crop needs is essential.

NITROGEN DYNAMICS IN MAIZE AGROECOSYSTEM

Mineral Nitrogen Distribution

Knowledge about distribution of mineral N in the soil profile is important. Nitrogen distribution is actually affected by inherent characteristics of soil, as well as amount and nature of fertilizer-N supplied to maize fields. Mineralization/immobilization patterns, recovery by crop, leaching, seepage, gaseous losses, and volatilization, all of these factors also affect soil-N distribution. Influence of above factors on soil-N distribution may after all be different based on location of maize belt and soil types. Nyamangara (2007) studied effects of fertilizer-N and organic manure application on temporal variations in soil-N distribution in a Sub-Saharan location. On sandy soils, mineral N was highest in surface layer if both fertilizer-N and organic manure were supplied. It reached 60 mg kg^{-1} soil in the top most layers. Soil-N accumulation was least in fields given only organic manure or control fields. It was 20 mg kg^{-1} soil. Mineral N concentrations in the entire profile decreased as seasons progressed. It was attributable to recovery by maize plants. Partial immobilization as organic-N was another factor affecting N distribution. Actually, immobilization suppressed undue leaching of mineral-N. Yet, there was clear evidence for movement of mineral N from surface to subsurface horizons during early stages of the crop. Hence, it was advisable to apply a small fraction of N in organic form. In fact, Nyamanagara suggested that only 30% of fertilizer-N stipulated be added at planting (i.e., 10–20 kg N ha^{-1}). The remaining N could be split in 2 or 3 portions and applied as season progresses. This procedure may allow us to attain better temporal distribution of soil-N. Further, soil-N may localize closer to maize roots. Shepherd et al. (2001) studied NO_3-N distribution within Ferralsols in Kenya. Here, maize is cultivated under subsistence farming and without extraneous inorganic fertilizer supply. They reported that soil NO_3-N measured till a depth of 2 m varied depending on physiographic location (e.g., ridge top, slopes, etc.) parent material and cropping pattern. Soil NO_3-N ranged from 0.1 to 10 mg N kg^{-1} in the top 0.25 m; and from 1.0 to 16 mg N kg^{-1} in subsoil (0.5–2.0 m). Sub-Soil NO_3-N was not related to anion exchange capacity, plant N status or mineralization patterns. Instead, effective cation exchange capacity (CEC) could be utilized to predict subsoil NO_3-N status of soils. The subsoil NO_3-N seems important while predicting N requirements of a maize crop. Let us consider an example from Thailand. Sipaseuth et al. (2007) found that their prediction of fertilizer needs based on most of the crop models did not match actual situation. Instead, often prescriptions were much higher than that needed by the crop. As a result NO_3-N tended to accumulate in the subsoil layers. No doubt, maize roots were prolific at 40–60 cm depth and matched N distribution in upper layers. However, maize roots also spread sufficiently deep and explored subsoil-N present in layers beyond 1.2–1.5 m depth. In fact, Sipasueth et al. (2007) found that subsoil NO_3-N declined significantly after each crop of maize. Roots grew into subsoil and explored NO_3-N perceptibly. They concluded that within the maize growing regions, it is necessary to have a prior knowledge about N distribution in both surface and subsurface layers. Most importantly, quantum of subsoil-N needs to be included while devising fertilizer-N supply schedules.

Field trails in the Indus plains have shown that irrigation affects NO_3-N distribution significantly during maize production. Usually, NO_3-N accumulation is relatively low (35 mg N kg^{-1}soil) in the surface layer and it progressively increases with depth. Greatest accumulation of NO_3-N (57 mg NO_3-Nkg^{-1}soil) was traced in deeper layers of soil profile at 90–120 cm depth (Niaz et al., 2007). Long term trial extending for 16 cropping cycles, on Ustochrepts in Northwest India has shown that N distribution in soil profile is significantly affected by fertilizer supply schedules. Annual supply of 320 kg N ha^{-1} without P and K resulted in peak N stratification at 135 cm depth. However, if N is applied with P and K, recovery was higher and it did not result in peak accumulation or stratification of N. Repeated inputs of FYM did induce N accumulation in the subsoil. The amount of N accumulation in subsoil decreased as crop harvests tended to be higher (Benbi et al., 1991).

Nitrogen Inputs into Maize Agroecosystem
Nitrogen derived from Atmosphere
The atmospheric sources of N derived mainly through dust, storm or precipitation events could range from negligible to almost 50–60 kg N ha^{-1} yr^{-1} depending on geographic locations, closeness to industries, sea shore and farming practices *in situ*. Generally, atmospheric N sources are small. Quantitatively, its share in maize cropping is not comparable to those derived via fertilizers, organic manures, mineralization or symbiotic N fixation. In the 'Corn Belt of USA', a maize crop may derive some amount of N via atmosphere that needs to be considered while devising fertilizer-N schedules. Atmospheric sources may contribute nil to 25–30 kg N ha^{-1} yr^{-1}.

Plate 3.1. *Top:* An Imminent Rain Storm near Dosso/Gaya, Niger, a location in the Transition zone between Guinea-Savanna and Sahel in West Africa. Maize and Sorghum become conspicuous in the Guinea-Savanna and vegetation gets denser. The Rain storms that occur intensely bring in massive amount of dust and sand prior to down pour. Nutrient transfer from a location to other is indeed massive. Further, sand deposits may drown maize seedlings.

Bottom: A sand/dust storm in progress in the Southern Niger/Nigeria border region of West Africa. Large amounts of sand along with nutrients shift from one field to other. They induce loss/additions of nutrients via atmospheric deposits. *Harmattan,* is a massive sand storm common to Sahel and Guinea-Savanna regions in West Africa. It again causes large deposition/loss of nutrients from cropped zones.

Source: Dr. K.R. Krishna, ICRISAT, Hyderabad, India.

In the European Plains, sizeable quantity of soil-N is lost via NH_3 volatilization, also as NO_2, N_2O, and N_2 emissions. It is said that a share of it is re-deposited into crop land in a few days. As such atmospheric N depositions in major maize growing regions of Western and Eastern Europe is relatively high and ranges from 25 to 65 kg ha^{-1} annually. In Germany, airborne N input was reported to be 20 kg N ha^{-1} yr^{-1}. The ^{15}N based analyses at long term trial locations such as Broadbalk in England suggest that up to 65 kg N ha^{-1}yr^{-1} could be accrued into maize forage fields due to atmospheric deposition. In other locations of Europe such as Prague in Czeckoslvakia, Rothamsted in England, Bad Lauchstad in Germany, atmospheric N deposits into farming areas ranged from 45 to 58 kg N ha^{-1} yr^{-1} (see Krishna, 2002).

In the Guinea-Savanna regions of West Africa, atmospheric dusts, precipitation and sand storms like *Harmattan* may add 2.2 to 5 kg N ha^{-1} in a season (Plate 3.1). Of course, N deposited depends on enrichment ratios of sand particles. We should note that N is also removed from a field due to sandstorms. Therefore, N input reported is actually a difference between N deposited in a given time minus that lost due to storms. Timely application of mulches (crop residues) to trap *Harmattan* dusts is important. It improves nutrient trapping capacity of a maize field. In Semi-arid plains of South India about 6–8 kg N ha^{-1} could be added to maize fields via atmospheric deposits and precipitation (Murthy et al., 2000).

Nitrogen Supplied through Chemical Fertilizers

During recent years, N dynamics in almost every major cropping zone has experienced a definite change. Nitrogen supply has increased compared to previous levels. The N turnover in maize fields indeed has enhanced remarkably with the advent of chemical fertilizers. Urea, Ammonium nitrate, Diammonium phosphate are some of the common N fertilizers used by maize farmers worldwide. Maize belts in particular were impinged with higher amounts of N in order to enhance productivity. In the "Corn Belt of USA", N inputs were originally derived predominantly from residue recycling and FYM. Legumes grown in rotation added small amounts to soil-N status. However, N supply into the agroecosystem increased gradually initially in 1940s and rather sharply since 1950s as chemical fertilizer technology got standardized. Nitrogen inputs were progressively revised upwards to suit the fertilizer responsive high yielding composites and hybrids. Fertilizer-N inputs were also increased to support the higher yield goals. Several types of soil and plant tests were used to ascertain crop's need for N more exactly. In fact, N has been considered as one of the best inputs in the "Corn Belt of USA". It is estimated that about 3.6 Tg of N fertilizer was impinged to "US Corn Belt" at a cost of over 800 m US $ during 2008 (Stanger and Lauer, 2008). During recent years, computer–based programs that simulate N requirements versus yield goals have been generously utilized to recommend N inputs and predict grain yield. A step further, N input has also been guided by economic returns. Fluctuations in cost of N fertilizer and returns from grain yield has been given due importance while deciding fertilizer-N inputs. Following is an example:

Cropping System	Economic Return			
	(Price Ratio = $ kg^{-1} N / $ kg^{-1}grain)			
	0.05	0.10	0.15	0.20
Soybean-Corn (kg N ha^{-1})	171	135	121	106
Corn-Corn-Corn (kg N ha^{-1})	225	197	170	157

Source: Sawyer, 2007.

Note: Yield goals were 10–11 t grains ha^{-1} plus forage.

It should be emphasized that until now, factors like soil, crop genotype, environment, and agronomic procedures dictated N inputs and crop productivity equations.

However, with the advent of computer-based simulations, nutrient dynamics in maize belts have also been influenced by economic considerations rather routinely. Field scale evaluations at Iowa State University farms during 2000 to 2006 has revealed that economically optimum nitrogen rates (EONRs) for maize was 135–145 kg N ha[-1] if it is soybean-corn rotation and 190–210 kg N ha[-1] if it is corn mono-crop grown year after year (Sawyer, 2007). The EONR is a good example of how economic benefits affect nutrient dynamics and other natural processes within maize agroecosystems.

The fertilizer N rate in US Corn Belt has varied with inherent soil fertility status and yield goals. For example in Central Iowa, low yielding maize plot is provided with 67 kg N ha[-1], medium rate is 135 kg N ha[-1] and high yielding plot receives at least 172–240 kg N ha[-1] (Jaynes et al., 2001). Much of the fertilizer is supplied as Urea or Ammonium Nitrate with or sometimes without urea and/nitrification inhibitors. Halvorson et al. (2006) have reported that in Great Plains area, optimum N input for maize ranges from 168 g N ha[-1] to 228 kg N ha[-1]. Maximum grain yield was obtained at 260–280 kg N ha[-1] depending on tillage system adopted. A maize silage crop in the Appalachians receives between 57 and 224 kg N ha[-1] depending on yield goals. However, on the piedmonts, best forage harvest could be obtained by supplying 170 kg N ha[-1] (Staley and Perry, 1995).

In the Rolling Pampas, maize mono-crops planted under no-till (NT) systems are provided with 50 kg N ha[-1]. The side dress N dosages are dependent on soil fertility and yield goals (Plate 3.2).

In the Nigerian rainforests, maize farmers supply 60–70 kg N ha[-1]. However, in Savanna, 100–120 kg N ha[-1] seems economically optimum. In the Southern Guinea Savanna, a maize crop that yields 3 t ha grains is provided with 50 kg Nha[-1] (Kogbe and Adediran, 2003). It is said that N inputs >50 kg ha[-1] removes N dearth if any during hybrid maize production. Split application of fertilizer-N is also in vogue in African maize belts (Plate 3.3). In Kenya, maize responds to fertilizer-N inputs. On Red and Brown soils, farmers supply at least 50 kg N ha[-1] to achieve optimum grain harvests of 2–5 t grains ha[-1] (Smaling et al., 2006).

In the Indus Plains, a maize silage crop with a yield goal of 20 t ha[-1] is supplied with 75 kg N, 60 kg P, and 60 kg K ha[-1]. Nitrogen input is split and the second split is usually supplied at 45 DAP (Niaz et al., 2007). Whereas P and K are applied at sowing.

Nitrogen input to maize fields in Southern Indian plains varies enormously depending on soil type, irrigation facilities, genotype, and yield goals. Maize cultivated as dry land crop under rain fed conditions may receive 20–40 kg N ha[-1] plus a top dress of 20 kg N ha[-1] at knee high stage. Rainy season crop is supplied with 40–80 kg N ha[-1] depending on soil tests. Irrigated maize with relatively higher yield goal of 4–5 t grain is supplied with 80–120 kg N ha[-1]. Often N is supplied as both inorganic fertilizers and organic manures. Usually 30–50% of total N envisaged is supplied as organic manures. Maize is often rotated with legumes. In that case, it is preferable to calculate N requirements of entire sequence and apply N to fields. Legumes may contribute a certain amount of N.

Plate 3.2 Nitrogen fertilizer application to no-tillage fields in the Rolling Pampas of Argentina.
Source: Dr. Carina R. Alvarez Facultad de Agronomia, Universidad de Buenos Aires, Argentina.
Note: Nitrogen fertilizer is applied in 3 to 4 side dresses.

Plate 3.3 Nitrogen Top Dressing done manually in East Africa.
Source: CIMMYT, Mexico.

On typic Kandiudults found in Malaysia, optimum N rate recommended for maize production fluctuates between 240 and 275 kg N ha^{-1} (Ahmed et al., 2009).

In Queensland, maize is cultivated in most parts of the province. Soils often get exhausted of nutrients through various rotations. Nitrogen is replenished based on rainfall pattern and irrigation facilities. In dry lands, maize is initially supplied with a starter dose of 5 kg N ha^{-1} and rest 40–75 kg ha^{-1} is applied in split doses at various stages of crop. The irrigated crop may receive up to 20 kg N ha^{-1} as starter and rest 160 kg N ha^{-1} in splits at 25–30 day intervals.

Nitrogen is also supplied to a maize crop using foliar sprays. Most commonly, a solution of Urea (0.2%) or KNO_3 (1–2.5%) in water is sprayed at seedling/tasselling stage. In a foliar spray, N sources are absorbed through leaf tissue and it avoids contact with soil phase of the agroecosystem. Many of the soil related transformations that fertilizer-N undergoes belowground in soil is absent with foliar fertilization. Nitrogen is required in relatively lower dosages if sprayed on foliage. Therefore, fertilize-N efficiency is generally high. Foliar sprays are also common on maize grown for forage.

The chemical N fertilizers are also channeled into maize belts using irrigation water. It is sometimes referred as fertigation. Fertigation is a process where in fertilizer-N is dissolved and allowed to spread across the field in a dissolved state along with water. Fertilizer-N required is relatively small compared to soil application (broadcast or placement). A study in the Central Platt Valley of Nebraska, has shown that utilization of N supplied using irrigation depended on crop's need and amount of N supplied using other systems. The number of side dress N envisaged and concentration of N in irrigation water also affected N recovery. Due care is needed to avoid excessive leaching or seepage to ground water (Francis and Schepers, 2005). It is interesting to note that generally, irrigation source may itself contain a certain amount of dissolved N in it. The N content in irrigation water may be dependent on source and intensity of farming practiced in the area. For example, in western parts of Corn Belt of USA, Halvorson et al. (2006) have reported that irrigation water contained 2.8–3.6 mg NO_3-N L^{-1}. It amounted to supply of 12–15 kg N ha^{-1}.

In some farms of USA, Europe, and Australia, extraneous N is also supplied into maize fields using liquid NH_3 dissolved in irrigation water. Pressurized NH_3 is dissolved into irrigation water source and allowed to mix with soil. Due care is needed to avoid undue loss of NH_3 through volatilization. Liquid ammonia is usually added in cool conditions and during night to avoid evaporation. Fertilizer-N efficiency is higher if liquid NH_3 is utilized compared to broadcasting or banding of powder or granules.

Nitrogen Supplied through FYM, Crop Residue and Green Manures
Farm Yard Manure has been the main organic source of N in most of the maize farming belts. As stated earlier, before the advent of inorganic source, rates of FYM applied almost decided N supply to maize crop. Nutrient dynamics that ensued in the ecosystem was regulated by FYM inputs. The rapidity of FYM decomposition actually dictated the release of N and other nutrients into soil solution. However, higher yield goal, availability of inorganic fertilizer, high yielding maize genotypes, and high yield goals have necessitated higher inputs of N. Maize farmers have used a wide range of organic manures derived mostly from crop residues, vegetation, animals, industrial wastes, and effluents. Often, it is preferable to apply organic manure a couple of weeks ahead of planting to allow proper mineralization. The nature of organic manure, its

physical form, N concentration and ratio of other nutrients are crucial. The extent of N derived also depends on soil traits such as texture, moisture content, microbial load, temperature, season, and timing of organic manure supply. Following are few examples of organic manures utilized during maize farming and their percentage N contents on dry matter basis:

Percent N in Organic Manures

Crop and Tree residues: Paddy straw = 0.36; Rice hulls = 0.3–0.5; Sorghum straw = 0.41; Pearl Millet straw = 0.65; *Cassia auriculata* = 0.98; *Careya arborea* = 1.67; *Terminalia chebula* = .46; *Terminalia tomentosa* = 1.39; Cowpea = 0.71; Dhaincha = 0.62; Guar = 0.34; Kulthi = 0.33; Mungbean = 0.72; Black gram = 0.85; Sun hemp = 0.75; Groundnut husks = 1.0–1.8; Farm Yard Manure (general) = 0.4–1.5. Compost= 0.4–0.8; Groundnut cake = 7; Coconut cake = 3

Animal Manures: Cattle dung = 0.3–0.4; Sheep Dung = 0.5; Rural Compost = 0.5–0.76; Vermi-Compost = 1.6; Piggery Slurry = 0.8-1.2; Guano deposits = 0.80–1.4; Poultry manure =1.08–1.2.

Industrial Byproducts, Effluents, and Wastes: Coir Pith = 0.26-0.35; Coir Pith Compost = 1.3; Press mud =1.12; Distillery Yeast Sludge = 1.45; Paper Industry wastes = 1.25

Reports from CIMMYT in Mexico suggest that maize grown on recently cleared tropical forest may derive about 45–55 kg N via organic residues. Maize grown in high rainfall tropical savanna may derive up to 15 kg N ha^{-1} from residues if planted after long fallow and about 5 kg after a short fallow. Maize planted in low rainfall tropical savanna may derive 2–4 kg N ha^{-1} from the organic residues.

Supply of N via organic manures is common to all regions where ever maize is cultivated. The extent to which N is supplied via FYM or other organic manures depends on inherent soil fertility status, availability of organic manures, timing, agronomic procedures, and yield goals. Organic farming may stipulate supply of entire N via FYM or animal manures. The N supplied to soil depends on quality of organic manure, its C:N ratio, succulence, microbial load of soil, soil temperature, and other environmental parameters.

The amount of FYM or other organic manures applied varies widely depending on region, season, cropping intensity, soil type, its inherent fertility, cropping systems, and yield goals. In some of the low input subsistence farming zones, availability of crop residues or other organic sources decides extent of N derived. For example, FYM inputs are low and range between 2 and 5 t ha^{-1} in Sahelian West Africa. In this region, crop residue is precious and utilized to feed farm animals. Animal manures recycled are also small. Natural bird dropping is known to add 2.2 kg N ha^{-1} annually. In the dry lands, farmers again recycle only small quantities of crop residues ranging from 2 to 5 t FYM. Nitrogen derived may range 10–20 kg N ha^{-1}. Hence, grain yields are low at 0.8–1.5 t ha^{-1} in such subsistence belts. In the moderately intense farming zones of Argentina, Mexico or South India where crop yields range from 4 to 5 t grains ha^{-1} plus forage, maize farmers apply commensurately more, at least 10 t FYM or animal manure ha^{-1}. It improves soil quality, induces soil nutrient transformation at optimum rates and supplies N at higher rates to maize roots. In the intensive maize farming belts of USA, Europe, and Northeast China grain yields range from 7 to 11.5 t ha^{-1} and a sole forage crop may produce 30–40 t biomass ha^{-1}. In such areas, farmers supply at least 15 t FYM ha^{-1} plus inorganic fertilizers. The ratio of organic and inorganic source of N is still held optimum. It stabilizes soil quality, SOM and supplies 50–60 kg N ha^{-1} in organic form.

The proportion of N supplied via inorganic fertilizers and organic residues seems important. For example, in maize cropping zones of Michigan, in USA, it is common to add inorganic fertilizer-N [(NH$_4$)]$_2$SO$_4$ plus organic residue derived from alfalfa or soybean. Maize derives N from soil, fertilizer, and legume residues at different proportions (Hestermann et al., 1987). The ^{15}N based analysis has shown that proportion of N derived from legume residue decreased from 44 to 19% in the whole plant and from 57 to 23% in grains, if inorganic fertilizers impinged into soil increased from 56 to 168 kg N ha^{-1}. Therefore, ratio of N supplied through chemical fertilizers and organic residues is important. In Northern Italy, maize grown for forage is supplied with large doses of N both in inorganic and organic forms. Totally they may supply 215–285 kg N ha^{-1}. Organic manures such as FYM or cattle slurry are incorporated ahead of planting at high rates. Since extraction of N supplied is around 48–53%, this procedure induces N accumulation in soil. Such organic manures also improve C accumulation in soil. It has been reported that FYM or cattle slurry created a 45% surplus N and 18% surplus C in soil. It improves soil quality and residual N could be utilized by succeeding crops (Griganani et al., 2007).

In Kenya and other East African nations, it is a normal practice to grow perennial agro-forestry species and hedge trees around maize fields. These trees offer foliage rich in nutrients that could be recycled onto maize fields. Upon decomposition, such organic manures could supply sufficiently large amounts of N (150–200 kg N ha^{-1}) and other nutrients. Dry leaves and twigs from trees such as *Tithonia, Senna,* or *Calliandra* supply on an average 3.2–4.7% N upon incorporation and decomposition in maize fields. Such organic input becomes vital in subsistence farming zones (Kimetu et al., 2001) because there is no other source of N.

Incorporation of weeds prior to flowering/seed set is a common practice in maize farming belts. It is a very useful procedure in subsistence farming areas, since it adds to SOC and nutrients, especially N. Field trials at Bad Lauchstadt, in Germany have shown that weed incorporation into maize fields prior to sowing is a useful procedure. Nitrogen present in weeds is effectively utilized by maize crop. The ^{15}N tagging of weed residues has revealed that during first season after incorporation of weed residue, maize recovers about 2–23% N present in weed residue. In the following year about 7–32% N still available in weed residues is exploited by maize roots. In all, weed residues contributed 7–16% of total N garnered by maize crop during first year and 4–11% in the second year (Merbauch, 1998).

Nitrogen Input via Atmospheric N fixation by Soil Microbes

The extent of N derived from asymbiotic N fixation in the rhizosphere and surrounding root zone soil, depends on factors related to microbial inoculant, crop genotype, soil type, and environmental parameters. Worldwide, there are innumerable field trails conducted to ascertain extent of N derived by maize or other cereals from asymbiotic N-fixers such as *Azospirillum, Azotobacter,* and several other bacterial species. On an average, a maize crop may derive 20–40 kg N ha^{-1} through asymbiotic N fixation. On a field scale, N contributed by asymbiotic N-fixers may be relatively small in quantity. Often crop responses to inoculation with asymbiotic N-fixers could be insignificant. However, given the size of maize belts in different continents, blanket application of microbial inoculants does add large quantity of N. It definitely improves N dynamics in the cropping ecosystem. Nitrogen inputs to maize agroecosystem could be improved via adoption of appropriate maize genotypes, microbial inoculant mixtures and agronomic procedures. In the Sudano-Sahelian zones and Guinea Savanna regions of West Africa, N fixation during fallows may add 5–10 kg N ha^{-1} to cereal fields (see Krishna, 2008). In Southeast Asia, maize derives moderate amounts of N through asymbiotic N fixation by soil bacteria. In addition to it blue green algal patches in maize fields also add to soil-N.

Nitrogen Credit from Preceding Legume Crop

There are indeed several reports that depict the benefits of growing legumes before maize. This aspect of N benefits through symbiotic N fixation to succeeding cereals or even agroecosystem as a whole has been reviewed periodically. Leguminous crops add to soil-N status through symbiotic-N fixation. Legume residues are rich in N content. Therefore, its incorporation into maize fields adds to soil-N upon mineralization. The quantum of N benefit derived from legume species varies based on factors related to soil, crop species, and environment. Let us consider a few examples. Sawyer et al. (2007) report that in the "US Corn Belt", especially in Iowa, corn after corn sequence is generally supplied with 240 kg N ha^{-1}. It is about 55–60 kg N ha^{-1} more than that supplied to a soybean-corn sequence (160–180 kg N ha^{-1}). In this case soybean, a nitrogen fixing legume crop seems to contribute >55 kg N ha^{-1}. In the Northeast, especially in New York State, farmers are advised to apply a small starter N of 34 kg N ha^{-1} and side-dressing from 0 to 168 kg N ha^{-1} depending on EONR. The silage maize is rotated with forage legume-grass (FLG) sod. Lawrence et al. (2008) suggest that N inputs could be regulated to just starter dose depending on composition of FLG and N released upon its decomposition. The optimum grain/forage yield was achieved at 150 kg N ha^{-1} side dress. Fertilizer-N inputs depended on N credits derived from previous FLG sod. The N credits derived from sods depended on extent of legume in the mixture. The N turnover attributable to sod incorporation and decomposition ranged from 93 to 185 kg N ha^{-1}. In Nebraska, corn grown after soybean may derive up to 65 kg N ha^{-1} advantage over a field with continuous corn (Pikul et al., 2005).

In Northern Guinea Savanna of Nigeria, farmers consistently practice legume-cereal rotation because they wish to take advantage of N credits possible from legume. Again the quantum of N credits derived may vary depending on legume species and its genotypes, soil type, and environmental parameters (Obogun et al., 2005; Sanginga, 2003). Obogun et al. (2005) have reported that depending on soybean varieties, the N credit derived by succeeding maize ranged from 57 to 95 kg N ha^{-1}. In a later study, Obogun et al. (2007) attempted to ascertain N shortfall during maize production and to see if it could be corrected either partly or entirely using legume N credits. The N credit from soybean tended to be low. It ranged from 5 to 22 kg N ha^{-1} depending on soybean variety. Yet, succeeding maize did grow better and yielded high because soybean shed leaves and the N derived from leaf fall was significant. However, Obogun et al. (2007) have cautioned that sometimes, enhanced productivity of maize after a legume crop could also be due to several other factors in addition to N credits. During lablab-maize rotation, N credit derived from legume was significant and it got reflected in higher maize yield (Obogun et al., 2007).

Plate 3.4. Cowpea (top) and Pigeonpea (bottom) are rotated with maize in South India.
Source: Dr. K.R. Krishna, Bangalore, India.
Note: Legumes grown in rotation with maize contribute significant amounts of N to soil through their ability to form symbiotic association with *Rhizobium*. The extent of N fixed varies. The contribution of legumes to soil-N ranges from nil to 120 kg N ha^{-1} depending on legume species, *Rhizobium* isolate and environment.

In Southern Africa, particularly on sandy soils of Zimbabwe, maize is rotated with legumes like soybean, cowpea, or macuna. Chickowo et al. (2004) report that in the first season, soybean accumulated 82 kg N ha^{-1} and macuna 87 kg N ha^{-1} in the biomass. Maize that succeeded soybean garnered 14.8 kg N ha^{-1}; 16.4 kg N ha^{-1} if it succeeded macuna and 10 kg N ha^{-1} if grown after *Crotalaria* or *Cajanus*. Clearly, legume cultivation either as intercrop or in rotation is important in subsistence farming zones. Nitrogen derived from legume has direct impact on N dynamics. It could even be the main source of N in many areas where farmers adopt subsistence farming techniques. Over all, N supply into maize belts could be perceptibly affected by legume species. Therefore, choice of legume species planted prior to maize seems important.

In the Indian plains, maize is often rotated with legumes such as pigeonpea, cowpea (Plate 3.4), and groundnut. There are indeed several reports that pertain to extent of N benefits that accrue to maize due to preceding legume crop. The N credit derived from legumes may vary depending on crops species, its genotype, season, rhizobial isolate, soil, and environmental characteristics. On an average, 20–65 kg N ha^{-1} could be derived from preceding legume (Kumar Rao, 1996; Krishna, 2003; Krishna, 2008; 2010; Venkateswarulu, 2004). In the general course, N credits derived from legumes should be included while devising N input schedules to succeeding maize crop. In dry lands, N credits play a vital role during maize production. Resource poor farmers tend to supply very low quantities of chemical fertilizers, only about 20 kg N ha^{-1}. Rest of N has to be derived from residual-N (RSN) from preceding legume crop.

Peanut-maize rotations are common in Thailand. Peanut contributes significant amounts of RSN to succeeding maize crop. Estimates indicate that about 100–130 kg N is stocked in peanut stover that is generally incorporated to be utilized by following maize. The ^{15}N analysis has shown that 16–27 kg N ha^{-1} was derived from peanut residue (McDonagh, 1993). Nitrogen contributed by peanuts to following maize needs to be considered carefully during intensive cropping.

Nitrogen Mineralization

Nitrogen mineralization adds to available-N pool of soil that supports maize production. Nitrogen mineralization is a spatially variable biochemical process. Heterogeneity in SOM and microbial activity are prime causes. Soil environmental factors too affect mineralization rates. In addition, variable recovery of N by plants may also influence N mineralization. On a wider scale, vegetation that soil phase of the ecosystem supports may be an important factor. Brye et al. (2003) aimed at quantifying differences between natural prairies and maize cropping zones with regard to N mineralization patterns. They found that in 4 years, N mineralization rates were significantly higher in maize fields compared to tall grass prairies. They found that tillage made a significant difference. Generally, chisel ploughed maize fields supported higher rates of N mineralization compared to NT systems. Nitrogen derived from mineralization of SOM could provide for nearly 40–50% of total N uptake by maize crop. On humid and sub-humid silt about 1.0 kg mineral N could be derived for each ton of SOM mineralized. Oberle and Keeney (2006) report that irrespective of soil type encountered in the maize farms within Wisconsin, about 3.5% of SOM could be mineralized in a crop season. The resultant mineral-N could be absorbed by maize roots. Field evaluation in Western Nebraska has shown that potentially, a maize crop may derive substantial amount of N from mineralization. It could reach up to 56 kg N ha^{-1} through mineralization during the crop season (Pikul et al., 2005). On an average, 5.5–6.5 kg N ha^{-1} per week could get mineralized in the soils of

Great Plains region. Fluctuations are caused by the cropping system adopted. This mineralized N is available for maize, wheat or other sequences adopted (Weinhold and Halvorson, 1999).

Reports pertaining to farming locations in Pampas of Argentina indicate that soil-N mineralization during crop season averages at 50–58 kg N ha^{-1}, but may fluctuate between 5 and 125 kg N ha^{-1}. Hence, N derived via soil mineralization could satisfy a certain share of total N requirement. In some locations, up to 38% crop's demand for N could be derived through mineralization process. Gonzalez-Montaner et al. (1997) suggest that N mineralization rates are site specific and generally regulated depending on temperature, moisture, SOM, C:N ratios, and timing of fertilizer-N inputs.

In West Africa, N fluxes caused by mineralization of organic matter is an important soil phenomenon that has direct bearing on soil-N availability to maize roots. Different types of organic manures are applied to fields during maize production. The N mineralization rates and N derived (N$_{min}$) may vary widely depending on factors related to soil type, organic source, moisture, cropping season, etc. For example, an evaluation of various organic manures on sandy Oxisol found in Savannas of Nigeria has shown that N$_{min}$ derived ranged from 38 kg N ha^{-1} to 78.1 kg N ha^{-1}. Nitrogen recovery by maize was best under alley cropping with legumes, where in, legume prunings were regularly added to soil. Nitrogen mineralization was highest with prunings from *Gliricidia* > *Leucana* > *Cajanus* > Fertilizer > Control in that order. Among legumes, residue from *Cajanus cajan* supported relatively lower levels of mineralization and release of N (Okonkwo, 2008).

On sandy loams found in the maize cropping zones of Zimbabwe, Kamukondiwa et al. (1998) found that mineralization of recycled maize residue was improved, if fields were already primed with small amounts of inorganic fertilizers. Available-N pool increased proportionately. As a consequence, N recovery and grain yield improved. Zingore et al. (2003) state that, green manures like tree leaves, twigs, and prunings have been promoted briskly in the semi-arid regions of Zimbabwe. It is aimed at supplementing soil-N and sequestering C into sandy soils. However, synchrony between mineralization, actual release of N into available pool and maize root activity is important. Hence, they evaluated leaf material from a few different tree species with regard to quality, mineralization pattern, N release and productivity. Prunings with better C:N ratio and succulence may get decomposed quickly. It leads to release of N$_{min}$ rapidly. Following is an example:

Tree Species	Quality of Tree Prunings	Mineralization (%)	
		C	N
Tithonia diversifolia	High	70	30
Calliandria calothyrsus	Medium	40	10
Accacia angustissima			
Flemingia macrophylla	Low	} 25	5

Source: Zingore et al., 2003.

At a different location in Zimbabwe, Mtambanengwe, and Mapfumo (2006) found that N mineralization again depended on organic source. On soil with <100 mg clay kg^{-1}, *Crotalaria juncea* released 24% N and *Calliandra calothyrsus* released only 13% of N applied as organic manure. However, sizeable fraction of N released was vulnerable for leaching and immobilization processes. Immobilization of N was of higher order >30%, if maize residues or saw dust was applied. However, if C:N ratios are altered the immobilized N in soil does get mineralized and become available to maize roots.

It is interesting to note that in dry lands of South Africa, maize growth and productivity is generally higher immediately under tree canopy than in areas away and in open field. Soil nutrient dynamics, especially N and C mineralization patterns seem to influence this phenomenon rather significantly. Rhoades (2004) reported that N mineralization rates under the *Faidherbia* canopies were consistently higher than in open fields. It clearly improved N nutrition of maize under the canopy. The initial N pulse under the canopy was 60 μg N g^{-1} in the top soil (0–15 cm). During 4 month cropping period, N release into soil was 112μg N g^{-1} but it was a mere 42 μg Ng^{-1} in the open fields beyond canopy. On long term basis, N availability due to mineralization of organic matter under canopy improved appreciably compared with open fields.

Soil tillage and compaction that results due to repeated movement of tillage equipment also induce changes in N mineralization potential of soil. Long term trials and simulations have shown that N mineralization rates were dependent on soil characteristics like bulk density, soil porosity, moisture status, and cropping sequences. Mineralization rates are low in soils with high water content and low porosity (Jensen et al., 1996). Further, mineralization of manures seems to be affected by timing of its application and cropping season. In the Eastern European plains, mineralization is an important process that adds to soil-N pool during maize production. On Chernozems, mineralization of organic fraction and NH_4/NO_3 release was greater during spring (April–June) and summer (August) when soil temperature and moisture conditions were optimum for microbial activity (Maly, 2002). Whenever, NH_4-based fertilizers were added NH_4^+ were immediately transformed into NO_3 through nitrification. The rate of mineralization, especially nitrification was directly influenced by fertilizer-N inputs.

Anaerobically mineralized N also adds to soil-N during maize production. Field tests at INTA, Balcarce, in Argentina have shown that anaerobic mineralization differs between locations. Knowledge about quantity of N derived from anaerobic mineralization could improve reliability of fertilizer-N recommendations. Rozas et al. (2008) have reported that critical pre-plant N test values for 94% relative yield (RY) was 75 kg N ha^{-1} in sites with higher anaerobic mineralized-N and 90 kg N ha^{-1} for sites with low anaerobic mineralized-N. Demarcation of sites/fields based on extent of anaerobic mineralization improved accuracy of pre-plant fertilizer-N recommendation. Clearly, a certain amount of N is derived through anaerobic mineralization during maize production.

Soil and Plant Tests for Appropriate Nitrogen Recommendations
Soil and/or Plant based tests have played key role in regulating N dynamics in most of the farming belts wherever extraneous nutrients are impinged into soils. Techniques to

asses N status of soil/plant and assess exact needs of maize crop have been devised and refined rather consistently during past 4–5 decades. In fact, indirectly, such tests have guided and regulated dynamics of many nutrient elements, not just N in maize belts.

The problems related to estimation of soil-N status, N supply, N recovery rates by maize and productivity may differ based on geographic location. Wherever it is, decision regarding N supply to maize fields has been crucial. Nitrogen supply to maize cropping zones has been vital with regard to nutrient dynamics and productivity. Firstly, N inputs have direct influence on maize productivity. During recent years, N dynamics has been bestowed greater priority both in research and maize production fields for yet another reason. Insufficient N supply limits grain/forage production. However, excess N supply does occur during maize cultivation, especially if fertilizer inputs are based on yield goals or maximum yield. It has resulted in deterioration of soils, aqueducts, ground water, and even atmosphere through emissions (NH_4, NO_2, and N_2). Hence, researchers and farmers employ different soil and plant test procedures and adopt strategies to maximize nutrient use efficiency. Firstly, they aim at supplying quantities of N that matches its absorption rates. It then avoids undue accumulation and loss of N from the agroecosystem. During past 4–5 decades, maize farmers have been exposed to several types of soil and plant tests. Such tests have helped them in arriving at appropriate N inputs. For example, on sandy soils of West Africa, ^{15}N tests have shown that total plant-N and N derived from soil are important indicators of soil-N fertility. Estimation of N available using Warring-Bremner test helps in assessment of soil-N fertility (Bonzi et al., 2003). Overall, N diagnostic tests could be classified into those dependent on analysis of soil and /or plant.

Soil-based Nitrogen Evaluations
Nitrogen occurs in different chemical forms in soil. Mainly, they include inorganic and organic forms. The total N present in soil may vary widely. Often, total N found in soil may be optimum or even more than that needed for crop production and well distributed along the profile. However, total N content and concentration are less directly related to maize growth and productivity. It is the N forms available to maize roots that matter. Therefore, most soil tests aim at deciphering N forms easily available to roots such as soil NO_3 and NH_4. Actually, soil NO_3 level in the root zone is crucial for maize cultivation. Therefore, soil NO_3 based tests have been used worldwide in different maize belts. The critical soil-NO_3 required for optimum growth and grain formation may differ based on soil type, its texture, fertility status in general, agronomic procedure, and crop genotype sown. The critical soil-NO_3 level is defined as that amount of soil-NO_3 which supports formation of at least 95% of maximum grain/forage possible in a given location or environment. In case of a field with low soil-NO_3 and less than stipulated critical level for the soil type, extraneous source as either inorganic or organic N is applied. This step corrects the available-N in soil. The extent of fertilizer-N or organic manure needed to correct the soil-N status and bring it to levels above critical limits may vary enormously. Generally, soil with <10 ppm N (DTPA extractable) is termed low in N fertility. It may need large doze of fertilizer-N to raise available-N level to optimum or above critical limits. Soil with 11–20 ppm N is classified as medium N fertility soil and those with >21 ppm N as high N soils. This

system of estimating soil-NO_3 status, critical limits and fixing fertilizer-N inputs is in vogue in the maize belts since past 5 decades. Obviously, such diagnostic tests have played a key role in regulating N dynamics in the maize belts.

As stated earlier, soil NO_3 based tests have been used in most of maize growing regions of the world. The pre-plant or pre-sidedress estimation of soil-NO_3 has problems with regard to accurate estimations of soil-N status. In the high rainfall belts prone to excessive leaching, NO_3 measurements at the start of crop season may not reflect actual N availability during the entire crop period. Warmer climate in tropics tends to support loss of NO_3 through de-nitrification. Again, soil-NO_3 estimations may not account for such loss via de-nitrification. In addition, soil-NO_3 estimates at planting do not consider N that could be added to available pool via mineralization that occurs during the crop season. Perhaps, estimating or obtaining data about potentially mineralizable-N in soil is more useful. Such factors bring in inaccuracies into soil-test crop response data. Often, fertilizer-N supplied may exceed actual requirement of the crop at different growth stages.

In the "Corn Belt of USA", farmers have adopted several techniques to assess soil NO_3 levels and predict maize grain yield. Most common soil-N tests used during recent years are Illinois Soil N Test (ISNT), the Gas Pressure Test (GPT) and Incubation and Residual N Test (IRNT). The Incubation and RSN test combines residual NO_3, residual NH_4, and NH_4 mineralized from short term incubations. The IRNT seems to have limited success in predicting N requirements and response of maize crop (Williams et al., 2007a). The GPT was originally devised to assess N in manures. The GPT used on soils from Northern Plains and other regions of USA was well related to other plant available-N tests. The correlation value between GPT and other tests suggests that it could be employed to assess soil-N and crop responsiveness (Williams et al., 2007a). The GPT is rapid and needs only 25 min to arrive at soil-N value for a sample (Picone et al., 2002).

Illinois Soil Nitrogen Test (ISNT)

Williams et al. (2007b) state that fraction of potentially mineralizable-N that becomes available to maize crop during the course of a crop season is difficult to estimate accurately. It is attributed to variations in soil moisture, temperature, and microbial activity. Actually, most of the soil mineralizable-N estimations could be grouped into biological and chemical methods. Biological methods involve series of incubations of soil samples and they try to mimic soil conditions and mineralization rates that occur in maize fields. The incubation period is often stretched and it is difficult to use the data in the same crop season to decide N inputs (Bundy and Meisinger, 1994). The chemical N mineralization tests quantify N in specific fractions of soil. They may often show low correlations with available-N and consequent grain yield response. These chemical N mineralization tests are quick and could be used at various stages of the crop easily. During recent years, chemical tests based on organic-N liberated through acid hydrolysis have become popular, both with researchers as well as maize growers in USA and other areas. It is based on a study of estimation of N in organic fractions like amino sugar-N (Mulvaney et al., 2001; 2006; University of Illinois Urbana-Champaign, 2004). The ISNT was developed as a quick method that allows us to judge soil-N and predict maize grain productivity. Actually determining hydro-lysable amino sugar N using distillation method is tedious and not conducive during practical maize production. The ISNT was developed as a quicker alternative for de-termining amino sugar-N by performing alkalization directly on the soil, rather than using the soil hydrolysate and measuring alkali-hydrolyzable amino sugar-N. Both, alkali hydrolyzable amino-N and acid hydrolysable amino sugar-N are well related to corn grain yield response. However, alkali hydrolysable amino sugar-N correlates better with corn grain response (Williams, 2007a). The ISNT could also be used to estimate EONR and prescribe N inputs accordingly (Williams et al., 2007b).

PLANT-BASED NITROGEN EVALUATIONS

Critical-N Concentration in Maize Tissue

Maize forage production is wide spread and occurs at different intensities based on geographical location, soil fertility, and demand for fresh fodder. Fertilizer-N supply is one of the key factors during forage production. In low input subsistence farms, fertilizer-N supply is either nil or confined to that obtained via FYM or residue recy-cling. In high input zones, like European plains or Corn Belt of USA, forage maize receives high amounts of fertilizer-N. The critical N concentration (CNC) in the leaf at a given stage is indicative of, firstly the N status of forage crop. It could also help in judging deficiency or excess of N supplied to forage. The CNC is also used effec-tively to predict maize grain/forage productivity. The critical N in stover and/or leaf varies with stage of the crop and soil fertility level. Hermann and Taube (2005) report that under high input intensive farming conditions of Western European plains, CNC in leaf tissue is 10.5 g kg DM. It is in upper range of soil fertility status and a forage yield of 25–40 t fresh biomass ha^{-1}. Other reports specify CNC at 7–8 kg N kg^{-1} DM (Lemaire and Gastal, 1997; Planet and Cruz, 1997). Based on data from Great Plains of USA and standardizations using CERES-Maize model, Jones and Kiniry (1986) reported 4.5 g N kg DM in stover from silking to physiological maturity. In this case, maize forage production was only of moderate intensity with moderate yield goals of 15–20 t biomass ha^{-1}. In low input forage fields common to Guinea Savanna or Sub-Sahara of West Africa, maize for forage may yield 10 t biomass ha^{-1} and CNC may fluctuate between 4.7 and 6.0 g N kg^{-1} DM. In any region, if forage maize is cultivated on low fertility soil without N supplements, the CNC fluctuates between 2.1 and 4.7 g N kg^{-1} DM (Hermann and Taube, 2005).

The CNC could be used to decipher minimum fertilizer-N to be supplied to achieve 95% of maximum grain/forage yield in a given environment. As a corollary, estimates of leaf/stover tissue N concentration may allow us to extrapolate and arrive at prob-able grain/forage yield level. From the above examples, it is clear that CNC in maize tissue could help in adjusting fertilizer-N supply to fields. In fact, it is being practiced in many farms all over the world. Further, we should realize that leaf/stover tissue N is an important indicator of N status of above-ground portion of maize farms. It could play a vital role in regulating the nutrient dynamics of the field/cropping zone. Since forage-N is a crucial aspect of cattle nutrition and mixed farms in general, it has far reaching influence on nutrient dynamics that ensues in farm/village.

Chlorophyll Meter Readings (SPAD) and Nitrogen Dynamics

Chlorophyll based plant tests to judge Fertilizer-N requirement has its share of influence on net N dynamics in the maize agroecosystem. The chlorophyll meter readings (CMR) are useful in diagnoses of the leaf-N status. Presently, CMRs are being regularly used to ascertain N deficiency during maize production. In the Corn Belt of USA, chlorophyll meters are being routinely used to manage N dynamics. The concept depends on relationship between CMRs and yield response to N inputs. Zhang et al. (2008a) pooled data from maize fields at Dayton in Ohio that were supplied with fertilizer-N at different rates. Grain yields obtained were then related to CMRs. They state that firstly, fluctuations in CMRs were directly indicative of N status in leaves. Nitrogen deficiencies were easily discernable early during the crop season. However, small or medium intensity N deficiency could not be detected authentically. Chlorophyll meters could detect severe N deficiencies rather perfectly. Further, it was suggested that CMRs were useful in diagnosing N status of the crop and fertilizer-N requirement of maize fields. However, its utility in research fields was constrained due to several lacunae including low sensitivity in detecting small variations in leaf N status.

Let us consider few examples wherein CMRs have been used efficiently to judge N requirements and predict grain yield. The CMRs are used to manage N status of the maize crop as crop seasons progress. There are indeed different strategies to manage in-season N needs of maize crop. In Iowa, Ruiz Diaz et al. (2008) evaluated about 30 fields for 2 years with chlorophyll meters to determine N stress and adjust in-season N needs. Among the six treatments imposed, N stress detection was successful and accurate in 70–80% of instances. Mean grain yield improved by 0.5 t ha^{-1} across all sites that was supplied with post-silking N rates (POST-N) compared to 67 kg Nha^{-1} PRE-N (Pre-plant or Pre-side dress-N) rate. The POST-N application tended to suffer from poor N recovery and loss of grain yield potential. Economic analysis indicated that agronomic 134 kg PRE-N rate and 134 kg N+POST-N in-season strategy were similar, with regard to returns, but higher than 67 kg N ha^{-1} + POST-N strategy. It was concluded that agronomic PRE-N rate with confirmation of N stress and determination of additional N need through CMRs was most effective. It was better than starting the crop season with low N rates. A set of field evaluations at Dayton in Ohio has shown that during luxury production, CMRs may not be affected if plants were supplied with above optimal levels of N (Zhang et al., 2008b). The time at which N deficiency first appeared seems to affect N requirements. Also the measured N deficiency using CMRs changed with time and temporal patterns were affected by pre-plant N inputs. In-season N inputs made to crops at the start of N deficiency symptoms and those supplied with adequate N showed similar CMRs. It was concluded that luxurious N inputs may induce chlorophyll formation that is analogous to luxury N absorption.

Accurate determination of N demand by maize during various stages may be tedious. Plant-based monitoring of sufficiency and deficiency is accomplished much easily. Most of the systems that determine sidedress N for maize depend on estimation of leaf chlorophyll content using hand held Chlorophyll meters (Piekielek and Fox, 1992; Scheppers et al., 1992). However, chlorophyll meters do not identify N deficiency accurately at V6–V12 stage of the crop. The mid-season fertilizer-N requirements may not get assessed accurately (Jaynes and Colvin, 2006).

Scharf et al. (2006) found that CMR were significantly related to EONR. It is rapid and can be used to decide early season sidedress N on to maize fields. It predicts grain yield response to N inputs on maize crops grown in North Central USA. A different study involving 102 sites in Iowa has shown that validity of CMRs improves, if they are Normalized and Relative Chlorophyll Meter Reading (RCM) is used. The RCMs could be utilized to decide in season N rate (Hawkins et al., 2007).

There are indeed several plant based diagnostic tests employed during maize production. Comparative studies on their relationships could be of value to agronomists. Zaidi et al. (2008) compared chlorophyll meter based N recommendations and those based on N Nutrition Index (NNI). Evaluation across different sites in Northeast Canada has shown that generally CMRs and NNI increased with increasing N inputs to a maize crop. Such relations between CMRs and NNI were site specific. The CMRs and NNI were directly related to RY. Hence, both the tests and their relationship could be adopted to ascertain N deficiency in a maize crop, and based on it, fertilizer inputs could be recommended

Maize crops may suffer due to N paucity during late stages of the crop. Nitrogen insufficiency during late stage could be corrected by applying suitable N sources and placing it close to root system. Foliar spray that is quick and very efficient in supplying N to maize crop could also be adopted. There are several late-season Diagnostic tests that assess N status of maize crop and help in predicting N requirements, if any. Following are few examples:

- NO_3-N concentration of stalk sections at black layer;
- NO_3-N concentration of stalk sections at one-fourth milk line growth stage (MLGS);
- Chlorophyll meter test at one-fourth MGLS
- A relative CM test (normalized values) at one-fourth MGLS;
- A visual test based on number of green leaves below and including the ear leaf at the one-fourth MGLS;
- A relative visual test with normalized values at one-fourth MGLS.

Generally, stalk NO_3-N test helped in separating N-sufficient and N-insufficient maize crops (Fox et al., 2001). Such N diagnostic tests could be conveniently adopted by farmers to avoid excessive N supply and ground water contamination. In fact, in Pennsylvania, these diagnostic tests are being applied to reduce the number of drinking water wells that are getting contaminated with NO_3-N beyond permissible levels (Fox et al., 2001).

Active Sensor Reflectance, Corn Nitrogen Status, and Grain Yield

Remote sensing could also be used to assess variations in N status of a crop. We can devise fertilizer-N supply schedules based on reflectance measurements that are indicative of N status of crop.

In Argentina, active sensor reflectance assessment of maize canopy has been adopted to direct variable N applications. Solari et al. (2008) suggest it improves accuracy of in-season N distribution in a given maize field. The sensor data was converted to Normalized Difference Vegetative Index ($NDVI_{590}$) and Chlorophyll index (CI_{590}).

Grain yields were also determined. Their assessments indicate that sensor indices were directly related to CMRs. Also, CIs were more sensitive and accurate in assessing greenness of maize canopy. They conclude that Sensor CI_{590} values are best suited to judge variable N needs during vegetative growth.

A recent report by Schmidt et al. (2009) states that sensor reflectance index based on green normalized vegetation index (GNDV) measurement on corn at 6th or 7th mature leaf stage was as good a indication as optimum N rates or Pre-side dress N test (PSNT) or that derived using CMR. Further, it has been stressed that deciphering within field temporal and spatial variability for N content is crucial. Soil reflectance and soil fertility/cropping history of fields together can help us in arriving at better judgment regarding N supply to maize fields. Such remote sensing based techniques are yet to gain ground in maize belts of Africa and Asia.

Nitrogen Recovery by Maize Crops

Nitrogen recovery by roots and its translocation to above-ground parts has a major impact on nutrient distribution in the agroecosystem. Actually, it is the N recovery rates and accumulation pattern in plants that affects the above-ground N dynamics. As a corollary, it has an influence on nutrient depletion patterns in soil. As a thumb rule, a maize crop, within physiological limits, depletes at least to 6–8 kg N ha^{-1} from soil to produce 1.0 ton forage, if it is grown exclusively for silage and about 18–23 kg N ha^{-1} to produce a ton grain depending on harvest index. As stated above, N recovery by maize is dependent on geographic location, soil fertility status, especially available-N, crop genotype, agronomic procedures adopted, environmental parameters, and yield goals.

In the US Corn Belt, N recovery has improved from a mere 20 to 25 kg N ha^{-1} during early 1900s to as much as 235 kg N ha^{-1} per season at present. Obviously, it is directly related to intensity of cropping and yield goals. Incidentally, yield goal in the Corn Belt has gradually increased from a mere 1.2 t grain ha^{-1} in 1930s or 1940s to 10.5 to 11 t grains ha^{-1} in 2005. However, within physiological limits, maize crop grown in any region seems to recover 15–24 kg N t^{-1} grain. Small variations are generally attributable to agronomic procedures, soil type, moisture status, season and environmental parameters. Let us consider few examples from Corn Belt of USA and maize cropping zones from other continents. Based on field trials across several locations, Williams et al. (2007a) have reported that a maize crop in North Carolina and Coastal Plains absorbs on an average about 159–180 kg N to produce 8.1–9.5 t grains. Nutrient recovery per t grain ranged from 18.4 to 22.3 kg N. Long term (5 years) field evaluation under NT system has shown that maize absorbed 19 kg N and that under conventional tillage (CT) was 20 kg N to produce 1.0 t grains (Halvorson et al., 2006). In Wisconsin, maize crops are normally supplied with 210–265 kg N ha^{-1} in splits. During a season, maize may absorb as much as 50–67% of fertilizer applied to loamy soil compared to only 25–50% on silt loams (Oberle and Keeney, 2006).

Reports from Maryland suggest that about 16 kg N is recovered from soil to produce 1.0 t grains. From the above examples in North America, it is clear that between

15.5 and 22 kg N could be garnered by maize to produce 1.0 t grain. Over all, fluctuations in N recovery that occur are easily attributable to geographic location within Corn Belt, soils, agronomic procedures, genotype and environment. Reports from Mississippi Delta region indicate that 139–265 kg N ha^{-1} is removed by crop maize crops that yield 6–10 t grains ha^{-1}. In case, farmers aim at grain yield above 10 t ha^{-1}, then N supply should be proportionately increased beyond the present 250–300 kg N ha^{-1} (Bruns and Ebelhar, 2006). In the Appalachian region, maize silage crop recovers about 10.1–10.7 kg N t fresh forage (Staley and Perry, 1995). Total N recovery improved as N inputs were enhanced from 57 kg N ha^{-1} to 224 kg N ha^{-1}. However, fertilizer-N use efficiency fluctuated between 49 and 55% irrespective of N dosages and tillage systems. Tillage does not seem to affect N uptake, its accumulation in tissue nor productivity of the crop.

Examples from Argentinean plains indicate that at physiological maturity maize crop that yields 10.4–11.0 t grain absorbs 155–167 kg N. It equates to 15.2 kg N t^{-1} grain (Sainz Rozas et al., 1999). Nitrogen uptake below 17 kg N per t grains could be considered as lower range.

Maize is an important forage crop in European plains. The productivity of maize grown for forage is generally held high by applying manures. In Netherlands, maize grown on light clay loam recovers about 80 kg N ha^{-1} to produce 10.3 t silage. It is equivalent to 7.5 kg N for each t of forage. However, N recovery by forage crop increases as fertilizer inputs are enhanced. Nitrogen recovery improves from 7.4 kg N to 11.2 kg N, if fertilizer N input is enhanced from 0 to 200 kg N ha^{-1} (Van Dijk and Brouwer, 1998).

In the Indus plains, forage maize is supplied with varying levels of fertilizer-N depending on inherent soil fertility, cropping systems, and yield goals. Irrigation seems to affect the NO_3 distribution in the soil profile and extent of N recovered by the crop. On an average, maize crop removed 6.2–6.4 kg N t^{-1} fresh biomass (Niaz et al., 2007).

Nitrogen Loss in Maize Belts
Soil Erosion, Leaching, Seepage, and Runoff

In the "Corn Belt of USA", N loss via leaching assumes greater importance in sandy soils, than in medium or fine textured Mollisols. Nitrate is the form in which much of N is lost via leaching. Leaching occurs during and immediately after a heavy rainfall or irrigation event. Leaching physically removes NO_3-N away from the root zone. Hence, it reduces fertilizer-N use efficiency. Nitrogen loss via leaching is also dependent on amount of fertilizer-N or NO_3 present in soils. For example in Central Iowa, Jaynes et al. (2001) have reported that in low fertilizer-N (67 kg N ha^{-1}) plots, NO_3 loss was 29 kg N ha^{-1}; in medium input fields NO_3 loss was 35 kg N ha^{-1} and in high fertilizer-N input zones NO_3 loss was 48 kg N ha^{-1}. Next, period for which NH_4 or NO_3 form of N accumulates and stays in soil without being transformed into organic forms is also important. Further, Urea-based fertilizers are highly soluble. If rainfall occurs well before urea is transformed into NH_4 then much of fertilizer-N as urea is prone to leaching. However, if urea applied gets hydrolyzed to NH_4 form then sizeable amount of fertilizer-N is still held in upper horizon for maize crop to absorb (Laboski, 2009).

Essentially, timing and intensity of irrigation and/or rainfall pattern are important. Fertilizer-N inputs and NO_3 or NH_4 concentration in soils are prime factors that affect leaching. Maize crop genotype may also be important in thwarting undue leaching. Maize genotypes that typically absorb NO_3 forms quickly and from deeper layers of soil could be preferred. A genotype efficient in NO_3 absorption basically avoids leaching loss of N. A genotype with larger and deeper root system intercepts and absorbs NO_3-N otherwise prone to leach or percolate to lower horizon.

Nitrogen inputs into maize field should be regulated in order to avoid undue loss to lower horizons and contamination of groundwater. Pikul et al. (2005) have reported that under high N input, maize grown on Calcic Hapludolls may not explore soil-N efficiently. It causes soil-N to accumulate in subsurface soil up to 3 m depth. For example under high N, top 3 m of soil profile accumulated 200 kg N ha^{-1}, but in no-N fields only 73 kg N ha^{-1} got accumulated. The extent of N vulnerable to erosion and percolation proportionately increased and there was every reason to suspect it would percolate further into groundwater and contaminate it. Carefully structured rotations and N input schedule seems necessary to avoid N loss via erosion and percolation.

Basically, there is a strong need to understand trends in transport of NO_3-N and NH_4-N from maize fields. Proper selection of fertilizer formulations may be important in reducing runoff loss. For example, cumulative NO_3-N recovered in runoff expressed as a proportion of total was 0.37 for ammonium nitrate, 0.25 for sulfur-coated urea, 0.10 for composted dairy manure and 0.07 for poultry manure during a 10 week period from sowing (King and Torbert, 2007).

Maize cropping zones in the European plains are also prone to significant loss of fertilizer-N via NO_3 leaching process. The Cambisols region of Western Europe, Loess region, and Chernozem belts of Eastern Europe, all suffer loss of fertilizer-N due to NO_3 leaching and other chemical transformations such as nitrification, ammonia volatilization, etc. In Southern Spain, fertilizer N applied to irrigated maize tends to be converted rapidly into organic fraction. If not a large fraction of NO_3-N is leached out from maize field via drainage, seepage, and leaching into lower horizons. The intensity of maize cropping and fertilizer N inputs regulate the NO_3 leaching loss. Overall, NO_3 loss from fields reached 150 kg N ha^{-1} in plots given high N inputs and it reduced to 43 kg N ha^{-1} in low input plots (Moreno et al., 1996).

Maize farming in Southern Africa occurs predominantly on coarse and sandy soils prone to nitrate leaching. Nitrate loss from field is accentuated depending on distribution and intensity of precipitation during the crop season and fallow. Nyamangara et al. (2003) report that high N input at 120 kg N ha^{-1} plus manure results in proportionately higher concentration of NO_3 (up to 34 mg N L^{-1}) in drainage. Hence, it is common to experience NO_3 loss from maize fields. Soil NO_3-N loss could even be 56 kg N ha^{-1} yr^{-1}. Other reports by Kamukondiwa and Bergstrom (1994) suggested that 34 kg N ha^{-1} could be lost from the maize fields via NO_3 leaching. There are reports that depending on sand fraction of soil and intensity of rainfall, 40–54% of fertilizer-N applied could be lost through NO_3 leaching (Twomlow, 1994). The NO_3 leaching was greatest in the top 0–50 cm layer of soil. Drainage and percolation of water also accentuated NO_3 loss. The NO_3 concentration in the leachate was proportional to N inputs. In control plots, NO_3 was <10 mg N L^{-1}. It was 15 mg N L^{-1} in fields amended with organic manures and above 22.5 mg N L^{-1} in fields given organic manure plus fertilizer-N. Leaching loss via NO_3 was low during second year and much less in the 3rd cropping season.

In the West African cereal farming zones, for example in Mali, major avenues of N loss were soil erosion and de-nitrification. About 17% of fertilizer-N supplied was lost via soil erosion and 22% because of N_2 and N_2O emissions (Vander Pol and Traore, 1993). Nitrogen loss from the agroecosystem generally gets accentuated with higher dosages of fertilizers. Wind erosion via dust storms such as "*Harmattan*" could be severe in the transition zones and even in Savannas. The extent of loss depends on storm height, sand particles and nutrient enrichment. Actually, N loss due to wind erosion is the difference between that received from neighboring regions and that lost to other regions due to storm. About 18 kg N ha^{-1} could be lost in a cropping season via soil erosion mediated by winds and *Harmattan* (see Krishna, 2008). A maize-peanut rotation may suffer substantial loss of N due to wind erosion. Hence, farmers adopt mulching and *Zai* system of planting. Irrigation is not common but percolation after rainfall event could induce a certain amount of N loss from the upper layers of sandy soils.

Around Kilimanjaro in Eastern Africa, soil erosion and resulting loss of N is rampant. It ranges from low or moderate to severe depending on soil texture and slope of the field. Loss of N occurs due to soil erosion, runoff, leaching, seepage, and of course due to uptake by maize crop. All of these factors contribute to mining of soil-N. During a period of 2 years, N loss due to erosion (100 kg N ha^{-1}) and leaching (200 kg N ha^{-1}) amounted to 300 kg N ha^{-1}. In comparison, N loss from the field ecosystem due to uptake by the crop was 180 kg ha^{-1}. Soil analysis indicated a net depletion of 125 kg N ha^{-1} during first year and 272 kg N ha^{-1} during next year (Vaje et al., 1999).

Long term estimates of soil erosion and nutrient loss from Alfisols found in the maize cropping zones of Indonesia, indicate that about 19 kg N ha^{-1} could be lost from the top soil during peak rainfall period. In addition to N, soil erosion induced loss of SOC, P, K, and micronutrients (Hermiyanto et al., 2007). We should note that extent of soil erosion and N loss is also site specific.

In the Northwest China, maize fields may suffer excessive loss of inherent soil-N and that applied via fertilizer. Both, Urea and NO_3-N applied to soil are prone to loss via soil erosion and leaching. Zhou et al. (2006) have reported that about 5.7–9.6% N of the total applied was lost through leaching from clay loam that supported maize production. Whereas, on sandy soils, N lost via leaching was greater and it ranged from 16.2 to 30.4% of total fertilizer-N applied via fertigation. On a sandy soil of Inner Mongolia, maize fields tend to loose larger fraction of fertilizer-N, about 32–61% applied due to leaching (Hu et al., 2008). Generally, leaching losses were greater if NO_3-N was applied. Li et al. (2007; 2008) observed that NO_3-N and NH_4-N leaching was higher than N loss through volatilization. Both, N leaching and volatilization increased with fertilizer inputs. Mostly, leaching loss coincided with precipitation pattern. Incidentally, spurts of intense rainfall events do occur in the Huanghuaihai region of China. However, top dressing was a good method to curtail N leaching.

Studies on N leaching have shown that in Japan precipitation pattern, irrigation and N supply trends are key factors that control N loss via leaching. On Andosols and sandy soils, two split applications instead of a single lumped application reduced N leaching significantly, almost by 33% compared to control. A three way split further reduced N leaching, but it was efficient only in sandy soils. Nitrogen fertilizer application in six splits did not give any further advantage regarding N leaching (Nakamura et al., 2004).

Ammonia (NH$_3$) Volatilization

Maize is grown intensively in the "Corn Belt of USA" in order to achieve high yield goals. Nitrogen inputs are high ranging from 180 to 240 kg N ha^{-1}. Nitrogen is supplied using different fertilizers such as granulated urea, urea–urea-phosphate (UUP), urea-ammonium nitrate (UAN), and prilled ammonium nitrate (AN). The extent to which N is lost as NH$_3$ from these sources is variable. Evaluations on sandy soils of Lyles in Indiana has shown that 51 kg N ha^{-1} of the 168 kg N ha^{-1} applied as granular urea was lost through NH$_3$ volatilization, in 5 days (Keller and Mengel, 1986). Peak loss of N as NH$_3$ occurred 23 hr after application and it was 3.0 kg N ha^{-1} h^{-1}. Volatilization of NH$_3$ was not significant with AN. About 26 kg N ha^{-1} was lost from UUP and 14.6 kg N ha^{-1} from UAN. Peak rates of N loss via NH$_3$ volatilization was 1.83 kg N ha^{-1} h^{-1} for UUP and 0.37 kg N ha^{-1} h^{-1} for UAN. Precipitation or application of water subsided NH$_3$ evolution. Therefore, it was advisable to apply fertilizer-N just before irrigation event. In other wards, in the "Corn Belt" selection of fertilizer-N source, timing and soil moisture status are crucial, if N loss as NH$_3$ is to be curtailed.

Sainz Rozas et al. (1999) state that maize grown under no-tillage system is often provided with fertilizer-N by surface application. It is prone to significant loss of N through NH$_3$ volatilization. Therefore, N inputs are held high to overcome N loss plus to satisfy crop's need. Fertilizer-N loss due to ammonia volatilization ranged from 2.6 to 13.3% of applied N. Ammonia volatilization increased with N rates. Application of Urea-N along with urease inhibitors like nBTPT [N-(n-butyl) thiophosphoric triamide] had little effect, but delaying fertilizer-N inputs and matching it with plant's demand at later stages (6th leaf stage) did improve N uptake and fertilizer-N use efficiency. It seems action of nBTPT in reducing urea hydrolysis depends on soil pH and temperature.

In Pampas of Argentina, N loss via NH$_3$ volatilization is relatively low during early stages of crop. However, NH$_3$ evolution increases with temperature around V6 stage (18–22°C) of maize crop. The NH$_3$-N loss increases if N inputs to fields are enhanced and more so if it is broadcasted on the soil surface. Models developed using data from maize crop grown in Rolling Pampas and actual measurements indicate that about 5–6% N is lost as NH$_3$ during early stages of the crop. The NH$_3$N loss may increase to 10–12% of total N inputs by the time maize crop reaches grain fill (Alvarez and Steinbach, 2006).

On fertilized soils of Southern France, total N loss to atmosphere reaches 3–11 kg N ha^{-1}. About <1% N loss occurs via NH$_3$ volatilization, 40% as NO, 14% as N$_2$O, and 46% as N$_2$ (Jambert et al., 1997). Obviously, nitrification/de-nitrification processes cause significant amount of N loss from maize fields.

In South Africa, Yerokun (1997) compared different fertilizer-sources with regard to extent of N lost through volatilization as NH_3. Among them, NH_3 loss via volatilization was similar for Co-granulated Urea-Urea-Phosphate and Urea. Nitrogen lost as NH_3 in 14 days ranged from 7.2 to 7.8% of fertilizer applied to soil. Urea-phosphate and ammonium nitrate induced relatively lower levels of NH_3 volatilization. Nitrogen loss through volatilization increased with fertilizer-N dosage. Soil moisture too affected the extent of NH_3 volatilization from different fertilizer-N sources.

In South India, N loss from fields due to NH_3 volatilization ranges from 5 to 20% depending on fertilizer formulation, placement techniques, quantity, soil types, soil moisture conditions, and weather parameters (see Krishna, 2010).

In the Northern Plains of China, maize-wheat rotation is a preferred cropping system. Fertilizer-N inputs are held high in order to achieve higher yield goals. However, fertilizer-use efficiency gets depreciated due to loss of N via volatilization and other modes. Several techniques such as Micrometeorological Integrated Horizontal Flux (IHF), Calibrated Drager-Tube method (TDM), and ^{15}N balance methods using ^{15}N urea have been utilized to estimate the loss of N via NH_3 volatilization (Pacholski et al., 2008). According to Wang et al. (2008), NH_3 volatilization is an important pathway by which both soil-N and fertilizer-N are lost from the system. It reduces N availability to maize crop plus causes environmental pollution in the Chinese maize belt. Measurement made using both closed chamber and modified vented chamber methods indicated that N loss via NH_3 volatilization is significant. During maize phase (summer) of crop sequence, NH_3 loss from urea treated fields ranged from 4.8 to 5.6% of the total fertilizer-N applied (320 kg N ha^{-1}). Ammonia volatilization was relatively higher if urea was surface-broadcasted. Ammonia volatilization peaked rather quickly within 2–3 days after surface application of urea. Zhang et al. (1992) had earlier shown that deep placement (5–10 cm) of fertilizer-N in soil reduced NH_3 volatilization to 12% of that applied, but if broadcasted NH_3 volatilization was over 32%. Nitrogen loss increased from 2.5 kg to 24 kg N ha^{-1} if fertilizer-N supply was enhanced from 40 to 80 kg N ha^{-1}. About 30–32% of applied N was lost in 188 to 280 hr after application. Splitting fertilizer-N dosage reduced loss via NH_3 volatilization. In the drier regions of Northwest China, maize is grown with relatively low N inputs. Therefore, minimization of N loss via NH_3 volatilization or emissions as NO_2 and N_2 is important. Estimates indicate that about 5% of fertilizer-N applied to sandy soils of Inner Mangolia was lost as NH_3 (Hu et al., 2008). The semi-arid climate seems to induce rapid loss of N. It is said that a larger fraction of fertilizer-N is lost because of leaching and through volatilization.

In a different study reported by Tsinghua (2002), ammonia volatilization from maize plots was again found to be an important avenue of N loss from the wheat-maize cropping systems. On calcareous soils of North China plains, NH_3 volatilization was dependent on factors like soil pH, temperature, relative humidity, and wind velocity. The amount of fertilizer-N supplied directly affected N loss as NH_3. During maize phase, NH_3 evolution occurred rapidly after fertilizer-N inputs. The NH_3 volatilization peaked within 2 days and it ranged from 3.3 to 5.41 kg N ha^{-1} day^{-1} depending on fertilizer supply and inherent soil-N status. However, N volatilization as NH_3 did not last long. It subsided to negligible levels within 7 days after fertilizer-N application.

Let us compare maize and rice fields. Maize is produced on arable soils held in oxidized state; rice is grown on flooded soils held under anaerobic conditions. The nitrogen transformations stop at NH_4 formation. Generally, NH_3 volatilization from flooded rice fields is much higher compared to those recorded for dry land cereals like maize or millets. For example, NH_3 volatilization accounts for 5.1 to 10.6% N loss depending on timing of fertilizer-N inputs. Ammonia volatilization from flooded paddy fields may range from 11 to 60% depending on timing and method of urea disbursement (Fillery et al., 2005).

Field trials at Puchong in Malaysia have shown that we can reduce loss of fertilizer-N via ammonia volatilization by mixing the fertilizer formulation with zeolites. The clinoptilolite zeolites used actually entrap NH_4 and NO_3 in its matrix and reduces chances for mineralization by microbes. As a consequence, fertilizer formulation releases NH_4 ions slowly into soil and avoids loss via emission or transformation reaction in soil (Ahmed et al., 2009).

Farmers adopt different methods in order to deliver fertilizer-N more efficiently and in close proximity to root systems of maize plants. Usually, granulated or powdered forms are broadcasted or band placed in furrows. Deep placement reduces loss via volatilization and leaching. Foliar application of fertilizer-N is also practiced, especially when top dressing a maize crop that is already at 6–8 leaf stage. Foliar fertilizer application avoids many of the N transformations that otherwise occur in soil. It also avoids loss via erosion and leaching to a great extent. Fertigation is yet another procedure where in N fertilizers are dissolved into irrigation water and applied to field. It again improves fertilizer-N efficiency. In Australia, maize farmers adopt yet another method wherein N is channeled into maize fields by dissolving NH_3 gas into water that is meant for irrigation. It is an efficient system in provided technology is available and precautions are carefully adopted. According to Denmead et al. (1982), this system of N application is subject to loss via NH_3 volatilization. In fact, NH_3 volatilization is slightly accentuated. On an average, about 7% of NH_3-N was lost from the fertilizer source during application. Sometimes uneven loss of NH_3-N results in variability of soil-N available to maize crop. The NH_3-N loss was greater if temperature and wind velocity was higher. Hence, it is better to apply NH_3 gas dissolved irrigation water in the night and as rapidly as possible using furrow irrigation.

Crops like maize are prone to emit NH_3 into atmosphere. In fact, most crop species are know to emit nitrogen either in reduced or oxidized form through various above ground organs. Discussions in the above paragraphs relate to NH_3 volatilization from soil or soil plus crop if micrometeorological methods were adopted. Now let us consider the extent of N lost to atmosphere from above ground portion of maize crop exclusively and its significance to N cycling nature. Firstly, N emitted from the maize crop seems to depend on genotype, growth stage of the crop and its physiological status. For example, reports suggest that soybean crop may release up to 45 kg N ha^{-1} directly as NH_3 into atmosphere. In case of maize, NH_3 emissions range from 4 to 23 kg N ha^{-1} (Francis et al., 1993; Harper and Sharpe, 1995). Previous records indicate that based on crop species and environment, N emitted directly from plants ranges from 0 to 80 kg N ha^{-1} (Kastori, 2004). The intensity of NH_3 volatilization from plants depends on ambient temperature. The NH_3 volatilization was relatively higher under warm conditions with temperature ranging from 30 to 35°C. Amount of N supplied to maize crop also affected extent of N emissions from above ground parts. Over all, we should note that in addition to loss of N from soil that is often rampant, crop plants too emit NH_3 and other N containing gases directly into atmosphere. This portion too needs to be accounted while calculation fertilizer requirements and N balance in a given maize field or a larger expanse.

Nitrification/De-nitrification and N-emissions

Soil nitrification/denitrification is a phenomenon that reduces N availability to maize crop. It induces formation of NO_2 and N_2 and accentuates N loss via emissions. A certain amount of fertilizer-N applied to maize crop escapes soil as emissions and reduces fertilizer-N use efficiency. Nitrification-denitrification processes in soil induce N emissions. Together, N-emissions as N_2, N_2O, or NO_2 accounts for 5–10 kg N ha^{-1} loss from soil ecosystem in the maize cropping zones. Tillage intensity and cropping systems are factors that affect N_2O emissions. Such N fluxes may often range from 2 to 4 kg N_2O-N yr^{-1}. Loss of N due to N_2O fluxes is greater (2.7–4.8 kg N ha^{-1}); if fertilizer-N is supplied in one stretch compared to starter and split inputs (Tan et al., 2009). According to Del Grosso et al. (2008), vast expanses of maize/wheat are major causes of N_2O emission in the Corn Belt and Western USA. In fact, there are extensive reports on effects of tillage and fertilizer inputs on N_2O emissions. There are also computer models like DAYCENT that simulate the effects of tillage and N inputs on N_2O emissions. Forecasts indicate that N emissions increase with enhanced fertilizer-N supply, but N emissions are relatively lower under No-tillage systems compared to conventional tillage. According to Jarecki et al. (2008), DAYCENT is an ecosystem model that simulates and estimates N_2O emissions accurately. Using extensive measurements of soil characteristics and N_2O fluxes from maize fields in Central Iowa they concluded that DAYCENT could be adopted to understand N-emission pattern during maize production. Venterea and Stanenas (2008) have stated that it is not easy to predict N_2O emission and minimize its fluxes during farming. They suggest use of models that estimate N_2O production, diffusion rates in soil, soil microbial load, soil enzyme activity, and chemical conditions that prevail during crop growth. According to them, stratification of microbes and their activity may affect N_2O emissions significantly. Under nitrification-dominated conditions N_2O emissions were greater, especially if NT systems were adopted. Whereas, under denitrification dominated conditions, presence of NO_3, higher bulk density and soil moisture in no-till plots promoted denitrification rates more than conventional tillage. Such effects were offset by higher soluble organic-C, temperature and lower N_2O reduction rates under conventional tillage. Use of fertilizer form that does not promote excessive N_2O emissions ahs been suggested. The oxygen concentration prevalent in soil affects N emissions both due to nitrification and denitrification (Khalil et al., 2004).

Farmers in the Corn Belt adopt different crop rotations. The tillage and fertilizer supply levels too differ based on a range of factors related to geography, soil fertility status, genotype, season, yield goals, and economic advantages. Halvorson et al. (2008) have made interesting reports regarding maize-based crop rotations and

N-emissions. Generally, they found linear increase in N_2O emission with fertilizer N supply. However, rotations such as maize-*Phaseolus* bean induced greater N_2O emissions than several others such as continuous maize or maize-barley. Importantly, crop rotation and N supply influenced N emissions much and almost masked that caused by tillage systems. Nitrification inhibitor is widely used in the Corn Belt (Plate 3.5) to suppress emissions.

It has been observed that excessive fertilizer-N and FYM inputs may induce loss of N via NO_3 leaching and cause fluxes of N_2O emissions. In order to suppress nitrification processes that generate NO_3, inhibitors of nitrification such as Nitrapyrin are used. Nitrapyrin (N-Serve: 2-chloro-6-(trichloromethylepyridine) may help in suppressing nitrifiers in the maize root zone. Nitrapyrin was actually effective in reducing nitrification thus enhancing NH_4-N accumulation and possibly reducing NO_3 leaching to lower horizons (Calderon et al., 2005; Plate 3.5). Let us consider an example from Maize belt of Pampas in Argentina. The no-tillage system practiced in this belt involves application of crop residues on the surface. It definitely reduces soil erosion and improves water use efficiency, but at times reduces N uptake by maize. Sainz Rozas et al. (2001) have opined that reduced N recovery could be due to N immobilization, NO_3 leaching and more importantly due to gaseous loss of N as NH_3, N_2O, or N_2. They have reported that out of the surface applied fertilizer-N about 4.7–7.4 % is lost due to denitrification processes that result in N emissions. The extent N lost via denitrification depended on time of N supply to maize crop. On Typic Arguidolls of Balcarce, about, 7.6–9.8 kg N ha^{-1} was lost via denitrification, if 210 kg N ha^{-1} was supplied to maize crop. The denitrification rate was dependent on soil properties like pH, temperature, C:N ratio, and water content. Denitrification rate increases rapidly if water filled pore space exceeds a critical value between 70 and 80%. We should expect denitrification losses to be more during the period after irrigation.

It is interesting to note that loss via N emissions could significantly affect N budgets during maize production. According to Sainz Rosas et al. (2004), application of N at planting resulted in higher N loss via emission and leaching. However, N supply at V6 stage avoided excessive N emissions (<13.8%). Grain yield increased from 10.5 to 11.5 t ha^{-1} if emissions were curtailed.

Incidentally, about 60–80% of N_2O emissions could be arising from agricultural expanses (Beuchamp, 1997). A review of reports on N_2O emissions from agricultural soils in North America indicates that build up of N in soil due to excessive chemical fertilizer supply is a major cause of N_2O emission. However, we should note that high dosages of organic manures and BNF may also contribute to N emissions. In Eastern Canada, maize-soybean rotations are applied with fertilizer-N at higher rates. The efficiency of fertilizer-N could be lessened due to several factors including emissions as N_2O. The denitrification rates are dependent on tillage system. Basically, higher N_2O emission was attributable to large fertilizer-N input, wet weather and NT systems. Wet weather and irrigation induced anaerobic conditions in soil that supported denitrification. Reduced loss of N noted during soybean phase of the sequence was attributable to N inputs accrued via BNF (Gregorich et al., 2008). Rochette (2008a) made a detailed analysis of published data regarding effect of soil aeration on N_2O

emission when maize fields are under NT systems. On an average, N_2O emission was lower by 0.06 kg N ha⁻¹ if soil was well aerated; but higher by 0.12 kg N ha⁻¹, if moderately aerated and 2.0 kg N ha⁻¹ higher if poorly aerated or stagnated due to irrigation. Soil type on which maize was cultivated too had its specific effects on N emissions. Reports from Eastern Canada suggest that on heavy clay loam, annual emissions of N_2O ranged from 12 to 45 kg N_2O-N ha⁻¹. Such emissions were not associated with fertilizer-N much, but were due to denitrification sustained by decomposition of high organic matter stocks (192 t C ha⁻¹) available in clayey soils (Rochette et al., 2008b). On an average, NT systems induced N_2O emissions twice that traced in conventional tillage. Nitrogen emissions were relatively small in sandy soils. A comparative study of N emissions from maize field given fertilizer-N or cattle slurry with grass land showed that in maize fields, annual gaseous loss was 4.9 kg N ha⁻¹ due to N_2O emissions. It was attributed to higher microbial biomass and activity under maize fields. The N emissions were relatively smaller by 20% in grass fallows (Mogge et al., 1999).

Plate 3.5. Field evaluation of Response of Maize to Fertilizer-N and Nitrification inhibitor –Nytrapyrin.
Source: Carrie Laboski and Tod Andraski, Soil Science Department University of Wisconsin, WI, USA.
Note: Above field investigation on Nytrapyrin was conducted at Arlington in Wisconsin, during 2008.

In Northeast China, maize is grown under high N input in order to achieve high (7–8 t grain) yield goal set by farmers. Fertilizer-N efficiency is an important criterion. Hence, they opt for all measures that decrease N loss via nitrification/denitrification reactions in soil. Estimates on soil denitrification by Zou et al. (2006) indicate that on Aquic Cambisols of North China, maize fields may experience loss of 4.7–9.7 kg N ha⁻¹ depending on fertilizer supply levels. Further, analysis has indicated that much of the N was lost as N_2O proving that nitrification induced greater loss of N.

Denitrification is a process accomplished mostly by soil microbes, wherein nitrate is converted to gaseous NO_2 or N_2. Denitrification is significant in medium and fine textured soils with optimum SOM. Denitrification is not a major avenue of N loss on sandy soils with low amounts of SOM contents. Nitrate is the substrate for denitrification. Hence, denitrification depends directly on NO_3 contents of soils that support maize production. Dentrification requires soils saturated with moisture. It is generally more during and after a precipitation or irrigation event when soils are held in flooded situation. The anaerobic microsites in soils too support denitrification. Soil temperatures above 75°F are congenial for dentrification. For example, if soil is held in saturated conditions for 3–5 days, say after heavy rain and at 13°–18°C, then about 10–25% of fertilizer-N applied will be lost as N_2O and N_2. Denitrification mediated N loss reaches 60–80% or above, if soil temperature is 24°–30°C. Soil pH below 5.5 may not support denitrification processes. The SOM that supports denitrifying bacteria is an important factor. Obviously, soils richer in organic matter are prone to loss of N through denitrification.

The formation of NO_3 and its accumulation in soil could be delayed in order to avoid denitrification process to occur. For example, if urease inhibitors are added along with Urea-based fertilizers, formation of NH_4 could be delayed. Usually it takes 10–14 days for urea to get completely transformed to NH_4 depending on urease inhibitor and its concentration. Conversion of NH_3 to NO_3 takes another 1–2 weeks. This process could be delayed by adding nitrification inhibitors. The time for which fertilizer-N stays in NH_4 or NO_3 form is crucial. It depends on fertilizer source and formulation used. Addition of urease and/or nitrification inhibitors delays accumulation of NO_3. Therefore, it avoids undue loss of N via denitrification for which NO_3 is the substrate (Laboski, 2009).

Nutrient loss due to removal of maize forage and residues has been discussed in other chapters dealing with organic matter and integrated nutrient management.

Fertilizer-N in Maize Fields
[15]N Fertilizer Recovery by Maize
Legume-corn rotations in the "Corn Belt of USA" are usually supplied with a mixture of organic residues and inorganic fertilizer. The idea is to enhance nutrient recovery into maize crop. The ^{15}N isotopic analyses have shown that fertilizer-N efficiency is 51% for whole plant estimates and 36% if only grain-N is accounted. However, N recovery from legume residues incorporated prior to sowing seems to depend on amount of N supplied via chemical sources. The ^{15}N recovered form legume residues decreased if fertilizer-N inputs were enhanced from 57 to 168 kg N ha[-1] (Hestermann et al., 1987). Clearly, N recovery by maize is influenced by the proportion of chemical fertilizer-N and organic-N sources supplied. Micro-plot assessments using ^{15}N labeled fertilizers on Kandiudults of Iowa has shown that crop removal accounted for 20–34%, while 20% got accumulated in the subsurface at 90–120 cm in the soil profile (Karlen et al., 1996). About 40–50% of fertilizer was lost from the plots to depths below root zone through leaching, seepage, de-nitrification, and volatilization. Timing of fertilizer-N supply appropriately and placing it deep in soil reduced loss of fertilizer significantly.

Knowledge about N recovery from soil and fertilizer sources is important. Further, an idea about distribution of N recovered from fertilizer into various parts of plants is vital. The ratio of fertilizer-N that gets into seed formation may be an important physiological aspect that affects grain yield. Let us consider an example. Maize grown on tropical soils in Malaysia is supplied with 150 kg N ha^{-1} as $[(NH_4)]_2 SO_4$. The ^{15}N experiments have shown that out of 150 kg N ha^{-1} applied, 23% could be traced in above-ground plant parts (Abdelrahman et al., 2001). Total N recovered from fertilizer was 20.3%. More than 60% of labeled fertilizer-N was traceable in stalks and leaves, 11.6% in husks, 3.0% in empty cobs, and 23.7% in grains. Total depletion of ^{15}N in top 0–50 cm was 74.5%. Almost 40% of ^{15}N labeled fertilizer-N applied was unaccounted. In a different study from Malaysia, maize–groundnut rotation was assessed for N recovery using ^{15}N labeled fertilizer. The ^{15}N fertilizer recovered by the maize crop was 20–22% (Mubarak et al., 2003). The proportion retained in soil was relatively high at 44% during the first year and 24–34% during next year. Removal of crop residues reduced ^{15}N recovery.

Maize crop grown on sandy soils of Zimbabwe recovered 44–47% of ^{15}N fertilizer applied, during the first season. The ^{15}N recovered during second season declined to 26.3%. Application of inorganic fertilizer-N along with organic sources improved net N recovery. Mineral-N use efficiency improved if organic and inorganic N sources were applied together. Pilbeam and Warren (1995) studied the pattern of N recovery by maize/bean intercrops from a Ferralsol in Kenya. Under tropical conditions, N recovery from 120 kg N ha^{-1} was low at 7.5% during first season and 17.7% during second season. Precipitation pattern influenced N recovery. The fertilizer-N recovery was 5.3% into grains and 0.79% into straw, if rainfall was high and lasted for longer duration. Whereas, under short rainy period ^{15}N fertilizer in grains was 9.86% and 5.90% in straw. Major fraction of N recovered by maize was derived from soil. It was 32% from soil and 0.60% from fertilizer. The ^{15}N derived from soil was above 62%, if precipitation was low and soil erosion was within limits. Soil erosion, runoff, and leaching loss of fertilizer-N seemed to be major cause of low N recovery under long rains.

Generally, knowledge about ^{15}N uptake patterns and the factors that affect it is necessary, to account for N distribution in the maize agroecosystem. Fertilizer-N applied to maize fields gets distributed to plants, soil layers, drainage water, and a portion reaches atmosphere through emissions. According to Van Der Kruijs et al. (1988), about 70–93% of fertilizer-N applied to tropical soils at Ibadan in Nigeria was traceable in the crop, soil, and leachate. Depending on treatment, 10–30% of ^{15}N was immobilized in soils in the organic phase; 22–29% was lost via drainage channels and seepage; 7–30% ^{15}N was lost due to denitrification and emissions as N_2O or N_2. Maize crop itself removed only 31–38% of labeled fertilizer-N. These studies using ^{15}N fertilizer indicate that N loss via drainage and emissions should be reduced. Perhaps, timing fertilizer-N and its dosage more accurately helps in reducing N loss. Placing fertilizer-N at a depth avoids emissions to a large extent. Lehman et al. (2006) have shown that in Brazilean Cerrados, where maize-soybean rotation is a common practice, distribution of fertilizer-N, particularly its accumulation pattern in subsoil is important. They conclude that NH_4NO_3 applied is rapidly immobilized into SOM or nitrified. Large amount of NO_3-N is adsorbed into subsoil layers (2 m depth) amounting to 150–300 kg

N ha^{-1}. It is almost equivalent to uptake by maize crop which is 130 kg N ha^{-1}. The ^{15}N tracer tests showed that about 49–77% of N could be immobilized immediately into soil organic fraction. Most of the ^{15}N fertilizer applied, about 71–96% got accumulated in subsoil layers at a depth of 1.5–2.0 m. It automatically enhances the proportion of organic-N in subsoil. Hydrophyllic organic-N was the dominant fraction. Much of the subsoil-N was still prone to leaching to vadose zone and further into ground water. Seepage and loss of N via irrigation channels was also significant.

Reports on N dynamics of Australian maize cropping areas suggest that ^{15}N fertilizer recovery is only 64–68% (Edis et al., 2006). Loss due to NH$_3$ volatilization was less because fertilizer was placed deep as solution. Drainage and seepage loss accounted for 19–25% of ^{15}N fertilizer applied. Denitrification was estimated at 13% of fertilizer-N. Nitrous oxide emission was feeble but depended on rate of N input. Over all, it was suggested that positioning fertilizer-N deeper in soil and in solution form was necessary to avoid undue loss of fertilizer-N.

Investigation of barley-maize rotation in Japan has shown that ^{15}N recovery from plots treated with chemical fertilizer is influenced by the amount of barnyard manure supplied to maize field. Often, fields supplied with barnyard manure recovered lower levels of ^{15}N compared to those solely supplied with chemical fertilizers. The ^{15}N remaining in soil at the end of rotation ranged from 11 to 23%. Analysis of maize crop showed that 24–48% of plant-N was derived from chemical fertilizer. Mixing with barnyard resulted in better recovery and only 7–15% N remained in soil (Li et al., 2001).

Residual Nitrogen, Surface and Ground Water Contamination

The RSN is a fraction of fertilizer-N that is left unused in the soil profile. Maize crop fails to utilize this portion due to various factors related to soil profile, fertilizer-N formulation, its placement, more importantly timing, soil moisture, and nutrient uptake pattern of crop genotype. In the "US Corn Belt" and other areas of USA, among crops, maize garners most of the fertilizer-N applied. It is also the crop often grown under tile drained irrigation which offers an easy conduit for NO$_3$ leaching into surface and ground waters. We should note that RSN is often vulnerable to loss via leaching, percolation and seepage. Percolation easily removes fertilizer-N away from the root zone. Added to it subsurface seepage further affects RSN. During maize cropping, maximum amount of residual NO$_3$ is lost due to percolation and it leads to ground water contamination. There is no doubt that regulating fertilizer-N inputs and build up of RSN is crucial. During recent years, farmers adopt EONR during maize production. The EONR could lead to either higher or lower RSN. Field test by Hong et al. (2007) suggests that at EONR, RSN found in top 0.9 m soil profile within maize plots was 33 kg N ha^{-1} at the end of crop season. It is at least 12 kg N ha^{-1} less than that found in Producer's plots that receive higher levels of fertilizer-N. The RSN actually increased linearly in fields as fertilizer-N inputs increased from 65 kg Nha^{-1} below EONR to 20 kg N ha^{-1} above EONR. It is clear that N inputs should be regulated at or around EONR in order to lessen RSN. This step indeed has a major impact in regulating N dynamics within maize belts. It economizes on fertilizer-N, maintains optimum yield and most importantly avoids undue detriment to soil and water resources in the region.

Jaynes and Colvin (2006) suggest that, in the Corn Belt of USA, fertilizer-N inputs scheduled at early stages of the crop that is pre-plant and early sidedress-N may get absorbed into maize crop effectively. Whereas, mid-season sidedress and that applied at later stages may not be recovered effectively. A certain portion of fertilizer-N remains unutilized in the lower portion of soil profile. Analysis of RSN status immediately after harvest of maize suggested that generally soil-N gets depleted through out the soil profile. The RSN in the top 50 cm increased as fertilizer-N inputs increased from low (67 kg N ha^{-1}) to high (168 kg N ha^{-1}). Maize crop provided with mid season fertilizer-N as side dress had greater concentration of RSN. Jaynes and Colvin (2006) point out that mid season fertilizer-N applied (dribbled) to maize fields is mostly held in the soil profile as RSN or absorbed into plants. About 78% of fertilizer-N got distributed between plant system and soil profile. A small fraction could have been lost through ammonia volatilization despite its placement on soil surface. Jaynes and Colvin (2006) opine that mid-season fertilizer-N inputs allow insufficient time for the crop to effectively exploit NO_3-N pool in soil profile. Placing the midseason N nearer to root zone by injecting may avoid accumulation of RSN. We should also note that RSN could be effectively utilized by a succeeding crop, if it is not prone to loss via seepage or percolation.

Maize cultivation in Northwest Europe is intensive and meant mostly for forage. The intensification has necessitated supply of high amounts of fertilizer-N. The luxury supply of N and other nutrients has caused a certain degree of environmental problems. It is said that in Western Europe, maize cropping zones are primary sources of NO_3 contamination. The excess N inputs find their way into drainage channels, aquifers and ground water. Nearly 30% of fertilizer-N input into arable fields that support maize production is found in ground water (Hermann and Taube, 2005). It seems during past 15 years, there has been a decrease in NO_3 contamination of groundwater and aquifers due to cautious supply of N. Still, in western European plains, intensive forage production and dairy makes farmers supply a sizeable amount of excess N than required (or removed) to sustain productivity of maize fields. On an average, surplus N was 117 kg N ha^{-1} during 2000–2005. In fact, to avoid excess use of fertilizer-N in maize fields, there are suggestions that fertilizer subsidy should be linked to maintenance of environmental standards within a farm. A few other suggestions are to supply in-season fertilizer-N based on careful monitoring of leaf-N and comparing it with critical nitrogen concentration. Incidentally, CNC for maize forage may fluctuate between 4.5 g N kg^{-1} dry matter to 10.5 g N kg^{-1} dry matter, based on intensity of cropping and nutritional quality of forage (Hermann and Taube, 2005). Use of computer models to assess fertilizer-N needs of forage/grain crop as accurately as possible during various growth stages is a good idea. The percolation of N to vadose zone and contamination is rampant in areas with shallow, sandy soils, for example in the "Loess region of Europe". Some examples of the measures to curb movement of fertilizer-N to lower horizon are: mulching to avoid percolation of N to lower horizons; planting genotype with massive roots that trap as much fertilizer-N, planting deep-rooted crops during off season, etc. Overall, accurate matching of fertilizer-N demand with supply, effective exploitation of fertilizer-N, curbing undue accumulation in subsurface soil and Vadose zone are the key aspects in the intensively cultivated areas such as Corn Belt of USA, European plains, or Northeast China.

Now let us consider an example dealing with surface water contamination due to high fertilizer-N supply in the "Corn Belt of USA". Usually, high inputs result in massive accumulation of RSN. This fraction represents N inputs beyond actual requirements of the maize crop. It also represents the fraction that is easily lost via percolation, seepage, drainage and runoff, mostly into irrigation channels or canals nearby the maize fields. Subsurface seepage also results in loss of NO_3 from the maize fields (Zucker and Brown, 1998). It may cause NO_3 contamination in the surface waters and ground water. The normally accepted safe limits for NO_3 at 10 mg NO_3 L^{-1} is easily exceeded if fertilizer-N inputs to maize belt is of higher orders. Jaynes and Colvin (2006) report that a sizeable portion of NO_3 found in seepage and surface water actually arises from mid-season N inputs recommended based on yield goals. Whereas pre-plant N may get used effectively and in time. Numerous studies at field scale and larger areas have shown that surface waters of Midwest corn-soybean production zones are contaminated with NO_3 at levels beyond acceptable limits. According Jayne and Colvin (2006), a common strategy is to fine tune fertilizer-N applications. Optimum N inputs may depend on topography, specific field location, crop sequence, season, and yield goals. In addition, variation in mineralization rates, leaching loss and denitrification may also affect N demand. Splitting N supply into pre-plant, early season and smaller dose in mid season seems a better strategy to avoid loss of N through drainage. It is important to note that mid-season N application based on sensor readings may improve crop yield (9–29%), if maize crop suffers N insufficiency at that stage. However, a sizeable fraction of mid-season N is traceable in subsurface drainage and canals (Jaynes and Colvin, 2006).

Physical displacement of fertilizer-N is an important phenomenon in maize cropping zones. It affects N dynamics rather quickly. Kent Keller et al. (2008) found that fertilizer-N could be lost to vadose zone and ground water in perceptible quantities based on precipitation and drainage systems adopted. In tile drained maize fields, base flow of NO_3-N was 4 mg N L^{-1}. The NO_3-N fluxes were greater in early winter when 150 mm rainfall was received and soil profile was saturated with fertilizer-N. The greatest loss of N via lateral flow and percolation (20–30 mg N L^{-1}) coincided with high precipitation events. Clearly, fertilizer-N efficiency depreciates if its application is not split. Also due care is required to avoid undue accumulation just before rainy period.

In Spain, field trials with maize have shown that out of 150 kg N ha^{-1} applied as Urea or slow release-N along with 27 t FYM ha^{-1}, about 25 kg N ha^{-1} could be lost via NO_3-N leaching into drainage channels. Lateral flow of NO_3 could perceptibly affect N dynamics in the maize field and channels. Percolation towards vadose zone and ground water contamination could be avoided by N loading of soil profile and regulating drainage discharge.

Field evaluations of maize-alfalfa intercrops at Zaragoza in Spain have shown that regulation of irrigation and NO_3-N inputs is crucial, if NO_3-N accumulation is to be avoided. If irrigation and drainage flow is not regulated, then NO_3-N level in water reaches 18–49 mg NO_3-N L^{-1}. Cavero et al. (2003) have suggested that accurate irrigation and split application of fertilizer-N avoids undue loss of N to ground water.

It is interesting to note that, maize forage crops supplied with 75 kg N ha^{-1} and irrigated with 7.5–15 cm of water did not alter N loss due to leaching into drainage channels (Niaz et al., 2007). The NO$_3$-N did leach to lower horizons. NO$_3$-N concentration was relatively low (41 mg NO$_3$-N) at top 0–15 cm of soil profile, it increased to 47 mg N kg^{-1} at 30–60 cm depth, then to 53 mg NO$_3$-N kg^{-1} at 60–90 cm depth and 59 mg NO$_3$-N kg^{-1} at 120 cm depth. Nitrate did not leach into drainage channels and vadose zone because irrigation levels were accurately managed and water seepage to groundwater was almost nil. It could be inferred that by regulating irrigation quantity and timing, we can restrict NO$_3$-N movement away from soil profile into channels and ground water. However, a heavy rainfall event may induce NO$_3$-N loss into groundwater.

Whatever is the soil/plant test, computer model adopted, yield goal and economic returns planned, the bottom line nature stipulates is that N added to soil/plant systems should match recovery pattern by maize crop. The residual N, if any, should be least if we suspect it could be lost to atmosphere or may reach drains and ground water. Actually, in a cropping belt, pattern of N removal by entire crop rotations devised should match N supply.

Nitrogen Cycling Maize Belts

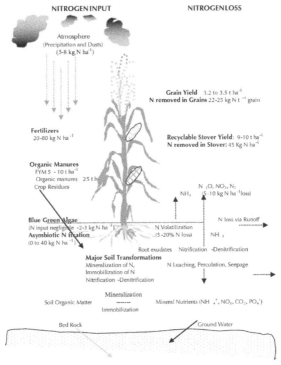

Figure 3.1. Nitrogen cycle in the Maize Cropping Zones of South India.
Source: Krishna, 2010.
Note: Nitrogen priming into the ecosystem is commensurate with moderately low grain/forage yield. Nitrogen removal too is low. Intensification of cropping requires higher N inputs. Regarding volatilization, N loss values in parenthesis refer to percentage of fertilizer-N lost.

The above discussions delineated various aspects of N cycling that occurs in maize fields or agreocosystem at large. A summarized view of N inputs, transformations possible and outputs or loss from the ecosystem has been shown in Figure 3.1. It relates to a low/moderate productivity region of South India. It is self explanatory regarding the quantum of N that is supplied or accrued, that transformed or lost in a moderately productive farm. The N turnover rate itself is highly dependent on geographic location, intensity of cropping, and grain/forage yield goals. It is obvious that N supply, turnover and output tend to be greater in intensive farming zones (e.g., Corn Belt of USA). There are advantages in summarizing N transformations and cycling as a diagram. It allows us to identify, and concentrate on improving specific steps of N transformations. For example, improving N-use efficiency helps in reducing N loading and undue accumulation of N in the agroecosystem. If a particular area is prone more to N loss via percolation or seepage, then mulches could be adopted. In case grain yield is low, improved genotype could be used. In case, forage requirement is high then genotypes that partition less of N to grains but more into vegetative portion could be selected. In case, emissions are pronounced, then, nitrification inhibitors could be added along with fertilizer-N. Overall, obtaining data about various aspects of N cycling in maize fields is helpful to farmers/researchers who intend to control N dynamics more efficiently.

PHOSPHORUS DYNAMICS IN MAIZE AGROECOSYSTEM

Maize agroecosystem extends into various geographic locations within each continent. The soil types encountered may vary enormously with regard to aspects such as inherent soil-P status, various forms of P, their distribution and availability to maize roots, chemical and biological transformation in the profile, interactions of P with other soil nutrients, loss of P from soil profile, its accumulation in the subsoil and ground water contamination. The above aspects of P dynamics may influence maize productivity either marginally or immensely.

Phosphorus Forms and Fractions in Soils of Maize Belts

Regarding soil-P, various fractions of soil P encountered in soil and their availability to plant roots seems to affect P recovery. Obviously, knowledge about various forms or fractions of soil P is necessary. Soil-P fractions can be characterized using Hedley's fractionation method (Hedley et al., 1982) or other suitable methods. It essentially involves extraction of readily available fractions using mild extractant first and then stable P forms using stronger extractants. This way, soil-P could be categorized into three forms. They are: (a) readily available forms, in other words biologically available and easily mineralizable P forms; (b) moderately resistant P forms are reversible forms; and (c) stable or resistant forms are sparingly available and highly resistant. Maize roots explore biologically available forms first. These fractions are exhausted quickly by maize roots. For example, resin-P is readily available for uptake by roots. Whereas, $NaHCO_3$ extractable-P is readily mineralizable. Soil-P extracted using NaOH is moderately available to plant roots. Sparingly available and highly resistant P forms include insoluble inorganic-P and organic-P forms. These fractions may become available in 3–4 crop cycles in a field (Basamba et al., 2006). Tillage systems adopted may affect the extent of P derived from each fraction. Also, maize-based rotations or intercrops seem to affect soil P fractions and P availability. The SOM content is also important. For example, Columbian Oxisols accumulated available-P in the top soil rather consistently. It was attributed to periodic P supply and enhanced storage capacity due to SOM. Further, it has been suggested that in highly weathered acid soils, P dynamics could be regulated by organic-P fraction.

In the Southwest European plains, long term trials on maize cropping have shown that tillage and other agronomic procedures affects individual soil-P fractions to different extents. Saavedra et al. (2007) have reported that after 21 years under NT systems, vertisols accumulated higher amounts of organic matter, organic-P, Olsen's-P and total-P compared with conventional or minimum tillage. Ratio of labile inorganic-P to sparingly soluble $Ca-PO_4$ in the top 0–5 cm was much greater with no-till (0.8) than under conservation (0.57) or conventional tillage (0.55). Increase of inorganic-P in surface soil makes this fraction of P vulnerable to loss via erosion, runoff, and percolation.

Let us consider an example from Northern Plains of India, where Ustochrepts are predominately utilized to produce maize or other cereals. Long term maize-wheat rotations, in general, cause massive alterations in P availability to crops. The individual P fractions may get depleted to different degrees. Sharma (2006) found that repeated P inputs between 17 and 35 kg P ha^{-1} each season for 22 years caused "Build up of soil P". Whereas, in the fields not provided with fertilizer-P there was "drawn down of soil-P" through out 22 years. The dynamics of not just available-P but many other P fractions were also affected due to cropping systems and fertilizer-P schedules. There was significant decrease in fluoride-P (24%), NaOH extractable-P (44%), and H_2SO_4-P (25%) but increase in chloride-P (41%) in fields provided with balanced fertilizers (120 N: 17.5 P: 33 kg K ha^{-1}). Among the P fractions, saloid-P contents served as a good index for soil-P availability to maize roots. Clearly, fertilizer input schedules and maize crop production systems both affect dynamics of various soil P fractions differently. Prior knowledge about soil P availability trends and P fractions that affect crop growth and productivity is necessary.

Phosphorus Distribution and its Movement/Diffusion in Soil

Garcia et al. (2007) have reported that stratification of nutrients, especially P is rampant in maize blocks kept under NT systems. Such stratification actually builds up P concentration at certain depth in soil profile, but it is vulnerable to runoff loss. A one-time chisel or disk tillage after 8–10 years of no-till sequence did not alter the situation regarding stratification and P accumulation. The Bray P1 was still 5–21 times more at 0–5 cm depth compared to 10–20 cm depth. Greater redistribution of P was possible only with compost plus a moldboard plowing. Otherwise, infrequent tillage seems to redistribute soil P, reduce undue stratification and reduce loss of P via runoff. Boem et al. (2008) have reported that NT system reduces P buffering capacity of surface horizons. It could be attributed to organic matter contents. However, variation in P sorption index of surface soils was related to clay contents instead of organic matter. Knowledge about subsurface P accumulation is also important.

Critical P in Soil for Maize Production

Maize production reduces soil P levels. Fertilizer-P inputs are required if P availability goes below critical limits envisaged. On the other hand, if soil-P levels are high then P inputs are to be restrained. Methods to achieve draw down of soil P to levels just around or above critical limits are needed. By definition, critical P in soil is quantity of plant available-P that results in 90–95% of maximum yield possible in a given soil or agro-environment. The critical P limits vary depending on soil types, their physico-chemical characteristics, cropping sequences followed, irrigation systems, and yield goals. During maize production, fertilizer-P inputs are almost always decided based on soil P test values. More appropriately, farmers often consider the critical soil-P levels before supplying P. The critical P level in the maize producing regions of Northeastern USA fluctuates between 12 and 20 mg P kg^{-1} using various extractants. In Pennsylvania, critical P ranges from 19 to 20 mg P kg^{-1} soil extracted using Bray-Kurtz P1 or Mehlich's-3 solution. In Iowa, critical P concentration in soil was found to be 13 mg P kg using Bray-P1 extractant and 12 mg P kg^{-1} using Mehlich-3 extractant. In Vermont and NewYork state, critical P is said to be around 4 mg P kg^{-1} based on modified Morgan extractant (see Heckman et al., 2006). Maize is cultivated on sandy loams in Nebraska. Again, knowledge about inherent soil P is crucial while deciding on fertilizer-P inputs and yield goals. The current critical P limits is 15mg kg^{-1} Bray-P1 for Haplustolls found in Nebrasks region.

Agriculturally, Pampas of Argentina is a productive belt. It contributes large amount of cereals and soybean. Yet, its productivity is dependent on optimum availability of two major nutrients, namely N and P. For major cereals like wheat and maize, critical P limits range from 15 to 20 ppm. Rubio et al. (2006) have reported that much of maize cropping zones in Pampas suffers due to paucity of P. The soil P levels are usually <12.0 ppm. In order to reclaim optimum P levels, these Mollisols of Pampas need to be refurbished with appropriate amount of fertilizer-P. Based on a simple linear equation, Rubio et al. (2006) devised methods to enhance soil P levels above critical limits (15 ppm). They have shown that, if soil depth is 0–20 cm and bulk density is 1.2 t m^{-3}, then about 2.4 kg P ha^{-1} is to be supplied in order enhance soil P availability by 1 mg kg^{-1}. Incessant cropping, at times may deplete available-P levels much below critical limits. In that case, proportionately larger P inputs are needed to achieve optimum soil P level.

The sandy Oxisols found in West African Savannas are low in P content. The available-P pools are low at 2–3 ppm. However, the critical P needed is above 10–12 ppm for a crop that produces 2–2.5 t grains plus forage ha^{-1}. Subsistence farming with low-P supply made via organic manure results in 0.8 t grain ha^{-1}. It is said supply of fertilizer-P and addition of organic manures is essential to retain available-P pools at optimum levels. Soluble fertilizer supplied at 40–60 kg P ha^{-1} often restores optimum P levels, at least for a season.

The loamy soils of South India are again low in P content. At times, total P may be optimum but available-P pools are well below critical limits of 10–15 ppm. The available-P in Red Alfisols of South India is about 5–8 ppm P. Consistent supply of soluble fertilizers and organic matter maintains available-P pools at levels above critical limits.

Phosphorus Inputs to Maize Belts

Phosphorus inputs into maize fields are mostly based on soil test values. Pre-planting soil test values usually guide the quantum of P supply to soil. Generally, soils are classified as low or high with regard to P, based on various soil-P tests. Soil tests that depict plant available-P accurately are preferred. Mehlich-3 P, Bray-1P, and Olsen's P are some of the examples. Following is a set of guide lines for fertilizer-P supply, available to maize farmers in Iowa (US Corn Belt). It is based on pre-plant soil-P status and it effectively considers the subsoil P availability to maize root system:

Soil P test	Phosphorus in Soil Sample (ppm)				
	Very Low	Low	Optimum	High	Very High
Bray P, Mehlich's-3 P					
Low Subsoil P	0–8	9–15	16–20	21–30	31+
High Subsoil P	0–5	6–10	11–15	16–20	21+
Olsen's P					
Low Subsoil P	0–5	6–10	11–15	16–20	21+
High Subsoil P	0–3	4–7	8–11	12–15	16+
Mehlich's-3 ICP*					
Low Subsoil P	0–15	16–25	26–35	36–45	46+
High Subsoil P	0–10	11–20	21–30	31–40	41+
Phosphorus Input (kg ha^{-1})	100	75	55	0	0

Source: Sawyer et al., 2008.

Note: Fertilizer-P supply suggested is based on a yield goal of 10 t grain ha^{-1}plus forage. Fertilizer-P input could be revised depending on yield goal and soil test values for each field and variations within it.

* ICP = Inductively Coupled Plasma.

As stated earlier, in Nebraska, current critical P level is 15 mg kg^{-1} Bray-1 P. Phosphorus supply has ranged from 40 to 80 kg P ha^{-1} and commensurate yield goals have fluctuated between 5.9 and 12.9 t grain ha^{-1} (Wortmann et al., 2009).

Regarding P inputs to crops like maize, Ebeling et al. (2002) opine that it is essential to balance inputs and removals of P on each field. Amount of P impinged via fertilizers should be, as far as possible, equal to that removed as grains, fodder or via erosion and leaching. If we follow the above rule, soil test values for P do not alter much despite cropping. Actually, managing soil-P stringently based on above principles also avoids soil-P accumulation and negative effects on ground water quality. In Wisconsin, soil test P is measured using Bray-Kurtz P1 extractant. Generally, soil-P >100 ppm is considered detrimental to environment. For agricultural cropping, especially maize production, "high P soil" is one where response to added P is low (<30%). It often has 20–30 ppm Bray-Kurtz P1. Major guidelines for P inputs in Wisconsin are as follows:

    (a) Soil with 50 ppm P or less—P application is allowed to meet crop's need;

    (b) Soil with 50-100 ppm P—P application should not exceed crop's need;

    (c) Soil with 100 ppm P or above—eliminate or restrict P inputs to very low levels, much below the crop's need. It then allows "draw down" of accumulated soil P.

Reports from CIMMYT, in Mexico, suggest that in the maize belts of Latin America, soil tests are extensively used to gauge the extent of response to fertilizer-P inputs. On acid soils, values < 7 ppm P (Bray-1) means a response to P supply is likely. However, if Bray-II test is adopted values <15 are indicative of crop response to P supply. If Olsen's Test is used, value between 5 and 10 ppm P suggests that crop response is possible.

Maize producing zones in Kenya spread into different soil types. Among them, Acrisols are known to fix sizeable amount of P applied as fertilizer. Therefore, P inputs are calibrated and enhanced in order to first satisfy P fixation capacity of soil and then to improve crop yield. For example, Nziguheba et al. (2004) reported that seasonal P inputs of 10 kg ha^{-1} improved crop yield but did not cause accumulation of P in soil. The labile pool of P was unaltered. Soil analysis indicated that only inorganic-P (NaHCO$_3$ extractable) fraction was most affected by P inputs. Its increase coincided with P supply and decreased gradually due to interruptions. Application of 25 kg P ha^{-1} improved maize grain production and improved soil P status gradually without causing undue excess.

Maize-wheat rotations are popular in intensive farming zones of Northeast China. Long term field trials have shown that application of 60–80 kg P ha^{-1} maintains maize grain productivity above 6 t ha^{-1}(Tang et al., 2008).

In Queensland, soil P is replenished using soluble P fertilizers such as TSP or SSP. About 20–35 kg P ha^{-1} is added as basal dose to support the crop for the entire season. Phosphorus is also channeled into maize belts through FYM and other organic manures. The extent of P supplied via organic matter depends on percentage P contents, amount supplied and mineralization rates possible with organic sources.

Laboski (2009) compared P supplying capacity of various organic manures in relation to inorganic fertilizer like KH$_2$PO$_4$. Dairy slurry contained 530 mg L^{-1} of total P with 79% of it as inorganic-P. Swine slurry had 1254 mg L^{-1} total P with 88% available-P. However, relative availability (RA) of P in manures could vary depending on several factors related to soil, crop and environmental parameters. The RA of manure P is calculated as change in soil test P when manure was applied divided by change in soil test value when KH$_2$PO$_4$ was supplied. The RA for dairy slurry ranged from 0.71 to 0.79 and that for swine slurry ranged from 0.88 to 0.95. Knowledge about RA of manures could be useful while deciding on ratio of inorganic and organic P fertilizers to be supplied to a particular field.

Atmospheric Inputs

A certain amount of P, even though small or at times negligible, is added into maize farming zones via atmospheric dusts, precipitation and storms. It depends on geographic location. For example, in Guinea-Savanna and Sahelian regions, dust storm (*Harmattan*) adds perceptible amount of P. Sand particles carry sizeable amount of P and deposit it to fields farther away. The extent of P that drifts from other regions is highly dependent on storm strength, quantity of sand/dust received in a field and more importantly P enrichment ratio of sand/dust. For example, Sterk et al. (1998) found that P enrichment ratio was about 0.83 for dust storms in West Africa.

In the Semi-arid India, generally, about 2–3 kg P ha^{-1} could be added annually through atmospheric dusts and precipitation (Murthy et al., 2000). The exact amount of P added may vary depending on location of fields. Maize crops close to sea or industries may obtain greater quantities of nutrient via atmosphere.

Phosphorus input into maize field also occurs via irrigation water. It could be negligible in a short run, but consistent irrigation does supply perceptible amounts of P. It could be adjusted while devising fertilizer-P schedules. Irrigation water may often contain 1–2 ppm P or higher based on source and parent material of soils found in the channels. However, in practical agriculture, P drawn into fields via irrigation is often over looked. Dissolved P in irrigation channels is considered only if it reaches levels good enough to cause eutrophication.

Phosphorus Recovery by Maize

Maize absorbs 2.5–3.0 kg P to produce a ton grain. Agriculturally, "Corn Belt of USA" has been intensified enormously during past few decades. Phosphorus inputs and recovery rates have played a key role in intensification of this maize belt. Phosphorus recovery by a corn crop in US Mid-West has increased from 40 kg P in 1900 to 50–60 kg P due to intensification of farming systems and higher grain yield goals (Abendroth and Elmore, 2007). In Iowa, maize-soybean rotations are most common. In this region, P recovered by maize grown on Hapludolls is around 4.5–6.4 kg t^{-1} grains if it is a crop intended for grains and 1.52–2.75 kg P t^{-1} fresh silage (65% H$_2$O) if it is a silage crop (Sawyer et al., 2008).

Maize genotypes grown in Brazil vary significantly with regard to P acquisition and utilization (Parentoni and Lopes de Souza, 2008). On an average, P is accumulated at 0.22% in seed and 0.07% in stover, if the crop is supplied with optimum levels of fertilizer-P. It is common to encounter P deficiency in dry lands. Maize genotypes may accumulate about 0.19% P in seed and 0.06% in stover, if planted on soils deficient in P.

Field investigations across different locations in Kenya suggest that P recovery by maize is influenced by yield goals, nutrient supply and soil types (Smaling and Janssen, 1993). Following is an example:

Soil type and Phosphorus supply	Grain Yield kg ha^{-1}	Phosphorus Uptake kg ha^{-1}	
		Total	t^{-1} Grains
Kisii Red Soils			
P 0	2100	5	2.4
P 22 kg ha^{-1}	4900	12	2.3
Homa Bay Black Soils			
P 0	4500	24	5.1
P 22 kg ha^{-1}	6300	5	4.8
Kwale Brown Sands			
P 0	2600	7	2.7
P 22 kg ha^{-1}	3700	16	3.7

Source: Smaling and Janssen, 1993.

In the Southern Indian plains, an irrigated maize crop that produces 6.1 t grains ha⁻¹ recovers about 52–57 kg P ha⁻¹, but the same cultivar under rain fed conditions yields only 2.1–2.5 t grains ha⁻¹ and recovers 8–10.5 kg P ha⁻¹ (see Krishna 2010). In addition to soil moisture, other factors such as geographic location, soil type, season, agronomic practices like intercropping or rotations also influence P recovery and grain yield.

Reports from Chinese maize growing regions indicate that to produce a ton grains about 3.1–3.8 kg P is absorbed (Wang et al., 2008). Reports based on long term trials in Northeast China indicate that maize crop recovers 3.1 kg P ha⁻¹ from soil to produce 1.0 t grain. Physiologically P-use efficiency is about 240 kg grain for each kg P absorbed from soil. Grains may contain 2.1g P kg⁻¹grain (Tang et al., 2008).

Phosphorus Loss from Maize Ecosystem (including fixation)

Phosphorus inputs into maize belts are generally not utilized efficiently by the crop. In most agroclimates and soil types only 20–23% of P inputs are used up by maize crop. This situation is similar with many other crop species and cropping belts. During long term no-till maize, normally P accumulates in the surface layers of soil. This portion of P is vulnerable to loss via runoff in addition to other modes. Usually, as time lapses, solution P in the runoff increases (Daverede et al., 2003). However, P concentration in the surface layers of soil could be reduced through plowing. A one-time moldboard plowing is said to reduce soil P levels in the upper layers of soil and consequently in the runoff rather significantly (Garcia et al., 2007). Actually, a one-time tillage may affect a bunch of soil traits like aggregate stability, water filtration, runoff volume, erosion and runoff P loss. Further, a one-time tillage with moldboard actually improved maize grain yield by 11–25% over a continuous no-tillage system (Pierce and Fortin, 1997). Quincke et al. (2007) reported that a one time plowing after several years of no-till affects runoff P. It effectively reduces dissolved reactive P in the runoff when compared with NT system. Yet, it may not get translated into crop growth or grain increases.

Phosphorus loss from cropping systems is affected strongly by soil moisture status and irrigation schedules. Roose and Fowler (2004) simulated water and nutrient loss from cropping systems. The model suggested that P loss as surface runoff is more when fertilizer inputs occur repeatedly. If it is replaced with band placement or spot application, risk of P loss via surface runoff is reduced. Regarding accumulation of P in the groundwater and contamination, it was found that repeated irrigation and continuous moist conditions in soil accentuated movement of water plus P into lower layers. Generally, if high nutrient input is practiced, soils tend to loose accumulated P in runoff water and through percolation. It is common to trace leached P in irrigation channels, subsurface layers and in places where seepage water collects in the farm.

Loss of soil and P via sand/dust storms is rampant in the Savanna and Sub-Sahara of West Africa. Annually, about 6 kg P ha⁻¹ could be lost via wind erosion. Field trials with maize have shown that in the Sudanian zone, maize-peanut rotations tend to loose appreciable quantities of nutrients from fields due to wind erosion. About 24–27 t ha⁻¹ soil could be lost in first few storms during crop season. The P enrichment ratio of sand particles eroded ranged from 1.5 to 1.8. Mulching and practicing *Zai* system of planting for maize or other cereals reduced loss of soil and nutrient appreciably. *Zai* system involves dibbling of seeds in a trough created on sandy surface and placing seeds along with small amounts of organic matter and mulch.

The soil erosion peaks usually with high rainfall period. In Indonesia, reports regarding soil erosion and P suggest that 14–15 kg P ha^{-1} could be lost from the top soil due to runoff (Hermiyanto et al., 2007). Total P loss is dependent on sediment enrichment. In the above study, sediment enrichment for P ranged from 10 to 20%.

Soil P could be lost via percolation and surface runoff if texture is highly sandy. On the other hand, if soils are rich in organic matter, erosion and loss via seepage water may be less, but immobilization of P into organic fraction of soil is a clear possibility. Fertilizer-P immobilized into organic fraction may after all be available at a later date if mineralization process takes over.

Phosphorus Diversion from Maize Crop

Soil phosphorus gets diverted to weeds or to intercrops in a mixed farming situation. Actually, P mobility and its availability in the vicinity of maize roots are important. In certain cases, weeds or intercrops may garner high amounts of soil-P. It causes paucity of P in the root zone of main crop that is maize. There could be many factors that affect P availability to roots of maize, competing weeds, and intercropped annuals or trees.

Plate 3.6. A Maize farmer in Africa (Kenya) uses hoe to inter-culture and reduce weed infestation.
Source: CIMMYT, Mexico.
Note: Weeding at early stages of the crop avoids loss of nutrients and water to weeds. It helps in development of crop canopy and avoids competition for photosynthetic radiation.

Trees interspersed within maize fields is a common feature in East African cropping zones. Let us consider an example of maize crop intercropped with trees on Ferralsols of Kenya (Redersma et al., 2005). In this case, P is not lost from the field ecosystem, yet it affects nutrient dynamics envisaged by the farmer. It reduces P recovered by maize crop and therefore gets translated as lowered grain yield. Phosphorus is a limitation to maize growth in Kenya. It is possible that trees that dot the landscape or those intercropped may primarily affect maize through competition for P. The rooting pattern, available-P pool and soil moisture status are major factors that affect P recovery by maize crop. Tree roots that are well distributed may out-compete and remove larger pool of P from soil thus leaving inadequate quantities to maize crop. Literally, soil-P gets diverted to trees or weeds. Weeding or herbicide application is essential (Plate 3.6). Redersma et al. (2005) have stated that maize grown nearer *Gravillea* tree lines suffered from P deficiency. Maize productivity got reduced by 30–40% in locations nearer to *Gravellia* tree lines. Phosphorus limitation was caused firstly due to exhaustion by tree roots. Next, tree lines removed large quantities of soil moisture hence hampered soil-P mobility to maize roots. In other words, interaction of soil moisture x available-P pool affected P dynamics in the maize fields. In addition, maize root growth may have been hampered by allelopathic effects. It obviously affects ability of root system to explore soil moisture and P.

Fertilizer-P Recovery Studies

Recovery of fertilizer-P is governed by several factors related to fertilizer formulation, crop genotype, soil, and environment. Overall, in most situations, fertilizer-P recovery has not exceeded 20–22% of that applied. Crop genotype, its demand for inorganic P, rooting pattern are some of the crop related factors. Soil texture, pH, P fixation capacity, P-buffering capacity, and available-P pool are few soil related factors that affect fertilizer-P recovery. Phosphorus is relatively an immobile nutrient element in soil. It has to be absorbed in dissolved state. Therefore, soil moisture status immensely affects soil-P/fertilizer-P recovery. In the Guinea-Savanna of Western Nigeria, maize is supplied with slightly higher amounts of fertilizer-P, because the sandy soils are low in P. They are also prone to loss of P via erosion, runoff, and percolation. Soil-P fixation also affects fertilizer-P recovery. In this area, again maize crop absorbs only 20–21% of fertilizer-P added (Bromfield, 1969). Researchers at Federal University of Agriculture, Abeokuta, Nigeria examined influence of repeated application of fertilizer-P to maize-cowpea rotation, season after season. Cropping depleted soil organic-P by 42%. Obviously, organic-P was mineralized then absorbed by maize/cowpea roots. Yet, it was found that much of inorganic-P applied was transformed to insoluble-P such as Al-P and Fe-P. Every crop season only 17–31% of 30 kg P ha^{-1} applied at sowing was recovered into above ground portion. Cumulative P recovery never reached beyond 75% of total P input. Over all, it was suggested that application of 30 kg P ha^{-1} season^{-1} kept soil P levels above critical limits for maize or cowpea (Adetunji, 1994). Maize grown in the highlands of Cameroon and other locations in Central Africa is affected by soil acidity and related P fixation which is generally high. Phosphorus fixation reduces fertilizer-P efficiency and necessitates application of higher amounts of P. Resource poor farmers tend to adopt different procedures like mixing fertilizer-P with lime or organic manure like macuna etc. Yamoah et al. (1996) have reported that liming generally raised soil Ca level, pH, effective CEC, and lowered Al[+] levels. Macuna application behaved almost like lime with regard to P requirement. Over all suggestion was to supply <2 t lime ha[-1] or apply macuna green manure and <35 kg fertilizer-P ha[-1].

Field trails at Faisalabad in Pakistan, suggest that maize crop recovers greater share of fertilizer-P applied via fertigation compared to broadcasting. About 26–32% of 33 kg P ha[-1] supplied to maize fields were traced in above ground parts of fertigated plots, but only 16% of fertilizer-P reached stover and grains, if it was broadcasted (Iqbal et al., 2003). Availability of soluble P in proximity to roots is an important factor that improves fertilizer P efficiency under fertigation. Loss of P through fixation and leaching may be more if P is broadcast.

Residual-P during Maize Cropping

Evaluation of long term (15 years) maize–wheat rotations in Northwest China suggest that soil P gets depleted with incessant cropping. The plant available-P or Olsen's P (NaHCO$_3$ extractable) depleted until it reached a low point at 3 mg P kg[-1] soil. Where-as, in fields provide with fertilizer-P it accumulated progressively at an average rate of 1.21 mg kg[-1] yr[-1] (Figure 3.2). The Olsen's P increased at 1.55 mg kg[-1] yr[-1] soil for an input of 80 kg P ha[-1] yr[-1]. Many of these aspects on soil-P depletion and accumulation can be easily simulated and modeled. Soil-P status could be predicted accurately (Ma et al., 2009).

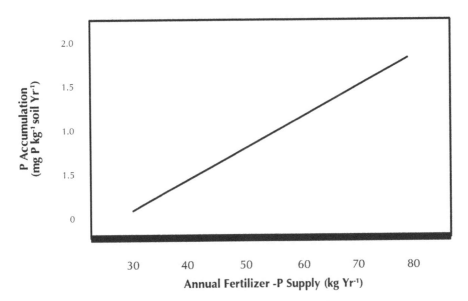

Figure 3.2. Phosphorus accumulation trend in soils used for Maize-Wheat rotation.
Source: Drawn based on Ma et al.. 2009.
Note: Phosphorus accumulation refers to increase in Olsen's or NaHCO$_3$ extractable P yr[-1.]

Residual-P could be a major problem in maize belts that thrive under high input systems. Repeated supply of P into agroecosystem allows rapid accumulation of P in the soil profile. Residual-P fraction may also be prominent in soils that are inherently high in total and available-P levels. For example, Mollisols in the US Corn Belt, Ustudolls of European plains or those in Northeast China may show high residual-P. Incessant P inputs plus inefficient extraction leads to this situation. Soil test for P after each crop is helpful in ascertaining residual-P pool. It is also preferable to supply P based on requirements of entire cropping sequence, instead of considering a single crop within a sequence. Residual-P could also be efficiently recycled by adopting green manure or catch crop during fallow season.

Fertilizer-P inputs are often calculated based on a single crop in a sequence. It could be more than that required for optimum yield. As such recovery of fertilizer-P is low. Usually, only 20–23% of fertilizer-P supplied into soil gets utilized by the first crop. It leaves a large fraction still in surface and subsurface horizons that could be slowly utilized by the succeeding crop. The localization of residual-P in the soil profile and its quantity needs to be ascertained. Knowledge about exact quantity of residual-P that could be garnered by succeeding maize crop is essential. It allows us to obtain better fertilizer-P efficiency during maize phase of the rotation. Residual-P could also be utilized by fallows that support pastures or green manure legumes. To quote an example, in Guinea-Savanna of Western Nigeria, soybean grown ahead of maize is supplied with slightly higher dose of fertilizer-P. It may range from 40 to 60 kg P ha^{-1} depending on soil test and yield goal envisaged. Therefore, during maize phase of a rotation, normally, available soil P levels are higher. Available-P levels reach 13.6–42.6 mg kg^{-1} soils from an initial level of 5.7 mg kg^{-1} for an input of 60 kg P ha^{-1}. Whereas, in no-P control plots available-P pool was only 7.7–13.8 mg kg^{-1} (Ogoke, 2006). Therefore, P inputs to crops that maize follows affects soil levels. Residual-P could be used efficiently by pastures, green manure, or catch crops that precede maize.

Studies in Southern Kenya have proved that P supplied at low rates each season was efficiently recovered by maize (Dirk et al., 2006). However, if large dosage of P is supplied at the beginning of a long term cycle, residual-P accumulation was high. The Olsen's P consistently remained above that required for maize. High P retention was due to mismatch between root growth and physical proximity of fertilizer-P applied a few seasons ago. Fertilizer placement improved P recovery hence residual P was low.

In the South Indian maize belt, where maize-sunflower rotations are common, farmers tend to supply a large share of nutrients via organic manures. Field trials using labeled ^{32}P manure have shown that extent of P derived from organic manures like FYM, bio-compost, press mud varies and it leaves a certain fraction of P as residual P in soil. The crops that follow maize in sequence could effectively utilize the residual-P. The amount of P derived by maize from organic sources ranged from 17.5 to 43% and the rest of P found in maize was from soil. In the succeeding sunflower, P from organic manures ranged at 19.6–42% and that from soil ranged between 57 and 82% (Meena et al., 2007). Obviously, we can calibrate and supply organic manures and chemical fertilizer better by considering the P dynamics of the entire cropping sequence. Knowledge about P derived from organic manures, inorganic chemicals, and soil should help us in deciding their proportions of supply more accurately.

Long term trials (15 years) have shown that in Northeast China, farmers tend to supply high levels of fertilizers to support intensive cultivation of maize and wheat. Depending on recovery rates, there is every possibility for accumulation of major nutrients in the soil profile. Tang et al. (2008) have estimated that for every 1.0 kg P applied in excess of that removed by maize crop, the available-P pool in soil (Olsen's P) increased by 3.4 mg P kg^{-1} soil. They have suggested that in order to improve efficiency of maize production and maintain optimum P dynamics, there is a strong need to understand the long-term P input-output trends. If not, fertilizer-P efficiency may dwindle and soil may accumulate P that becomes vulnerable to loss via leaching. Details about P dynamics (cycle) in the field/agroecosystem seems highly pertinent (Figure 3.3).

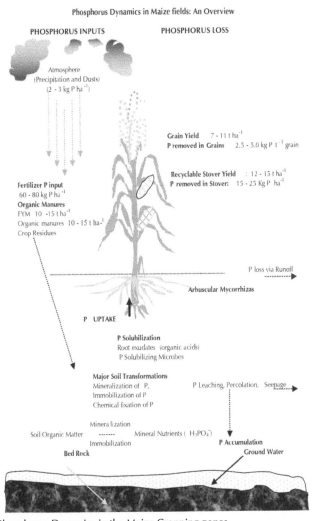

Figure 3.3. Phosphorus Dynamics in the Maize Cropping zones.

POTASSIUM DYNAMICS IN MAIZE CROPPING ZONES

Potassium in Soil

Potassium is an essential element required for growth and productivity of maize crop. It is classified as a major element required in larger quantities for optimum growth and grain formation. Maize is relatively a deep rooted cereal and it exploits K from both surface and subsurface layers of soil profile. Potassium is found in different fractions of soil. The extent of K traced in each fraction differs based on soil geographic location, parent material, weathering processes, cropping systems, and extraneous fertilizer-K sources supplied to maize fields. Major fraction of soil K is encountered in the structural component of soil minerals. As fixed or non-exchangeable-K, it is not easily extracted by maize, perhaps not at all. Such unavailable forms of K are held strongly in the lattice structures of micas, feldspars, and clay minerals. The exchangeable-K and solution-K are fractions that form available pool of K. The readily available-K is mostly held on the surface of clay and organic matter as exchangeable moieties. The available-K is perhaps most immediately important for crop growth and production. As stated earlier, it includes both exchangeable and soluble-K (Kemmler, 1980). However, more recent opinion is that, in addition to quantity of each fraction, it is the interrelationships and dynamic equilibrium between K fractions in soil that affects its extraction and utilization by maize crop (Krauss, 1997). Soil solution-K that is most easily utilized by maize roots is usually smallest of the K fractions. The solution-K is directly explored and absorbed by maize roots. Diffusion of soil K is primarily responsible for K availability surrounding the roots. Rapid extraction of K by roots creates a gradient of solution-K near the root surface. Steeper K^+ gradients obviously induce diffusive flux of K. The diffusive flux of K also depends on soil moisture content. The solution-K that gets depleted due to absorption by roots or natural factors like leaching, percolation, or seepage is replenished from exchangeable pool. The rate of replenishment of solution K from exchangeable pool is dependent on soil characteristics such as clay content, sand fraction, SOM, CEC, pH etc (Wolkowski, 2000). For example, sandy soils may release K rapidly from the exchangeable fraction. The release of K from non-exchangeable fraction to available pool is slow and small in quantity. It may not have an impact of K nutrition of maize roots immediately. Structural K is held in a chemically bound form. Release of K from this fraction is indeed feeble and negligible in terms of short term nutrient dynamics within maize fields. Potassium fixation is a phenomenon that affects availability of soluble-K fertilizers supplied to maize crop. Soils containing clay minerals such as bentonite, illite, montomorillonite have greater affinity to fix soil K. In comparison, soil with kaolinitic clay fixes much less of soluble-K applied. The soil-K fixation capacity in maize fields may fluctuate between 20 and 40%. No doubt, it reduces fertilizer-K efficiency. The maize crop should therefore be supplied with higher levels of K, first to satisfy fixation, then to enhance soluble-K levels and later result in enhanced recovery by maize.

Regarding K distribution in soil profile, large fraction of soil K is encountered in upper layers and in the region of maximum rooting. Nearly, 30–50% K could be localized in the upper 0–60 cm layer and rest may be found spread into layers well below 1.2–1.5 m depth. The subsurface K is a sizeable source for maize production. Hence, it is advisable to consider subsurface K and its variations during fertilizer-K supply. Soluble K is a mobile source to plant roots and is vulnerable to redistribution due to percolation, seepage, and erosion. Soil K may also be encountered at the vadose and ground water layer in appreciable quantities. It is attributable to rapid percolation. Actually, K distribution in soil is specific to each field, hence it is better to obtain an idea regarding it before basal K inputs. Potassium stratification is also encountered in many of the maize farming zones. It may be pronounced in regions that follow NT systems. Soil compaction and nutrient addition or removal pattern also cause stratification. Potassium stratification is common in South India both on Alfisols and Vertisols. Vertisols in South Indian plains are rich in montomorillonite clay and they impart high K-buffering capacity to soil. Moisture holding capacity is also high. In Australia, for example the Vertisol regions in New South Wales are prone to K stratification. Soils with smectite clays are prone more to K stratification. Sub-soil layers are also prone to fix K leading to stratified accumulation of K (Taylor et al., 2006). In Australia, maize is grown on Vertisols that possess high K buffering capacity due to smectite clay component (Carter and Singh, 2004). Generally, it is believed that soils with high K buffering have adequate K supplying capacity and those with low K buffering capacity need frequent additions of fertilizer-K. There is still a problem when high K buffering Vertisols fix high amounts of fertilizer-K applied.

Potassium Input to Maize Ecosystem

Firstly, there is a strong need to regulate K inputs into soil. The bottom line is that K supply to maize fields should match the removals and offset any loss via natural causes. This suggestion holds, despite the fact that soils in most maize growing regions are sufficiently endowed with K. Knowledge about inherent status of soil-K and its availability pattern to maize crop during the season is necessary before fixing-up the fertilizer-K supply schedules. Researchers use different soil test methods to assess soil-K available to maize crops. Measurement of available and exchangeable-K fractions using 1 N NH_4OAc is the most common method. Based on this test, soils for maize production can be categorized into low, medium, and high in K. Soils found in maize belts vary widely for different fractions of K. Some caution is also required while interpreting data from soil K tests. It is often remarked that result from a single measurement of exchangeable-K may not suffice. In fact, there are also suggestions that K release patterns from exchangeable and non-exchangeable K fractions may vary enormously even during a single crop season. Plant available-K content may be immensely affected by K release from exchangeable and non-exchangeable fractions. A recent study pertaining to soils from maize producing zones of Thailand suggest that there are in fact, two fractions of non-exchangeable-K that are released at distinctly different rates into available pool. Nilawonk et al. (2008) have argued that depending on nature of clay and their contents, simple NH_4OAc extraction for K may be suitable for Kaolinitic soils, but in certain cases with Smectite clays, available K data from both Ca-resin and NH_4OAc extraction may be needed.

In many of the maize farming zones of the world, soils are endowed with sufficient levels of total, non-exchangeable, exchangeable, and soluble K. Yet, it is wiser to refurbish the soil with K, at least to the extent that it gets depleted in a season. Otherwise, it may lead to soil K mining and appearance of K deficiency. During practical farming, K supply to maize farms varies widely based on soils, cropping systems, genotypes, agronomic practices, especially residue recycling practices, etc. State Agency recommendation for entire region is practiced in some maize belts, but it may not be common, since it could induce undue K accumulation in soil. Field scale K inputs are based on State Agencies Recommendations, Maximum Yield goals, and previous Soil-test Crop response data. Site-Specific Nutrient tests have also been conducted in the maize cropping zones. Let us discuss a few examples. In the maize farming zones of Iowa, K supply is mostly based on soil test values and yield goals envisaged by farmers. Soils are actually categorized based on soil-K tests covering both surface and subsurface layers. Soil characteristics such as sand or clay contents are also considered. Following is an example from Iowa in the US Corn Belt:

| Soil Test | Potassium Content in Soil Sample | | | | |
	(Ammonium Acetate or Mehlich-3 Extractable K in ppm)				
	Very Low	Low	Optimum	High	Very High
Low Subsoil K	0-90	91-130	131-170	171-200	201+
High Subsoil K	0-70	71-110	111-150	151-180	181+
K$_2$O Input Recommended (kg ha^{-1})					
Fine Texture Soil	143	99	50	Nil	Nil
Sandy Soil	122	77	50	Nil	Nil

Source: Sawyer et al., 2008.

Note: Fertilizer-K supply suggested is based on a yield goal of 10 t grain ha^{-1}plus forage. Fertilizer-K input could be revised depending on yield goal and soil test values for each field and variations within it.

According to Riedell et al. (2000), maize response to fertilizer-K inputs is highly dependent on inherent soil K levels. For example in the northern fringes of US Corn Belt, response of maize crop to K inputs were nil or feeble because Haplustolls in South and North Dakota were generally bestowed with high levels of K. Added K , if any, improved surface and subsurface K levels, but without much effect on maize productivity. Banding improved K recovery by crop. Reports from Mississippi delta suggest that K supply to maize fluctuates from 45 to 90 kg K ha^{-1} depending on yield goals. Potassium inputs enhanced soluble-K pool. At higher soil test values, K supply had minimal effects on maize growth and grain formation. Maize–cotton rotation is common in this region. It is said that K supply may not have strong impact on maize yield, but it is of immense value to succeeding cotton crop. Supply of K for the entire crop rotation has its advantages (Bruns and Ebelhar, 2006). In Michigan, soils are generally endowed with optimum levels of K. Yet, it is replenished based on yield goals. On soils testing <100 kg K ha^{-1}, 60–75 kg K ha^{-1} is supplied using chemical and organic manures (Christenson, 2008). Overall, it is interesting to note that during past 35 years, K recovery by corn grown in USA has increased steadily at a rate of 2.3 kg ha^{-1} yr^{-1}, considering that grain yield improved at 109 kg ha^{-1} yr^{-1}.

The average K supply differs enormously among the Corn states of USA. It is a mere 3.5 kg ha^{-1} in Nebraska to as high as 115 kg K ha^{-1} in Indiana. Since K removal per ton grain yield is almost uniform in the entire corn producing region, it generates imbalances in soil K distribution. The input-out put equations are obviously different in each of the corn states. There is a negative K balance in states like Nebraska, Kansas, and South Dakota. Such a situation is easily encountered in many of the corn producing countries. There is a common perception that soils in many of the maize cropping zones has sufficiently large inbuilt pool of K and this drives farmers to refrain from supplying K at optimum levels. According to Dobermann (2001) fertilizer-K supply has generally exceeded its removal by crop, but such surplus supply has declined during recent past. The soil K surplus has dwindled from 47 kg K ha^{-1} in 1980–1984 to 28 kg K ha^{-1} in 2005. Obviously, soil testing and appropriate replenishments have key role to play in maintaining optimum K dynamics in the corn ecosystem.

In the European plains, maize is predominantly a forage crop. As such, maize accumulates and partitions relatively larger amount of K into stem and leaves. Therefore, K supply should be proportionately high and it needs to be matched with periods or crop stages when maize absorbs K most rapidly. Reports suggest that maize crop sown in May absorbs greatest amount of K during July and August. In Great Britain and France, a maize crop yielding 40 t fodder ha^{-1} is supplied with 360 kg K$_2$O ha^{-1}. This is to match the high K demand of 8 kg ha^{-1} day during peak growth period. Forage yield in most parts of Eastern Europe ranges from 25 to 30 t ha^{-1}. In these areas about 200 kg K$_2$O is supplied.

Joshi et al. (2001) have compiled data on extent of fertilizers impinged into maize belts in Northern plains of India, where maize crop is supplied with low levels of inputs compared with Southern Indian Corn states such as Karnataka and Andhra Pradesh. The fertilizer (N, P, and K) supply to maize fields in Northern states of India ranges from 78 to 132 kg ha^{-1}. In Southern Indian Plains it ranges from 141 to 320 kg NPK ha^{-1}. According to Bala (2006), K supply into parts of maize growing regions of South India is less than sufficient considering net removals from the soil. Farmers tend to apply less mainly relying on the inherent K and buffering capacity. It has been pointed out that about 50 kg K is applied for crop that produces 2–2.5 t grains ha^{-1} during rainy season. However, during winter and summer cropping season maize still receives only 50 kg K ha^{-1}, although productivity is high at 4–6 t grain ha^{-1}. About 200 kg K ha^{-1} is required to support high grain yield aimed. Regarding K, based on an IFAD-CIMMYT survey, Joshi et al. (2001) have reported that in the dry lands of upper India, maize is supplied with 0.5–4.0 kg K ha^{-1}. This is mostly through crop residues. These are mostly subsistence crops or forage crops sown on fallows. Whereas, in non-traditional belts of Karnataka and Andhra Pradesh, maize productivity is relatively high at 4–5 t grains ha^{-1}. Nutrients are supplied at higher rates. For example, K is supplied at 5–10 kg ha^{-1} plus a certain amount of K is derived via mineralization of FYM. In South India, maize grown in post rainy season is supplied with 40 kg K ha^{-1} before sowing (Ikisan, 2000). Farmers in many regions rely excessively on inherent soil K and its buffering capacity. They refrain from adding fertilizer-K each season. Instead they supply only N and P and curtail K to once in 3 or 4 years. It may lead to soil K depletion. Hence, soil K mining is a definite possibility in many of the above situations.

In Queensland, maize crops receive at least 30–50 kg K ha^{-1} depending on rainfall pattern and yield goals. Usually, entire amount of K envisaged is supplied as basal dose. Potassium is also supplied using foliar fertilizers. They are quick to take effect. Plants absorb them directly through stomata and lenticells. Therefore, it avoids undue chemical fixation reactions and other biological interactions with soil components. Also, fertilizer-K is needed in appreciably small quantities when compared with dosages meant for soil application. To quote an example, maize grown in parts of Thailand is supplied with K using 2.5% KNO$_3$ solution as spray to a crop at tasselling stage. Sometimes, foliar sprays are done 2 or 3 times during growth to supply K as required (Suwanarit and Sestapukdee, 1989).

Potassium is also added to maize fields via FYM and other organic manures. The exact quantity of K supplied depends on percentage K concentration and release traits of manure and its interactive effects with soil. Manure characteristics such as source, succulence, K content, C:K ratio, SOM, microbial load, inherent soil K status, ambient temperature, and many other atmospheric factors may affect K supply to fields. In the intensive cropping zones of USA and Europe, organic K inputs are larger considering that 10–15 t FYM ha^{-1} is incorporated into maize fields prior to sowing. Whereas, in dry lands of Africa, India, and elsewhere in Southeast Asia, maize crops receive small amounts of K via organic manures. Basically, crop residues recycled and extraneous FYM applied are smaller ranging from 2–5 t ha^{-1}. Potassium is also supplied through green manures, or short season crops grown on fallows, crop stubbles that are incorporated, weeds if any, and crop residues left as mulch. The availability of K depends on ease and rapidity with which an organic source is decomposed. The exact amount of K derived from FYM and other organic sources could be calculated based on previous data and accounted in the fertilizer-K schedules prepared. Potassium salts are also supplied via drip irrigation. Potassium nitrate is pulsed at 15–20 g per dripper during irrigation. This induces accumulation of K salts around dripper. However, if maize crop has already been planted, then solution K levels decrease. The dynamics of K under fertigation may vary markedly. Firstly the distribution of fertilizer-K applied is itself affected. It is localized around dripper. Removal of K from surface is rapid.

Potassium supply to maize cropping zones via atmospheric deposits, wind, storms, precipitation and irrigation may appear small when calculated for a field or ha^{-1}. Its immediate impact on maize crop growth and grain formation could be marginal. However, it is a major K source, considering that larger patches of maize belts receive many essential elements including K via atmospheric processes. Measurements on sandstorms created in West Africa indicate that at least 4–12 kg K ha^{-1} could be received by maize fields during the 2 or 3 storms that a rainy season crop experiences. In the semi arid India, Murthy et al. (2000) have reported that 4–5 kg ha^{-1} could be added to maize belt via atmospheric sources.

Potassium Recovery by Maize

In the Northeast USA, Maryland in particular, maize crop grown for grains extracts 4.2 kg K t^{-1} dry matter. Whereas, a crop meant solely for forage removes at least 6.0 kg K t^{-1} dry matter. We should note that maize removes greater quantity of K into stover.

The maize crop in European plains recovers large quantities of K during seedling to knee-high stage. About 8 kg K_2O ha^{-1} day^{-1} is absorbed from soil during peak growth period. A crop that yields 40 t forage ha^{-1} in Western European plains recovers at least 175 kg K_2O ha^{-1}. In the Eastern European plains, maize yielding 30 t forage ha^{-1} removes 130 kg K_2O and that yielding 50 t forage ha^{-1} removes 220 kg K_2O ha^{-1}.

Reports from Zimbabwe, suggest that maize crop grown on sandy soil, accumulates 4.2 kg K in grains and 16 kg in stover in order to produce 1.0 t grains. Actual K recovery levels are slightly higher considering that roots too accumulate K. We should note that generally maize stover accumulates larger fraction of the total K absorbed. On the contrary, essential elements like N and P are accumulated in greater quantities in grains compared to stover. In the Eastern African tropics, total K uptake by maize clearly depended on soil type, nutrient supply levels, and yield goals. Smaling and Janssen (1993) have reported that grain yield and total K recovered on Homa Bay black soils was 6300 kg ha^{-1} and 126 kg K ha^{-1} respectively. It amounts to 20.4 kg K t^{-1} grain production. However, on Kisii Red soils grain yield was only 4900 kg ha^{-1} and K recovery 58 kg K ha^{-1} amounting to 12.2 kg K t^{-1} grain production. Overall, in Kenya, K recovery rate by maize crops fluctuated between 12 and 20 kg K t^{-1} grain productions.

Potassium requirement of maize is particularly high. Maize accumulates in large amount of K in its stem and leaves. The pattern of K absorption by maize varies at different stages. The K acquisition is high during knee-high stages till cob formation. During rapid K acquisition phase, a maize crop meant for forage may recover about 19 kg K ha^{-1} day^{-1}, whereas during rapid K uptake phase soybean recovers 12 kg K ha^{-1} day^{-1} and other cereals only 6–8 kg K ha^{-1} day^{-1} (see Krishna 2002). Hence, we need to adjust available-K pool to match rapid K-intake phase of maize crop. Potassium deficiency causes reduction in leaf area, therefore affects photosynthetic activity and grain yield (Pettigrew and Schorring, 2008). We should also note that maize genotypes vary with regard to tolerance to K deficiency. Cultivation of a genotype tolerant to K deficiency has its proportionate influence on K dynamics in the cropping belt.

Reports from maize growing regions of Queensland suggest that crop's biological demand for K is high. The K uptake peaks during vegetative phase and reaches 5.2 kg K ha^{-1} day^{-1} on Vertisols (White, 2000). The maize yields are optimum when leaf K content is 11.4–22.9 g kg dry weight basis.

Potassium Loss from Maize Fields

Soil K is fairly immobile, yet in the Northeast USA soils are prone to erosion and loss of K. On sandy or light textured soils, about 10–20% K held in upper layers of soil is vulnerable to loss. Soil K loss gets accentuated from fields situated on steep slopes and undulated terrain. Minimum tillage and mulching are often recommended to curtail loss of soil K. In order to avoid loss of soil K through leaching, erosion, and seepage and to derive higher fertilizer-K efficiency, band placement at a depth is preferable.

On Rhodic Ferralsols of Togo, it is said K loss via leaching could reach 7.5 kg K ha^{-1} with fertilization and 4.5 kg K ha^{-1} without fertilization (Poss et al., 2004). Large quantity of fertilizer-K is also lost due to chemical fixation. In some locations, about 58% of fertilizer-K applied has been lost through chemical fixation. Recycling organic-K held in crop residues is almost mandatory to maintain optimum levels of K in soil.

Erosion is a major cause of soil deterioration in Indonesia. Harmiyanto et al. (2009) report that peak loss of K via soil erosion, coincides with high rainfall event. It ranges from 0.34–2.2 kg K ha^{-1}. Net loss of K depends on sediment enrichment which ranges from 20 to 30% for K. Along with loss of other essential nutrients, K depletion leads to soil degradation and yield depreciation.

Potassium Diversion to Weeds

It is common to encounter weed growth during maize production. Weeds are especially problematic in no-till fields, if suitable herbicide treatments are not carried out. Hand weeding is essential (Plate 3.6). In order to obtain a decent maize crop a minimum of 40–60 weed-free days are essential. Weeding during early seeding stages is crucial, if a good canopy is to be established. Most importantly, both inherent soil K and fertilizer-K applied to crop as basal dose could be garnered appreciably by weeds if they are not eradicated in time. Weeds, no doubt, affect nutrient dynamics in the maize fields. Let us consider an example. *Imperrata* weed infestation is a serious problem in many of the Southeast Asian maize growing regions. It grows rapidly and affects growth and grain formation. Field trials by Lopez and Vlek (2006) suggest that despite fertilizer inputs, maize grain yield in fields infested with *Imperrata* was only 4.0 t ha^{-1}. Grain yield was 6.3 t ha^{-1} in field with low infestation of *Imperrata*. Regarding stover, *Imperrata* infestation impeded maize growth and fodder yield. It was restricted to 10.6 t ha^{-1} compared to 25–30 t ha^{-1} fodder achievable in weed-free fields. Detailed analysis for various minerals in crop tissue revealed that K was the most limiting in weed infested plots. Total K in stover from weed infested maize fields was very low compared to weed-free maize crop. Clearly, weeds divert a certain fraction of K meant for maize crop.

SUMMARY

Globally, maize cropping thrives in regions with different soil fertility conditions. Maize productivity is generally high (>8–10 t ha^{-1}) in intensive farming zones (e.g., Corn Belt of USA). Moderate level of fertilizer-based nutrient supply has resulted in maize belts that yield commensurately (2–4 t ha^{-1}) (e.g., Pampas of Argentina). Low input agricultural systems common to Savannas of West Africa has created subsistence maize farming zones. Aspects like nutrient supply, recovery rates, loss due to natural causes, and cropping intensity literally regulate grain/forage productivity. During recent years, nutrient supply into maize field is stringently guided by periodic soil tests, crop's demand at various stages and yield goals. There are innumerable soil tests and plant analysis methods that guide farmers in arriving at appropriate nutrient supply schedules. Rapid tests of both soil and plant have become popular. For example, ISNT that assesses soil-N rapidly or leaf color (chlorophyll) charts helps in regulating split applications of N. We ought to realize that soil/plant analysis procedures and crop models utilized to decide fertilizer inputs are indeed critical components that regulate both nutrient dynamics and productivity of maize crop. Perhaps, extra accuracy is needed in high input farming zones, because undue accumulation of nutrients can be deleterious to soil and ground water. The agronomic efficiency of fertilizer is an important aspect that has direct bearing on nutrient dynamics in the maize agroecosystem.

Nitrogen is a key nutrient required in relatively larger quantities. Knowledge about its distribution pattern in soil and physico-chemical transformations helps us in devising appropriate soil management procedures. Nitrogen supply into cropping zones occurs mainly via inorganic fertilizers, organic manures, atmospheric deposits, contribution by N-fixing legumes, crop residue recycling, and mineralization of immobilized-N. Computer models generally consider most, if not, all of the factors that affect N dynamics in soil and crop phase of the ecosystem. Nitrogen supply into maize field varies enormously. In the US Corn Belt, intensive farming requires supply of 180–240 kg N ha^{-1}. Nitrogen supply is usually split into 3 or 4 fractions in addition to basal inputs. This technique helps in regulating and matching N availability in root zone with N demand by maize crop at various stages. Interaction between soil-N and irrigation schedules needs due attention. Nitrogen supply into maize fields can be improved by careful selection of legumes. Legume N-credits to succeeding maize may range from 20 to 80 kg N ha^{-1} and this need to be maximized. In addition, asymbiotic N-fixing microbes that reside in rhizosphere and root zone soil, too add to soil-N. In parts of Africa and Asia, maize cropping systems include green manures and leguminous trees that add precious amounts of N to soil. Agronomic procedures like mulching, use of slow release fertilizers and nitrification inhibitors, deep placement of N, splitting, selection of N-efficient genotypes all help in improving fertilizer-N efficiency, and optimizing N dynamics in the soils. Methods that reduce loss of N from fields are equally important. Nitrogen loss via emissions and volatilization could be effectively reduced by adopting mulching, deep placement, splitting, etc. Maize genotype influences N recovery rates and accumulation pattern. Indirectly, it decides amount of N retained in soil or removed into seeds/forage.

Maize is a preferred cereal for grain and forage production. It is also preferred more for alcohol and biofuel production. Hence, in future, genotypes that produce both grains and biofuel may dominate the landscape and affect nutrient dynamics. It is generally accepted that N supply and planting density has played a crucial role in intensification of maize production. As yet we are not aware of limits to intensification of maize cropping zones. Maize belts that thrive at low N supply levels may also perpetuate for centuries to come. Techniques that enhance fertilizer-N efficiency and productivity in subsistence farming zones needs due attention, irrespective of cropping intensity.

Phosphorus dynamics in maize belts has direct bearing on grain/forage production. Maintenance of optimum levels of available-P is essential in most soils that support maize production. Phosphorus replenishments using inorganic soluble sources are usually done after soil tests. There are indeed several soil-P tests and standard recommendations based on yield goals. Fertilizer-P efficiency is generally very low at 18–22%. Soil factors like chemical P fixation, erosion, leaching, and seepage may affect P dynamics. Hence, appropriate measures such as supply of P at higher levels that satisfy fixation capacity are needed. In subsistence farming zones, much of the

P replenishment has to be augmented via efficient recycling of crop residues and organic manure inputs. Phosphorus dynamics in maize production zones could be further regulated by adopted efficient cropping systems. Excessive accumulation of P or percolation into irrigation channels and ground water needs to be avoided. Residual-P could be exploited by maize that succeeds a legume or any other species. Planting agroforestry trees helps in exploring P from subsurface. On sandy acid soils, maize farmers tend to supply P via rock phosphates. Application of partially acidulated rock phosphates and organic manures induces rapid growth and high grain yield. It also allows extended residual effects of rock phosphates supplied. Phosphorus dynamics is affected by maize genotype, especially its characteristics relevant to growth habit, duration, rooting pattern, foliage production, harvest index, and grain formation. In addition, there are genotypes that are P efficient. Therefore, selection of appropriate agronomically elite and P-efficient cultivars is required.

Maize belts are exposed to soils with rich, optimum, or low K contents. Maize roots exploit K from both surface and sub-surface layers. Therefore, knowledge about K distribution and physico-chemical transformation that affect K availability is important. Total soil-K could be high, but it is the dynamics of soluble-K and exchangeable-K that needs greater attention. Since, only these K fractions are exploited by maize roots. Potassium replenishment to soil is necessitated by incessant cropping and high amounts of recovery into forage and grains. Maize accumulates greater amount of K in stem and leaves. Potassium is supplied to maize fields based on soil tests. About 120–180 kg K ha^{-1} is supplied to high input intensive farming zones. Whereas, in subsistence farming belts, K is mostly recycled through crop residue or organic manure. Farmers in moderate productivity zones (4–5 t grain ha^{-1}) may supply 60–80 kg K ha^{-1}. Whatever, be the intensity of cropping or geographic region, a maize crop that puts forth a ton grain has to absorb 18–20 kg K ha^{-1}. Therefore, any intention to enhance productivity will mean excessive supply of K to support forage and grain formation in that proportion. This step will further alter K dynamics in the maize agroecosystem. Maize production for forage or biofuels again involves removal large quantity of this element from fields. Hence, repeated addition of K via fertilizers is necessary. Potassium is lost from soil phase through leaching and/or percolation. Further it reaches subsurface or groundwater. It is also exploited efficiently by weeds if they are not suppressed in time. Soil conservation techniques like contour bunding, mulching, planting agroforestry trees in strips, cover crops, periodic green manuring may allow us to retain K *in situ*.

In summary, it is necessary to manipulate soils, recycle and /or supply extraneous nutrients, cultivate appropriate crop genotype, and choose suitable cropping systems in order to optimize dynamics of major nutrients and stabilize productivity of maize agroecosystem. In some regions, soil nutrient dynamics and productivity of maize crop is being regulated using GPS guided systems and computer programs that decide fertilizer supply based on yield goals. Such improvements may help us in intensifying maize belts accurately and carefully avoiding disturbance to various ecosytematic functions. In future, the choice of soil/plant analysis tests, computer models/programs and GPS based site-specific techniques actually hold the key to nutrient dynamics that ensues in the maize agroecosystem.

KEYWORDS

- Biomass accumulation
- Cropping ecosystem
- Dernopodzolic soils
- Fertigation
- Micronutrients
- Volatilization

Chapter 4

Secondary and Micronutrients in Maize Agroecosystem

Maize agroecosystem has expanded into geographic regions that possess soil types rich, optimum, or deficient with regard to secondary and micronutrients. The intensity of cropping, grain/forage productivity, and yield goal set are among major factors that affects dynamics of secondary and micronutrients. In many of the maize belts, initially when maize production was subsistent, deficiencies of nutrient elements like calcium (Ca), sulfur (S), magnesium (Mg), or micronutrients were not felt. Their turn-over in soil and above ground portion was augmented satisfactorily by organic residue inputs and recycling. As such, in many of the maize cropping regions, fertilizer supply schedules may not prescribe or at best may suggest application of secondary and micronutrients once every 3 years to soil. However, incessant cropping of maize and other cereals does affect nutrient dynamics adversely. Rapid depletion of secondary and micronutrients necessitates periodic or yearly supply of these nutrients. Further, we should replenish secondary and micronutrients appropriately because their dearth can hinder maize crop's response to major nutrients, irrigation and several other agronomic procedures. The law of minimum as enunciated by Liebig may take over and impair dynamics of major nutrients. In addition, secondary and micronutrients may chemically interact in soil leading to specific antagonistic and positive influence on nutrient availability. Therefore, knowledge about physico-chemical transformations of secondary and micronutrients that occur in the soils of maize belt are important. In case of maize belts that thrive on acid soils it is customary to apply gypsum (Ca) to enhance soil pH. The critical limits of each secondary and micronutrient in soil is again an important factor that affects productivity of maize agroecosystem. Soil tests and fertilizer formulations are available that aid farmers in correcting nutrient deficiencies. Lastly, influence of secondary and micronutrients on maize crop growth and productivity needs to be understood accurately. With regard to intensification of maize cropping in any type of soil, it is said optimum supply and availability of major and micronutrients and at proper ratios is essential.

In the following paragraphs, dynamics of three secondary nutrients namely, Ca, Mg, and S has been discussed in greater detail. Discussions on micronutrients and their dynamics encompasses zinc (Zn), iron (Fe), copper (Cu), molybdenum (Mo), and boron (B).

SECONDARY NUTRIENTS IN MAIZE BELTS

Calcium in Maize Cropping Zones

In the Corn Belt of USA, soils with acidic reaction are treated with lime. Lime inputs are essential to raise soil pH from around 6.0 to 6.5 or 7.0 that is required for maize cropping. In general, lime recommendations are influenced by soil type, its depth, extent of pH increase envisaged, aluminum (Al) and manganese (Mn) content, source of lime and its calcium carbonate equivalence. It is said that depth of tillage should also be considered because it decides the volume of soil that needs pH correction. In Iowa, maize farmers tend to supply at least 3–5 t lime ha^{-1} to correct soil pH and supply Ca. Following is a table depicting lime requirements depending on soil depth and pH increase envisaged:

Soil pH Increase	Lime Input to Maize Fields (t ha^{-1})			
	Depth of Soil (cm)			
	5	7.5	15	20
5.7 to 6.5	3.2	4.7	9.5	13.4
5.7 to 6.9	4.3	6.5	14.2	17.6

Source: Excerpted from Sawyer, 2008.

In the Northeast (Maryland), maize farms are supplied with lime on a regular basis. Firstly, soils are inherently low in pH ranging from 6.0 to 6.3. This needs to be corrected and enhanced to at least above 6.5 or 6.9. Next, repeated application of nitrogen (N) as NH_3, NH_4NO_3, or urea induces acidity. The fertilizer-N induced soil acidity too needs to be neutralized. It has been reported that, if soil acidity problem is severe, about 4 kg limestone is needed for every 1.0 kg of fertilizer-N applied to maize fields. Hence, it is advisable to apply N both in organic and inorganic form. It reduces inorganic fertilizer-N inputs to soil. If not, repeated lime inputs are a necessity.

Soil texture influences $CaCO_3$ inputs into the Corn Belt. In Michigan, lime requirements of soils are judged based on extent of pH to be increased and soil texture. Soils with pH 5.5 are provided with lime to raise the pH to 6.8 which is conducive to maize crop. Lime recommendations are usually made based on 20 cm plow layer.

Texture of Plow Layer	Limestone Inputs (t ha^{-1})	
	Soil pH Increase	
	5.5 to 5.9	6.0 to 6.5
Clay and Silty Clay	5.50	3.50
Clay Loams	4.00	2.75
Sandy Loams	3.50	2.00
Loamy Sands	2.71	1.50
Sands	2.00	0.75

Source: Christenson et al., 1993.

The neutralizing value of lime stone source is also important. Considering neutralizing value of pure $CaCO_3$ as 100, neutralizing value of other sources are as follows: Pure $MgCO_3$-119; Pure Calcium hydrate-135; Pure Magnesium hydrate-172; Calcic Limestone-<100; Dolomitic Limestone-<106; Calcic Hydrated lime <135; and Dolomitic hydrated lime <172 (Christenson et al., 1993).

We should note that lime requirement during maize production is also calculated based on exchangeable Al^+ indices. This is in addition to usual method based on soil pH. In some regions, exchangeable acid saturation of the effective cation exchange capacity (CEC) is popular and is readily used to decide on $CaCO_3$ inputs. Evaluations on Haplorthox and Plinthic Paleudults of Southern Asian maize farming zones suggests that lime requirement based on soil pH is less accurate compared to those derived using Al^+ activity in soil. The relative maize grain/forage yield from fields where lime inputs ere decided based on Al^+ activity was consistently higher than that based on soil pH (Farina and Channon, 2006).

Calcium Recovery by Maize

On an average, maize crop that produces 1.0 t grain extracts 3.5–4.0 kg Ca from soil. It partitions about 1.1–1.3 kg Ca into grains and rest 2.4–2.7 kg gets accumulated in stem and leaf tissue. Calcium deficiency has not been observed in soils with pH greater than 5.5. In case of deficiency it is corrected using lime stone. The Ca concentration in the tissue varies based on crop growth stage and fertility status of soil. Seedlings have 0.30–0.70% Ca in the tissue. At tasseling, leaf below the whorl contains 0.25–0.50% Ca and at silking the ear leaf contains 0.21–1.00% Ca (Johnston and Dowbenko, 2009).

Loss of Calcium from Maize Fields

Loss of Ca from maize ecosystem could be important, especially if replenishment schedules are not appropriate. Firstly, a certain amount of Ca is removed via grains and forage. If crop residues are not recycled by incorporation it may increase loss of Ca from fields. Calcium is also prone to loss via natural processes such as soil erosion through wind and water, seepage through irrigation channels, percolation, runoff, chemical fixation into soil crystal lattice, and immobilization into organic phase of the soil. Most of the above factors that operate at different intensities may alter Ca dynamics in soil.

The experimental estimate from different locations in Indonesia suggests that maize fields may loose significant amount of soil due to erosion. Soil erosion peaks around maximum rainfall period. The composition of top soil and nutrient enrichment ratios are important factors that affect loss of secondary nutrients such as Ca, Mg, and S. Hermiyanto et al. (2007) have reported that in certain locations about 38 kg Ca ha^{-1} could be lost in a year from top soil. Soil processes like percolation, seepage, and runoff could also induce additional loss of Ca. Methods that restrict soil erosion are important. For example, contour bunding, mulching, band placement of fertilizers containing Ca, and other elements, selecting genotypes that extract both surface and subsurface Ca may all reduce loss of Ca from maize fields.

Sulfur Dynamics in Maize Belt

Sulfur Supply to Maize Belts

Sulfur deficiency is sporadic. Sandy soils with low organic matter are prone to S deficit. Most often, S status in the "Corn Belt of USA" is estimated prior to planting. Fields are then classified and quantum of S supply decided. In Illinois, soil with 0–12 kg S ha^{-1} is classified as low, that with 12–22 kg S ha^{-1} as moderate and those above

22 kg S ha⁻¹ as high. In soils with >22 kg S ha⁻¹ maize crops do not show response to extraneous S inputs.

In Central Europe, maize crop may often be provided with Mg and S. This is in addition to usual N, Phosphorus (P), and Potassium (K) recommendations. Since the soils are deficient for secondary nutrients, Mg and S inputs often result in improved grain formation. For example, around Poznan in Poland, maize crop is grown on gray–brown podzolic soils. It is provided with 25 kg MgO and 20 kg sulfate-S in order to satisfy crop needs (Szulc et al., 2008). Elsewhere, in other European locations, S inputs are based on soil test values. However, a certain amount of S is supplied as single super phosphate (SSP) or gypsum that are sources of P and Ca respectively. Crop residue recycling may also add to soil S. A small quantity of S is often derived from atmosphere depending on location of the maize fields. Those situated near industries or seashore may receive slightly higher quantities of S through precipitation and dust.

In India, S deficiency is wide spread in the maize belt. It has accentuated because of excessive use of S-free fertilizer formulations and inappropriate levels of residue recycling. The subsoil S status too is low in many locations. Hence, replacement of S through inorganic and organic sources is necessary. Singh (2001) has reported that in the four southern states of India, S deficiency is experienced in 29–36% soils. Based on soil tests in maize cropping zones it was concluded that critical limit is 13 mg S kg⁻¹ soil. The available-S in soil is usually extracted using 0.15% $CaCl_2$. Several other reports from Indian states suggest that critical level of S in soil is 0.3%. In other words, soils with <10 mg S ha⁻¹ are considered deficient. In many of the locations in South India, maize culture proceeds on soils with 7–8 mg available S kg⁻¹ soil, which is less than threshold. The available-S levels should be raised above critical limit by supplying it through chemical fertilizers or organic manures. Sulfur supply to maize crop in South India is often achieved via fertilizer meant to satisfy P needs. For example, SSP, TSP or gypsum used basically to supply P also adds S to soils. On an average, if S deficiency is suspected, 20–30 kg S ha⁻¹ is supplied as gypsum. It is incorporated before seeding. We should note that in most soil types, S tends to accumulate in the subsurface layers. Therefore, subsurface levels should be considered while deciding S supply to fields. As such, maize produces relatively deeper root system compared to other cereals like wheat or rice; hence, its roots actively scavenge a certain fraction of subsurface S. Often, maize may not respond to S supply, despite low soil test values. Such low soil test values are based on surface layers of soil. It excludes S content present in subsurface. Hence, crop response to S supply is either feeble or absent.

In addition to S derived from chemical fertilizer and organic manure application, a small amount of S enters maize belts via atmospheric deposits, dust storms, and precipitation. Sulfur derived from atmosphere can be significant if maize fields are located closer to industries, urban locations, or sea coast. About 0.5–2 kg S ha⁻¹ yr⁻¹ could be added on to maize fields. Sulfur containing pesticides also add to soil S status. In the Corn Belt of USA, about 8–10 kg S ha⁻¹yr⁻¹ could be supplied through precipitation, atmospheric dusts, and snow. The extent of S deposited depends on proximity of maize fields to industries. Specifically, in Wisconsin, Wolkowski (2003) reports that cropped fields may receive 4.5–9 kg S ha⁻¹ yr⁻¹ through atmospheric dusts and precipitation.

Sulfur Recovery from Soils

In general, maize crop grown on sandy loams of Northeastern USA extracts and accumulates about 3.5 kg S t^{-1} grain production. It partitions 2.8 kg S into grains and 0.6 kg S into silage. The percentage concentration of S in maize tissue varies depending on soil fertility regime and stage of the crop. Johnston and Dowbenko (2009) have reported that maize seedlings (12" tall) contain 0.15–0.50% S in the entire plant tissue; those at tasseling contain 0.15–0.50% in the leaf below the whorl and those at silking contain 0.21–0.50% S in the ear leaf. These values could be used as guidelines to verify deficiency/sufficiency of S in the maize crop. The ratio of S to N is also important. Generally, it ranges from 1:7 to 1:15 depending on soil fertility and crop stage. Wider ratios are indicative of S deficiency.

Regarding S, it is said that S requirements of cereals like maize is generally 9–15% of total N uptake. Maize may remove about 4–6 kg S ha^{-1} under rain fed conditions of South Indian plains. Whereas, in the same fields, pulses absorb greater amounts of S ranging from 8 to 10 kg S ha^{-1}. Maize grown under high input technology that yields between 9–10 t grains and 11 t forage ha^{-1} absorbs 31 kg S. It equates to 3.0–3.3 kg S t^{-1} grain. Reports from Southeast Asia suggest that, a maize crop, in general, extracts 2.8–3.0 t kg S for each t of grain formation (IPNI, 2007). Sulfur is partitioned into grains and stover in almost equal proportion. This is unlike N and P that are accumulated in higher quantities in grains and K that is found in greater quantities in stover compared to grains. Maize crop grown on Alfisols or Vertisol plains of South India may recover between 30 and 70 kg S ha^{-1} depending on grain and stover harvests. Maize recovers at least 22–30 kg S ha^{-1} to produce 6–7 t grain and 14 t stover. It equates to 2.5–3.0 kg S recovered for each t grain. Sulfur recovery from soils could be improved by supply of S fertilizers. The apparent sulfur recovery (ASR) is defined as S uptake in plots with S minus S uptake in plots without S divided by S added. The ASR was higher at S inputs of 15 and 30 kg S ha^{-1} but not at higher levels. Naturally, at higher levels of S input there was tendency to accumulate residual-S.

Sulfur is actually a secondary nutrient needed in small quantities. Yet, its insufficiency in soil or plant tissue can affect recovery of other nutrients, especially major nutrients. It is often reported that in addition to its own extraction, S input enhances uptake of all three major nutrients (Bharathi and Poongthai, 2008; Figure 4.1).

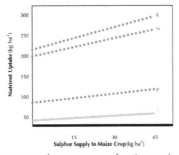

Figure 4.1. Effect of sulfur inputs to soil on recovery of major nutrients (N, P, and K) by maize crop grown on an Alfisol in South India.
Source: Based on Bharathi and Poongthai, 2008.

Residual-sulfur in Maize Belts

Residual effects of fertilizer-S are felt in many of the maize cropping zones. Sulfur accumulates in both surface and subsurface layers. Part of the S input may get converted to organic-S and become available to the crop at a later date through mineralization processes. Residual effects are conspicuous on the succeeding legumes. For example, in South India, residual-S still available after a maize crop is sufficient enough to cause grain yield increase of succeeding green gram (Bharathi and Poongthai, 2008). On Alfisols, available-S in soil after maize harvest ranged from 8.5 mg kg^{-1} S in control plots compared to 24 mg S kg^{-1} soil in plots given 45 kg ha^{-1} fertilizer-S during maize season. Fertilizer-S supply also induced higher recovery of N, P, and K both by maize and the succeeding legume. The residual-S present in the plots given 45kg S ha^{-1} during maize phase was sufficient to cause an improvement of grain harvest in the succeeding green gram to an extent of 300 kg ha^{-1}. Obviously, residual-S is an important aspect during maize–legume rotations. In fact, residual-S effects should be considered both while deciding S inputs and calculating fertilizer-S use efficiency.

Sulfur Loss from Maize Fields

Sulfur loss from maize fields is linked to soil erosion, seepage of irrigation water, and percolation. Sulfur may percolate and collect in subsurface layers. In this case it could be utilized efficiently by a deep rooted crop that succeeds maize. Sulfur could be immobilized into soil organic fraction. Organic-S becomes available to maize roots upon mineralization.

Fertilizer-sulfur in Maize Farms

Nitrogen deficiency is the most commonly felt nutrient dearth in the "Corn Belt of USA" or even elsewhere in maize growing regions of the world. However, during recent years deficiencies of many other elements have been traced. Sulfur deficiency has been noticed in many fields in Northern plains. Sulfur containing fertilizers are used to correct S requirements. Maize responds to both N and S inputs. For example, in Ohio, Chen et al. (2008) found positive N × S interaction with regard to maize crop yield. Sulfur supply at 33 kg ha^{-1} improved maize grain yield. Crop response to fertilizer-S inputs occurred at low to medium levels of N supply that is 0–113 kg N ha^{-1}. Hence, it was suggested that to derive better fertilizer use efficiency, a combination of N and S fertilizers should be supplied.

In Africa, sulfur requirement of maize crops is mostly satisfied through inherent soil-S. Incessant cropping may lead to S deficiency. Over all, fertilizer-S supply in Africa is sporadic. It is mostly achieved via SSP, gypsum, or organic matter recycling.

As stated earlier, S supply to maize fields are often achieved through SSP, gypsum, and other S containing fertilizers. Elemental S is also supplied in some areas. Fertilizer sources like SSP or TSP are primarily meant to correct P deficiency and not aimed at satisfying S requirements of maize. Maize does respond to S inputs. For example, in India, S supply improves its recovery, as well as grain and stover production. Grain yield improved from 6.4 to 7.3 t grains ha^{-1} for an increase in supply from nil to 45 kg S ha^{-1}. Optimum S input in Southern Indian maize belt depends on geographic location, soil types, cropping systems, genotype, and yield goals. The S use efficiency

(SUE) reported on Alfisols in South India ranges from 11.2 to 22.6 kg grain kg^{-1} S. In this case, SUE is defined as grain yield with S minus that without S divided by S added.

Magnesium in Maize Cropping Zones

Magnesium Supply to Maize Crops

Soils in the Canadian plains and Northern plains of USA may not exhibit Mg deficiency. Yet, its need may be felt, only if cropping is incessant and yield goals are kept consistently high. Soil tests values below 20 ppm are indicative of Mg exhaustion. On Mollisols of the "US Corn Belt", Mg inputs are mostly through $MgSO_4$ at 20–30 kg $MgSO_4$ ha^{-1} once in 2–3 years. In case soils are limed to correct soil pH or to replenish Ca, then it is easier to supply Mg using Dolomitic Lime. Dolomitic lime contains sizeable amounts of Mg salts (Omafra, 2002). Magnesium deficiency occurs in the European Corn Belt, but it is sporadic and confined to sandy soil with low organic matter. Accurate soil tests for Mg are available. Soils low in Mg could be treated with Dolomitic lime.

Magnesium Recovery

In the Northeastern USA, maize crop recovers 4.4–5.0 kg Mg t^{-1} grain formation. It partitions about 2.8 kg Mg into seeds and 1.6 kg Mg into silage. The concentration of Mg in the plant tissue varies with growth stage. Normally a seedling (12″ tall) has 0.15–0.45% Mg. At tasseling, leaf below the whorl contains 0.13–0.30% Mg and at tasseling, ear leaf contains 0.20–0.100% Mg on dry weight basis. The concentrations of Mg in tissue could be used as indicative of deficiency/sufficiency in the crop (Johnston and Dowbenk, 2009).

In the Savanna region of Ghana, maize is supplied with 25 kg Mg ha^{-1}. This is to correct the deficiency which is rampant. Magnesium supply is also dependent on N supply levels. On Haplic Lixisols found in Northern Ghana, Mg recovery by maize crop was less than 15% of 25 kg supplied ha^{-1}. Maize recovers about 2.8–3.0 kg Mg from soil for each t grain formation. It actually partitions about 0.88–1.0 kg Mg into grains and rest 1.85–2.0 kg Mg into stover. Reports by IPNI (2007) suggest that a maize crop grown on sandy soils of Southern Africa, extracts about 3.2–2.5 kg Mg to yield 1.0 t grains. It accumulates about 1.6 kg Mg in grains and 1.8–2 kg Mg in stover.

Magnesium uptake is dependent on various factors related to soil, crop genotype, and environment. Within the soil, interactions with major nutrient like N and micronutrient such as Zn could affect Mg recovery. A study at Tamale in Ghana suggests that Mg recovery from sandy soils was low (2.1–4.7 kg Mg ha^{-1}) despite its addition above optimum rates of 25 kg Mg ha^{-1}. It was attributed to antagonistic interactions of Zn^{+2} with Mg^{+2} at soil exchangeable sites (Abunyewa and Mercer-Quarshie, 2004). It seems Mg recovery was also affected by starch metabolism in the maize crop. In fact, on a long run, Mg contents in soil did not change significantly because of lack of proportionate recovery from soil. Therefore, it was suggested that Mg inputs may be necessary only in fields where intensive cropping is practiced and levels of other nutrients especially N, P, and K are held high and yield goals too are high.

Magnesium Loss from Maize Fields

Erosion, percolation, and seepage are usual factors that induce loss of Mg from soil phase. Crop harvest results in removal of a certain amount of Mg. Further, soil physico-chemical interactions with other nutrients may affect its availability to maize roots. In Southeast Asian maize farms, for example in Indonesia a certain amount of Mg is lost annually due to natural soil processes such as erosion due to wind and water, runoff during peak precipitation periods, seepage into irrigation channels percolation to Vadose and ground water, etc. Hermiyanto et al. (2007) estimated that in the watersheds about 9.5 kg Mg ha^{-1} yr^{-1} could be lost from maize farming zones. In addition, sizeable amount of Mg is lost via grain and forage removals from farming zones. Contour bunding, mulching, intercropping, band placement of dolomitic ores, appropriate genotypes, crop residue recycling are a few methods to restrict soil erosion and loss of Mg from maize fields.

MICRONUTRIENTS IN MAIZE AGROECOSYSTEMS

Micronutrient Status of Soils

Micronutrient deficiencies have been observed quite frequently in soils that support maize production in USA. The extent of deficiency experienced may vary depending on location within the Corn Belt. For example, in Nebraska, soils are optimum for most micronutrients except Zn and Fe. In Illinois, maize suffers Fe deficiency much more than other micronutrients. Soils in Indiana that support maize cropping are deficient for Zn. In Iowa, Zn and B deficiency are observed. Reports from most parts of Midwest of USA suggest that sensitivity of maize to micronutrient deficiency is especially low for B and Mo; medium for Cu, Fe, and Mn; and high for Zn (Johnston and Dowbenko, 2009).

Soils that support maize cropping are generally classified as low, medium, or high with regard to micronutrient contents/availability using various soil tests. Many of the soil types may contain optimum or even high levels of total micronutrients, yet they may be deficient in available forms. Therefore, micronutrient availability index or plant available micronutrient is considered while classifying soils. Most often, surface layer 0–20 cm depth is considered, but in some cases subsurface layer extending from 20 to 60 cm may also be sampled for estimating available micronutrient status. Let us consider a few examples. In the Corn Belt, particularly in Illinois, Iowa, and Canadian plains following classification is employed:

Micronutrient	Soil Classification		
	Deficient	Medium	Adequate
Fe (DTPA extractable ppm)	0–2	2–4.5	>4
Zn (0.1 NHCl extractable ppm)	0–2	2–7.0	>7
(DTPA extractable ppm)	0.05	0.5–1.0	>1
Mn (DTPA extractable ppm)	0.0–1.0	1–2.0	>2
(H$_3$PO$_4$ extractable ppm)	0.5	5–10	>10
Cu (DTPA extractable ppm)	0.0–0.2	0.2–1.0	>10

Sources: Mckenzie, 2001; Hoeft, 2004; Wortmann, 2008.

Micronutrient fertility status could also be gauged using plant tests. Micronutrient content of plant tissue could be indicative of its status in soil. Special care is needed while drawing tissue samples and interpreting tests. Normally, at least 25 plants are collected in each field to assess micronutrient status in plants. Following is an example depicting micronutrient status of maize plants and their classification:

Micronutrient	Micronutrient Status in Maize Plants (ppm)		
	Low	Marginal	Sufficient
Zn	<12	12–15	15
Fe	<15	15–20	20
Mn	<10	10–15	15
Cu	<2.3	2.3–3.7	3.7
B	<3	3–5	5

Source: McKenzie, 2001.

Micronutrient Availability

There are at least two forms of micronutrients in soil that are easily available to maize roots. They are micronutrients adsorbed on to soil colloids and that present in soil solution. Micronutrient availability in soil, like most other essential elements is influenced by the soil parent material, weathering, mineralization, and physico-chemical processes that involve micronutrients. It is generally agreed that, more than weathering or soil degradation it is mineralization and breakdown of soil organic material that contributes to available pool of micronutrients. Therefore, soils with low clay and organic matter may often be deficient in plant available micronutrients. Micronutrient availability to maize roots is also affected by soil pH. For example, Fe and Mn availability increases with acidity. Similarly, availability of Zn, Cu, and B gets enhanced as soil pH decreases. Whereas, availability of Mo increases if soil pH is higher at 7.0–7.5. Chelation is an important soil process that affects micronutrient availability. Formation of chelates containing organic fraction and metallic ligands is common. Chelation often increases solubility of micronutrients that occur in soil or that in fertilizer formulations (sprays). For example, it is preferable to apply Fe or Zn as chelates. Soil chelation may at times delay or cause deficiency of micronutrients. Nutrient interaction in soil may also affect micronutrient availability in soil. High levels of one nutrient may be detrimental to availability of others. For example, in most soils high P inputs curtail Zn availability and its absorption. Climatic conditions and crop growth rates also affect micronutrient availability and extraction. For example, slow root growth may hinder rapid absorption of micronutrients such as Zn, Fe, and Mn. Boron deficiency may get aggravated if dry conditions prevail for longer durations (Wortmann et al., 2008).

Micronutrient Deficiencies in Maize Belts

Micronutrient deficiency during maize production, firstly affects crop growth and grain formation. Next, it limits crop response to added major nutrients. It also alters nutrient ratios. Therefore, it affects a series of nutrient equations involving uptake, distribution, recycling, and harvest. In other words, micronutrient deficiency, a single nutrient or a couple of them can potentially disturb nutrient dynamics and affect crop productivity.

Zinc (Zn)

Zinc deficiency is quite frequently observed in many of the maize growing regions. It is common in areas with continuous cropping patterns. Zinc gets exhausted due to lack of proper replenishments. Zinc occurs both in organic and inorganic phase of soil. Zinc occurs as sulfides, silicates, carbonates, and hemimorphites in natural soils. Zinc found in the soil solution is more important in agricultural soils. Zinc occurs as Zn^{+2} and $Zn(OH)$ in soil solution. Inherent dearth for Zn is observed in many maize belts. Often, both surface and subsurface Zn levels reach below critical limits of 5–10 ppm.

In the "US Corn Belt", Zn deficiency is most likely due to high P supply on calcareous soils. According to Wortmann (2008), soils in Nebraska may show deficiency, if they are calcareous with pH greater than 7.3; or if the top soil has been eroded leading to Zn depletion; or if the land has been depleted of organic fraction and is sandy. Soil test guides are used to ascertain Zn deficiency. Soils with 0.8 ppm or above are termed adequate. Therefore, if it is less than 0.8 ppm, fertilizer-Zn is applied. In Illinois, Zn deficiency is more likely on soils with low organic matter and those with pH 7.3 or above. Soil with high P contents also exhibit Zn deficiency (Hoeft, 2004). Maize fields supplied with organic manures during each season are generally not deficient to Zn.

In the southern African Maize belt, especially in Natal, Zimbabwe, Zambia, and Mozambique, Zn deficiency is quite frequent and it limits maize growth and grain formation. Zinc deficiency actually curtails full expression of crop growth and its response to major nutrients. Soil test values in most areas range from 1 to 5 ppm, which is much below the critical limit (10 ppm Zn) required for optimum grain yield. In East Africa, maize crops experience deficiencies of micronutrients like Zn, Cu, and B. Greater details on extent of micronutrient deficiency and its impact on nutrient dynamics in general and grain yield is needed (Lisuma et al., 2006). Available literature indicates that maize grown on volcanic soil in Tanzania encounters micronutrient deficiencies, especially Cu. The grain yield reduces from 6.3 to 4.3 t ha^{-1} due to micronutrient deficiency. The volcanic soils may exhibit high Cu fixation. Hence, Cu replenishment is generally higher. It then helps to satisfy Cu fixation processes.

Regarding micronutrients in India, percentage soil samples deficient for Zn in the Deccan plateau and plains ranges from 51 to 57%. The Zn availability reaches below threshold that is <3–5 ppm. About 15% of soil samples from Coastal plains of Andhra Pradesh and Tamil Nadu exhibit Zn deficiency. Similarly surveys have shown that soils deficient for Fe ranges from 4.8 to 19% in North Karnataka and Andhra Pradesh, and about 4% in the Coastal plains of Andhra Pradesh and Tamil Nadu. Deficiency of Cu and Mn are feebly felt. It is perceptible in 3–4% soil samples drawn from different locations within maize belt of South India (Singh, 2001). Zinc deficiency has been reported from maize cropping areas of Chitwan valley in Nepal. Surveys indicated that DTPA extractable Zn in soil was below critical limits stipulated by agricultural agencies. Further, Zn content in leaf was deficient (Sherchan et al., 2004).

Iron (Fe)

Iron is an important micronutrient that affects maize productivity most regions of the world. Its requirements in terms of quantity may be relatively small, but its influence on growth and grain formation of maize, especially in intensive farming areas is significant. Iron deficiencies are wide spread across many geographic locations that support maize cropping. Iron is a relatively immobile element within plant system. Hence, younger leaves and tips may often express Fe deficiency. Sometimes, Fe deficiency occurs transitorily. Such transitory deficiency occurs when plant growth is rapid in comparison to Fe absorption and translocation. It actually leads to dilution of Fe in tissue and then deficiency. Iron deficiency is detected by examining leaves and stem at seedling stages. Younger leaves become chlorotic and show inter-veinal stripping.

Iron deficiency is wide spread in the "US Corn Belt". Soil tests using DTPA-Fe are used to classify soils for Fe status. Soils with DTPA-Fe <2.5 ppm are termed low, those with 2.5–4.0 ppm as marginal and values >4.5 are indicative of adequate Fe for maize cropping. Iron deficiency is more frequent on soils with pH >7.3. Maize grown in Canadian plains also suffers due to Fe deficiency. It is often related to soil pH and lime application schedules, if any. In Illinois, Fe deficiency is rare on maize. Yet, it may occur in soil with pH >7.3 (Hoeft, 2004). Soil tests are conducted to ascertain Fe levels. It allows us to judge Fe requirements properly. Symptoms on plants that show Fe deficiency could also be used to supply Fe. Foliar sprays of FeEDTA are common in large farms. Iron deficiencies are sporadic in the Pampas of Argentina. Iron deficiency is common on sandy soils low in organic matter. Incessant cropping and lack of organic matter replenishment also cause Fe deficiency.

Iron deficiency is wide spread on maize cultivated in the Southern plains of India. It is observed both in the Vertisol and Alfisol zones. The coastal sands that support maize also experience Fe dearth. In the dry lands, Fe deficiency occurs due to high soil pH, salinity, paucity of organic matter supplies, and interference due to high P supply. Iron deficiency is sometimes transitory and seen only at seedling stage, especially during rapid growth phase. Iron deficiency often curtails response to major nutrients and lowers fertilizer use efficiency. In South India, "Hidden Deficiency of Fe" that goes undetected has been amply reported.

Deficiencies of Other Micronutrients

Manganese deficiency has not been experienced by maize crops that flourish in the Corn Belt of USA. Maize crop may encounter Mn deficiency only in soils with pH >7.4. Foliar application of MnEDTA at 1.64 kg ha^{-1} after the symptoms appear rectifies the problem. We should note although Mn may not be required in large quantities, yet its dearth can have far reaching effects on nutrient dynamics in the cropping zone. The uptake and utilization of major elements too can be impaired due to law of minimum. Ultimately, maize growth is affected.

Copper deficiency is rare in the "Corn Belt of USA". It has not been observed in states like Illinois, Indiana, and Wisconsin. According to Hoeft (2004) Cu requirement is low. Traces of Cu found in fertilizer-K, usually suffices to correct if a deficiency occurs. Residue recycling and farm yard manure (FYM) inputs often satisfy Cu requirements of maize grown in the plains.

Boron deficiency has been reported from several areas within the US Corn Belt. Boron deficiency is common on sandy soils that are prone to leaching. Depletion of soil organic matter (SOM) also causes B deficiency. Drought and soil moisture dearth can accentuate B deficiency in most soils. Major sources of B are soil, irrigation water, SOM, and fertilizers containing B if applied periodically. Soil tests are conducted to ascertain B deficiency/sufficiency. Soil with <0.25 ppm B is termed low or deficient for B. Boron deficiency is also common in maize growing regions of European plains. Boron deficiency affects crop response to major nutrients. The calcareous soils of West Asia are prone to B deficiency. Droughts and low precipitation patterns accentuate B deficiency in maize and other crops. During recent years, B deficiency has been sporadically reported in the southern Indian plains where maize crop flourishes during rainy season. Boron deficiency occurs most commonly along with other micronutrients. It is attributed to consistent cropping of fields without appropriate replenishment schedules. The nutrient imbalances can be severe and it reduces grain productivity. As stated earlier, depletion of SOM, low crop residue recycling, and erosion may all contribute to B deficiency in South Indian plains. Soils with <0.2 ppm B are termed B deficient. However, in Southern African maize growing regions, deficiencies of micronutrients such as B, Mn, Cu, and Fe are less frequent.

Molybdenum deficiency is not a common occurrence in soils that support maize farming. In South Africa, especially in Natal province, maize yield could be limited due to Mo deficiency in soil. Similarly, Mo deficiency has been reported in parts of Zimbabwe and Zambian maize growing regions. It affects both maize and intercropped legumes. Legumes need Mo for optimum activity of symbiotic N fixers-Bradyrhizobium.

Chloride (Cl) deficiency has not been observed on maize grown in US Corn Belt. It is usually required in traces. Often, fertilizer formulation meant to supply major nutrients like KCl in case of K may contain sufficient Cl for the entire crop season plus a couple of them.

Micronutrient Supply

Soils are supplied with Zn using various types of formulations. Most commonly used products are grouped as follows:

Organic Chelates --- Zn EDTA, ZnHEDTA

Organic Non-chelates --- Natural Organic Complexes

Soluble Inorganic Compunds --- $ZnSO_4$, $Zn-NH_4$ Complex

Insoluble Inorganic compounds --- ZnO, $ZnCO_3$

Inorganic soluble sources of Zn have provided consistent correction of Zn deficiency in most of the maize farming zones (Wortmann, 2008). Insoluble sources are preferable when soil is slightly acidic with pH 6 or 6.5. The insoluble inorganic Zn sources need to be powdered for better contact with soils and broadcasted or thoroughly mixed with soil or incorporated. For example, ZnO or $ZnSO_4$ is effective if banded along with N and P fertilizer as basal dose. Zinc–Ammonium complex is often used as a starter fertilizer during maize farming. The chelated forms diffuse with soil water.

The organic forms are soluble in carriers and are not very mobile. Therefore, organic-Zn forms need to be placed near the root zone to achieve better contact.

Zinc deficiency seems to appear frequently in the Corn Belt. It is attributed to incessant cropping, leading to exhaustion of this element in available form. Also, during intensive cropping maize response to major nutrients immensely depends on availability of optimum levels of all micronutrients including Zn. Therefore, it is essential to maintain optimum availability of Zn to maize roots. Zinc deficiency can hamper proper response to major nutrients because of Liebig's Law of Minimum. In the Canadian plains, Zn deficiency is common. It is easily detectable by the appearance of white stripes at the bottom of maize leaves. Zinc is supplied at 4–14 kg ha^{-1} depending on extent of deficiency and yield goals. Higher rates of $ZnSO_4$ are possible but are spaced at once in 3 years. In some farms, Zn is also supplied as foliar spray. Foliar sprays avoid much of the soil transformation that occurs with soil application. Zinc is directly absorbed through lenticels on leaves and processed inside plants. In case of foliar spray, Zn is applied as soluble $ZnSO_4$ at 60g 100 L^{-1} (Omafra, 2002).

Zinc recommendations in the Corn Belt of USA (Iowa) are always based on soil tests. Soil with 0–0.4 ppm Zn is considered low in Zn content. About 10–12 kg Zn ha^{-1} is broadcasted or 2.2 kg Zn is banded along seedlings to correct Zn availability in soil. Soil with 0.5–0.8 ppm Zn is termed as marginal. Such marginal soils are supplied with 5 kg Zn ha^{-1} broadcasted or 1.0 kg Zn banded in rows. Soils with >0.9 ppm are termed as adequate for Zn and are not supplied with fertilizer-Zn (Sawyer, 2008).

Maize farmers may also prefer to correct micronutrient dynamics using a granular formulation containing 4–5 micronutrients. For example, TIGER Corn Mix (67% S, 10% Zn, 2.5% Mn, 2.5%, Fe, and 0.5% Cu) is a unique micronutrient mix that supplies at least 6 micronutrients to corn crop grown in USA. Maize crop is applied with 30–60 kg Tiger mix ha^{-1}. Such a mix could be incorporated into soil or banded.

In Canada, maize is usually provided with 2.2–5.5 kg $ZnSO_4$ ha^{-1} or 0.55–1.1 kg chelated Zn ha^{-1}. When Zn deficiency is suspected early at seedling stage, foliar spray of $ZnSO_4$ at 0.55 kg ha^{-1} is recommended. On eroded soils, 5.5 kg $ZnSO_4$ is broadcast or incorporated just before planting.

In the Spanish maize cropping areas, Zn deficiency has been corrected using a variety of sources. Lopez-Valdevia (2002) reports that on Haplo-Xeralfs, application of Zn-EDTA, Zn-lignosulphonate, or Zn-EDDHA resulted in high grain yield compared to other sources. Maize crop recovered Zn most efficiently and in high quantities when Zn was supplied at 20 kg ha^{-1} as Zn-EDTA or at 10 kg Zn ha^{-1} as Zn-lignofulphonate. Usually, soil tests are conducted prior to application of Zn. The Zn content in maize tissue seems to correlate best if water soluble-Zn plus exchangeable-Zn or available-Zn is estimated using diethylenetriaminepenta acetic acid or Mehlcih-3 extractants.

In the West African Savanna, maize production could be hampered by Zn levels and its availability to crops. Inherently low levels of Zn, incessant cropping, periodic droughts, and low amounts of residue recycling are some of the reasons. Abunyewa and Mercer-Quarshie (2004) have argued that maize crop in parts of Northern Ghana does not respond to supply of major nutrients, N, P, and K. It is attributed to dearth for Zn that actually curtails any response to major nutrients. Therefore, it is mandatory to maintain soil Zn at optimum or above threshold levels. Supply of Zn as $ZnSO_4$ might be crucial in enhancing maize productivity. Further, investigations have shown that $ZnSO_4$ at 5–7 kg results in best recovery of Zn and grain yield. Zinc inputs beyond 10 kg ha^{-1} did not improve its recovery. Hence, Zn accumulated in the subsurface layers. This fraction of Zn may become vulnerable to loss via leaching. In Zambia, maize cultivation spans both areas with subsistence and intensive farming techniques. The critical rate for Zn was 5 kg Zn ha^{-1} in most regions of Zambia. The critical Zn level in soil was reported as 2–5 kg ha^{-1} depending on soil type. Based on several field trials, it was concluded that if critical Zn level in soil is below threshold, then about 10 kg ha^{-1} $ZnSO_4$ should be incorporated in the intensively managed fields (ZARI, 2008).

Long-term trials on maize-finger millet rotations at Coimbatore in South India have shown that on calcareous or clayey soils with alkaline pH, micronutrient recovery may be affected, if appropriate quantities are not replenished. The availability of micronutrients such as Zn and Mn were sufficient, but Cu and Fe were deficient. Analysis of soils actually indicated that Zn and Fe levels had depreciated by almost 60–63% of initial level, due to 31 years of continuous cultivation. Hence, micronutrients had to be replenished after each crop or at least once in 3 years. The $ZnSO_4$ application improved soil Zn by 34%. The DTPA-Zn increased to 3.48 ppm (Selvi et al., 2006). Over all, accumulation pattern of micronutrients in maize fields was Fe > Mn > Zn > Cu.

In the Corn Belt, Fe deficiency is best controlled by systematic manure application. Manures are effective when Fe deficiency is due to low availability. Large amounts of FYM or organic manures usually restore Fe dynamics in the maize fields. In addition, farmers in the "Corn Belt" tend to apply $FeSO_4.7H_2O$ in furrows at 55–100 kg ha^{-1} to correct Fe dearth if any. Over-irrigation of calcareous soils could accentuate Fe deficiency. Therefore, it should be avoided. Maize genotypes tolerant to low Fe in soil are available. Foliar spray of Fe is also in vogue in many maize cropping zones. It firstly avoids undue soil chemical reactions that occur when Fe is applied on to soils. Foliar sprayed Fe is directly absorbed by leaves and hence is metabolized quickly in plant system. Foliar sprays are efficient in terms of fertilizer formulation requirements. Maize fields are sprayed with chelates of Fe at 1–1.5% (w/v) solution. Most importantly, Fe deficiency gets corrected rapidly upon foliar supply.

Maize cultivation in the Nigerian savannas and forested areas is affected by Fe dearth in soil. It is usually corrected by applying Fe salts once in 2–3 years to soil. Kayode (1984) states that maize cultivated in savannas of Nigeria, requires at least 3 kg Fe ha^{-1} more and it has to be supplied via fertilizers.

In the Northern plains of USA, maize crop is supplied with Cu as $CuSO_4$ or CuO. These sources of Cu are incorporated into soil before planting. On organic soils, broadcast, or incorporation rate for Cu is about 11–17 kg Cu ha^{-1}. Soil application of Cu is effective for 8–10 years. Foliar application could be in consistent. Foliar sprays are most effective if done at tillering stage. If Cu deficiency is severe two foliar sprays, one at mid tillering and other at boot stage is recommended. Chelated forms of Cu are also utilized to supply this micronutrient.

Manganese deficiency is rare in many of the maize belts. Manganese deficiency in corn is detected by discoloration of leaves to olive green. Manganese deficiency is corrected using foliar spray at 2 kg 200L^{-1} of water or 8 kg ha^{-1} soil application. It could be banded at seedling stage or incorporated into soil before planting (Omafra, 2002).

Boron is supplied into maize fields usually as borates or borax. Soil application rate for borax is 1.6 kg B ha^{-1}. In the Corn Belt of USA, particularly in Illinois, about 0.8–1.1 kg ha^{-1} B is recommended. Foliar sprays are done at much reduced rates at 0.11–0.55 kg ha^{-1}. In order to avoid B toxicity it is preferable to reduce B supply. Foliar application of B should not exceed 0.33 kg B ha^{-1}. Boron is usually not supplied until the crop has reached late seedling stage (Mengel, 2009).

There are many regions with in maize agroecosystems, where farmers supply a single or a composite of micronutrients using foliar sprays. Foliar sprays are advantageous when deficiencies have to be removed quickly avoiding interaction of fertilizers with soil components. A combination of micronutrients offers balanced micronutrient supply, especially if multiple deficiencies are suspected. There are indeed several micronutrient formulations that contain more than 2 or 3 micronutrients. Let us consider an example. Shabaan (2001) states that in Indus plains, soils are prone to rapid depletion of micronutrients. Soil deterioration and rampant use of $CaCO_3$ to correct pH have all caused imbalances in micronutrient availability. Micronutrients applied to soil may have to undergo series of physico-chemcial and biological transformations, before their dynamics get stabilized. Hence, supply of combination of micronutrients like 5.2% Mn, 0.65% Zn, 0.2% Fe, and 0.65% Cu is recommended. Foliar sprays of micronutrients are in fact suggested by many agricultural agencies all over the world.

Maize cultivated on Vertisols of South India may encounter micronutrient deficiencies, especially Zn and Fe. Micronutrient replenishment schedules are usually prepared based on soil tests. For example, in the Malaprabha project area of North Karnataka about 25 kg $ZnSO_4$ plus 25 kg $FeSO_4$ are applied prior to planting. It allows farmers to harvest 66–68 q grain ha^{-1} (Patil et al., 2006).

Micronutrient Recovery by Maize Crop

Reports from Nebraska and Indiana suggest that a maize crop that produces 5 t grain removes 2.7 kg Fe, 0.45 kg Mn, 0.45 kg Zn, 0.23 kg B, 0.1 kg Cu, and 0.01 kg Mo ha^{-1} (Wortmann et al., 2008). Mengel (2009) has reported that maize grown in Indiana, on an average recovers 2.1 kg Fe, 0.33 kg Mn, 0.33 kg Zn, 0.18 kg B, and 0.01 kg Mo ha^{-1}.

In the Canadian plains, maize grown on sandy loams yields 9–10 t grain and 12–14 t stover ha^{-1}. On an average, maize crop that yields 6.5 t grains and 12 t stover recovers 0.5 kg Zn, 1.2 kg Fe, 0.8 kg Mn, 0.2 kg Cu, and 0.76 kg B (McKenzie, 2001).

Field trials at Tamale in Ghana, suggest that on sandy soil of Savanna region, maize recovers only 1–10% of $ZnSO_4$ and 15% of $MgSO_4$ applied. Yet, it resulted in significant grain and fodder increase. The extent of benefit depended on the N input (Abunyewa and Mercer-Quarshie, 2004). Recovery of Zn and its effect on grain formation was best when soil Zn levels were 5 mg kg^{-1} soil or more. However, at 10 mg Zn kg^{-1} soil, grain yield stagnated at 2 t ha^{-1}. Soil Zn status affected both its recovery rates by roots and productivity. Zinc supply improves its recovery and grain yield. In addition, Zn interacts with N resulting better grain yield (Table 4.1). Maize grain yield generally peaked whenever soil Zn content ranged from 7 to 12 mg kg^{-1}.

Table 4.1. Influence of zinc supply and its interaction with nitrogen on maize grain yield.

Effect of Zinc alone			Effect of Zn x N Interaction on Grain Yield (t ha^{-1})	
Zinc Supply	Grain Yield	Zn Uptake	Nitrogen Inputs (kg ha^{-1})	
kg ha^{-1}	t ha^{-1}	g ha^{-1}		
			40	90
0	1.07	10.1	1.10	1.19
5	2.06	41.4	2.04	2.09
10	1.97	46.0	1.96	2.12
SE	0.11	4.48	0.11	

Source: Abunyewa and Mercer-Quarshie, 2004.

Yet another report based on several studies in the South Indian plains suggests that a crop that produces 6.27 t grains removes 3.0 kg Zn. It amounts to 505 g Zn to produce 1.0 t grains (IPI, 2000). Maize grown on fertile soil with high input technology that yields between 9–10 t grains and 11 t forage ha^{-1} absorbs 270 kg N, 50 kg P, 182 kg K, 31 kg S, 124 g Cu, and 520 g Zn ha^{-1}.

Micronutrient Loss

Micronutrient loss from maize fields/expanses is mostly mediated by natural factors that induce soil erosion, seepage, drainage, and percolation. Recovery into above ground parts of maize plant also induces loss of micronutrients from fields, if *in situ* recycling of crop residues is low or absent. Incessant cropping can definitely induce loss of micronutrient from soil phase. Micronutrients applied as chemical fertilizers could be lost to organic components through immobilization or fixation to organic components of the soil. This portion of micronutrient in organic components may become available to crops grown later in rotation. Soil conservation methods such as mulching, contouring, growing catch crops, and green manures may all help in conserving micronutrients and thwarting their rapid loss.

SUMMARY

Maize cropping zones spread into variety of soil types that are rich, optimum, or deficient in secondary nutrients. Generally, secondary nutrients are well distributed in both surface and subsurface horizons. However, there are several geographic locations with soil sporadically low or totally deficient with regard to secondary nutrients. Secondary nutrients are required in slightly lower quantities by a maize crop.

Calcium requirements could be judged by plant analysis and replenishment schedules planned accordingly. Maize crop may need about 3–4 kg Ca t^{-1} grain yield. Calcium supply is often achieved via lime, gypsum, or other sources. We should note that in Corn Belt of USA, Cerrados of Brazil and a few other locations, Ca is also supplied to correct soil pH. Calcium recommendations in most of Corn Belt States are meant to correct soil pH and raise it to reach 6.5 from 5.3. In Cerrados, knowledge about Ca, P, and K activity is important in order to overcome toxic effects of Al and Mn. In fact, it is often Ca + K versus Al ratio in soil that counts. Again, Ca input into such Al rich soil is meant to correct adverse effect of Al on root growth, its activity and nutrient absorption.

Knowledge about Mg availability and distribution is useful. Its deficiency seems sporadic in the maize belts. Yet, incessant cropping may deplete Mg concentration in soil to levels below threshold. Soils are replenished with Mg using $MgSO_4$ or dolomite. Crop residue recycling helps in containing Mg loss from the cropping zone. Soil erosion causes loss of Mg. It could be reduced through usual soil conservation methods.

Sulfur dynamics in soils needs due attention. Sulfur occurs in different forms both in surface and subsurface layers of soil. It is generally accepted that subsurface accumulation of S is significant in most soils used for maize production. Considering subsurface S has generally improved prediction of crop response. It also adds to accuracy in estimating S balance in soil. Sulfur supply is usually achieved partly when farmers apply SSP to correct P deficiency. Sulfur status of soil is also restored when farmers add gypsum or lime to correct soil pH. Crop residue recycling is an important method to restrict loss of S from the ecosystem. Soil erosion control measures also restrict loss S from fields. Generally, 4–6 kg S is required to support formation of 1.0 t grains. Hence, while replenishing S, we should consider various soil processes and distribution pattern plus crop's requirement based on yield goal.

Knowledge about micronutrient dynamics in maize farming zones is essential. Micronutrient deficiency has been reported in many regions of world. Micronutrients are required in small quantities. Yet, their deficiency can affect nutrient dynamics in general and restrict crop's response to major nutrients. Soil tests and generalized recommendations for all micronutrients namely Zn, Cu, Fe, Mo, and B are available. Micronutrients are supplied once in 3 years to field at 5–25 kg ha^{-1} depending on soil test values. In some cases, plant tests are conducted and foliar sprays are done to achieve rapid correction of micronutrient deficiency. Micronutrients undergo several types of transformation. They occur in both organic and inorganic fraction of soil. As such, organic matter recycling improves soil micronutrient status. Soil erosion, leaching and percolation may all result in loss of micronutrients. Again, soil erosion control measures restrict loss of micronutrients from farming belts.

Over all, knowledge about dynamics of secondary and micronutrients is essential. Details about critical limits in soil and plant tissue are important. Crop response to micronutrient supply is still a major concern in many areas. Methods to conserve micronutrients *in situ* without rampant loss are essential. It is said that, crop residue recycling and organic matter inputs may suffice to hold micronutrient levels above critical limits in subsistence faming systems. However, in intensive farming belts, consistent supply of micronutrient to soil or via foliar spray is a necessity to support high grain (10–11 t ha^{-1}) and forage (20–30 t ha^{-1}) yield.

KEYWORDS

- **Apparent sulfur recovery**
- **Corn belt**
- **Deficiencies**
- **Fertilizer-N**
- **Micronutrients**
- **Single super phosphate**
- **Soil organic matter**
- **Soil texture**

Chapter 5

Soil Organic Matter, Crop Residues, and Green Manures

Global carbon (C) cycling is intricately connected to cropping systems. Globally, soils contain 1500–1600 Pg C and it is second compared to C found in deep seas, but it is twice the amount traced in atmosphere. Soil organic-C fluctuations can have a major impact on atmospheric CO_2 concentration and C cycle in general. Excessive emissions can lead to deterioration of soil quality and induce green house effects. Maize agroecosystem with its expanse at 140 m ha is an important ecological entity that affects C cycle and crop productivity on earth. A sizeable share of soil C sequestration and emissions could be mediated via maize belts of different continents. Maize agroecosystem, individually may offset sizeable fraction of fossil fuel emission through its ability to sequester C into crop and soil phase of the ecosystem. As such, maize is a preferred crop in dry lands that are prone to massive loss of soil organic carbon (SOC) due to repeated tillage, resultant high microbial activity, and soil respiration. The SOC recycling procedures too could be meager in many locations within maize growing regions of the world, because it is preferred better as forage or utilized more frequently for biofuel production. Obviously, there is a strong need to enhance forage recycling and lessen C emissions. It has been aptly pointed out that an increase in 1.0 ton of soil-C pool in degraded or sandy low fertility soil can increase crop productivity by 20–40 kg ha^{-1}. In case of maize, about 10–20 kg grain ha^{-1} could be enhanced t^{-1} soil C sequestered (Lal, 2004).

SOIL ORGANIC CARBON AND ITS SEQUESTRATION

Organic-C status of soils utilized to produce maize varies based on geographic location, topography, soil types, cropping history, fertilizer and organic manure inputs, and harvest levels attained. In addition, environmental factors like temperature, relative humidity, precipitation pattern also influences the SOC status of field or cropping zone. A few examples of SOC status and pH found in experimental stations/fields found world wide where maize is cultivated has been depicted in Table 5.1.

Table 5.1. Maize cropping zones of the world, soil types encountered, soil organic carbon content, and pH.

Maize Cropping Zone	Soil Type	SOC %	pH	Reference
Corn Belt of USA				
Sterling, Nebraska	Cumulic Hapludoll	1.6	6.2	Lindquist et al., 2005
Fort Collins, Colorado	Aridic Haplustalfs	2.1	7.6	Halvorson et al., 2006
Dekalb, Illinois	Aquic Argidoll	2.4	7.2	Coulter and Nafziger, 2008

Table 5.1. *(Continued)*

Maize Cropping Zone	Soil Type	SOC %	pH	Reference
European Plains	Cambisols	2.6	6.3	Smalling et al., 2006
Rothamsted, England	Chernozems	2.2	7.4	
	Luvisols	1.8	6.2	
Sophia, Bulgaria	Pelic Vertisols	3.4	6.7	Alexieva and Stoimenova, 1998
South American Plains/Hills				
Santa Caterina, Brazil	Hapludox	0.4	5.9	Sangoi et al., 2001
Botucatu, Sao Paulo	Latosols	0.2	4.8–6.5	Theodora and Ferreira, 1995
Mossoro RN, Brazil	Red Yellow Argisol	0.2	6.8	Silva et al., 2004
Colombian Savannas	Acid Oxisol	2.4	5.5	Basamba et al., 2006
Balcarce, Argentinean Pampas	Arguidoll	0.6	5.8	Sainz Rozas et al., 2001
Eastern Colombia	Acidic Oxisol	2.3	4.8	Oberson et al., 2009
West African Savannas				
Togo	Ferralsols	0.07–1.00	5.2–6.8	Sogbedji et al., 2006
Ghana	Acrisol	0.97–1.56	5.5–5.6	Yeboah et al., 2009
Ghana	Haplic Lixisols	0.88–1.12	5.5–5.8	Abunyewa et al., 2004
Northern Nigeria	Arenic Haplustalfs	0.37–0.67	6.0–6.3	Kogbe and Adedira, 2003
Ibadan, Nigeria	Oxic Paleustalfs	1.11–1.51	6.5	Babalola et al., 2009
Borno State (Azir), Nigeria	Argilaceous Alfisols	0.82–0.94	6.1–6.5	Kamara et al., 2009
Ile-Ife, Nigeria	Oxic Tropudalfa	0.85–1.12	6.5	Tijani et al., 2008
Cameroon (Highlands)	Palehumults	1.32–1.45	5.3–5.7	Yamoah et al., 1996
Other Sub Saharan Locations	Aridisols	1.3	4.9	Smalling et al., 2006
	Acrisols	1.0	5.0	
	Luvisols	0.6	6.1	
South and East African Cropping Zones				
Mpangala, Tanzania	Volacanic soils	2.2	5.9	Lisuma et al., 2006
Chitedze, Malawi	Ferruginous soil	2.1	6.2	Sakala et al., 2009
Bembeke, Malawi	Ferraltic latosols	1.8	5.9	Sakala et al., 2009
Harare, Zimbabwe	Lixisols	0.4	4.6	Mapfumo and Mtambanengwe, 2008
Mediterrainian Region				
Hebron, El-Khalil Region	Clay Loam	2.1	8.2	Al-Bakeir, 2003
Central Valley, Jordan	Calcareous Loam	2.3	7.8	Khattari, 2000
South and South East Asia				
Dharwar, India	Vertisols	0.65.	7.2–7.6	Rajeshwari et al., 2007
Bangalore, India	Alfisols	0.35	6.5–6.8	

Table 5.1. *(Continued)*

Maize Cropping Zone	Soil Type	SOC %	pH	Reference
Coimbatore, India	Haplustalf	0.40	8.7	Prahraj et al., 2009
Ludhiana, India	Ustrochrept	0.42	8.0	Dhaliwal et al., 2010
Bikaner, India	Sandy Alfisols	0.24	8.2	Sharma, 2009
Faisalabad, Pakistan	Calciferous Inceptisols	0.85	7.7–7.8	Iqbal et al., 2003
Kampong Chang, Cambodia	Loamy Alfisol	2.4	6.7	Belfield and Brown, 2008
Berili, Philippines	Sandy Alfisols	2.1	7.8	Comia, 1999
Chinese and Fareast Maize Belts				
Pinglian, Gansu, China	Calcarid Regosol	0.9	8.2	Fan et al., 2005
Hainan, China	Ferralsols	0.9	4.8	Wu et al., 2002
Suwon, South Korea	Sandy soil	0.8	6.1	Hur, 2004
Kyoto, Japan	Alluvial soil	1.9	5.7	Li et al., 2002
Australian Maize Growing Regions				
Comet in Queensland	Vertisols	1.31	8.8	Carter and Singh, 2004
Quirindi in New South Wales	Vertisols	1.19	8.9	Carter and Singh, 2004

Historically, almost all agricultural zones have been derived from natural forests, savanna, or waste land. Conversion of natural vegetation like prairies and forested areas to cropping zones has immense influence on C sequestration in the ecosystem. Organic matter recycling is almost directly affected by the biomass production levels in a given agoecosystem. There are indeed innumerable long-term studies conducted across different continents, regarding the effect of various crop rotations on soil organic matter (SOM) (Beyer et al., 2002). There are several examples pertaining to the Great Plains of USA. David et al. (2009) have mentioned that most studies consider SOC and nitrogen (N) pools as good indicators of soil quality because they affect a series of physical, chemical, and biological properties of soil relevant to maize culture. Most studies show a decline in SOC due to conversion of virgin soils to maize cropping, but intensity and rate of decline may vary. Let us consider a few examples. In the Midwestern USA, C and N loss from virgin prairies and maize fields were 38% and 35% of initial levels over a period of 60 years (Jenny, 1941). Hass et al. (1957) reported 39–42% decline in SOC due to 40 years of cropping maize. Reeder et al. (1998) reported 16–28% decline in SOC due to 60 years of consistent cropping. On Mollisols, SOC in the surface layers (0–15 cm depth) declined by 33–38% of initial levels in a matter of 4 decades (Mann, 1986). There are also several reports regarding SOC decline due to conversion of natural vegetation to cereal cropping. For example, Mikhailova et al. (2000) state that on Chernozems and deep loess soils of Russia, SOC decline ranged from 38 to 45% of original levels in a matter of 50 years, by when the fields almost attained steady state of nutrient cycling. David et al. (2009) have cautioned that quite a number of studies on SOC or total N decline have considered only surface soils at 0–15 cm depth. The SOC changes could be different, if entire soil profile or at least rooting zone extending to 30–60 cm depth is considered. Crop rotations may influence C sequestration pattern. Evaluation during 1990–2006 at Brookings, Dakota in Northern Great Plains has shown that a well diversified cropping system improves SOC and reduces the burden of high N inputs. Addition of fertilizer-N at various stages in the rotations affects SOC accumulation. Fields under grass showed a 38% increase of SOC between 1996 and 2006. In fields with continuous corn, residue returned was 34% more than corn–alfalfa–wheat rotation, but it did not offset the SOC loss. Under high fertility conditions fields with continuous corn lost 2.3 Mg C ha^{-1} and those with corn–alfalfa–wheat lost 0.3 Mg C ha^{-1} (Pikul et al., 2008).

Maize is a highly preferred substrate for biofuel production. In most parts of the world, and more specifically in USA, demand for biofuel generated using maize crop residues seems to increase rather rapidly. In order to enhance maize acreage and productivity, natural vegetation, or grasslands that support smooth brome grass (*Bromus inermis*) are being converted to maize fields. Grass lands maintained under Conservation Reserve Program (CRP) may also be converted to maize for biofuel (Follet et al., 2009). It is well known that natural vegetation, such as grass lands sequester higher levels of SOC compared to maize fields under conventional or even no-tillage (NT) systems. Maize is a C$_4$ plant while brome grass is C$_3$ crop. Follet et al. (2009) have used this fact effectively in monitoring the changes (loss or gain) of SOC. They measured the replacement of C$_3$ derived SOC pertaining to brome grass by C$_4$ derived C from maize. Data from a 13 year long-term trial suggests that SOC sequestration is significantly reduced due to conversion of grass land to maize fields. Therefore, to offset SOC loss, at least partly, it was argued that it is useful to adopt NT corn that improves SOC accumulation. In addition to SOC, we should note that conversion of grass land to biofuel maize induces certain changes in soil aggregates and microbial load. The above situation is common to many regions of world, wherever, large scale and rapid conversion of natural vegetation/grass lands to maize occurs in order to enhance biofuel or grain production. Therefore, suggestion to adopt SOC sequestering techniques such as Nt and efficient residue recycling is applicable to most cropping zones. There are reports that corn stover-based bio-energy cropping systems could be managed efficiently to increase C sequestration rates and to control over all global warming potential by using NT systems along with a manure-based nutrient management system (Ultech, 2008). Integrating crop residue and livestock manure can further reduce loss of SOC by retarding CO$_2$ and N$_2$O emissions. There are several reports about possible improvements in C sequestration in European cropping zones. In England, adoption of suitable tillage systems and stringent recycling of crop residues is known to improve SOC status of arable soils perceptibly (King et al., 2004).

Soil Quality

Maize-based crop rotations may influence the soil C sequestration to different extents depending on crop species that precede or follow the maize crop. The soil quality indictors may actually get affected differently both in intensity and duration depending on crop sequences followed in a field. The crop residue recycling levels vary enormously in the maize agroecosystems of the world. In the subsistence farming zones, residue recycling is low compared to intensive cropping zones like Corn Belt of USA where corn–soybean rotations add large amounts of C to soil through residue recycling. Soil quality indices too differ depending on rotations practiced within maize belts. According to Karlen et al. (2006) extended maize-based rotations are common feature in the US Corn belt had positive effects on SOC sequestration and several of the soil quality indicators. Total organic-C was sensitive among soil quality indicators. Other soil quality traits such as bulk density, water stable macro-aggregations, total-N, microbial biomass C, extractable phosphorus (P) and potassium (K), and root penetration were all mildly improved due to residue recycling practices.

During a crop rotation, quantity of crop residue available for recycling seems most crucial factor that influences the SOC sequestration. In subsistence farming regions, fraction of residue recycled is relatively low because much of it is apportioned to feed farm animals. Whereas in intensive farming zones, forage yield is high and mulching is practiced routinely. Relatively larger quantity of SOC is recycled in the agroecosystem. Long-term evaluations in the Platte Valley of Nebraska have shown significant differences in forage recycling and SOC depending on maize-based rotation (Varvel and Wilhelm, 2008). The improvements in SOC due to maize sole crop or maize–soybean rotations were discernable up to 15 cm depth. Residue production greater than 6 t ha^{-1} was found to be sufficient to support recycling and biofuel generation systems in the farm.

There are several computer models and simulations that allow us to study the changes in C sequestration pattern in different maize belts of the world. Following are few examples—Century, Cquester, and CERES-Maize. In the US Corn belt, four long-term trials one each in Illinois, Missouri, Nebraska, and Iowa, ranging from 20 to 100 years have been simulated for consequences of cropping systems and residue removal pattern and their influence on soil C status (Liang et al., 2008). Carbon sequestration rate across the maize agroecosystem is influenced significantly by various agronomic procedures. Fertilizer-based inorganic nutrients and farm yard manure (FYM) supplied under integrated farming systems has immense effect on soil-C sequestration into maize ecosystem. Maize under intensive farming systems in Corn Belt of USA, Europe, or China accumulates large amounts of C in the above ground biomass and roots. It generates large amounts of recyclable C via residues. Whereas, in subsistence farms of West African Savanna or dry lands of Asia, C sequestered via biomass is relatively low, despite fertilizer inputs. Basically, biomass formation is of lower order in dry lands. Following is an example that depicts C sequestration trends in the maize growing regions of North Indian plains, where maize–wheat–cowpea is a predominant crop sequence followed by farmers:

Treatment (kg ha^{-1})	40 NPK	80 NPK	120 NPK	60 NP	40 N	80 NPK	Control	+ FYM
SOC (Mg ha^{-1})	1994	47.1	49.0	54.1	47.9	47.1	58.3	44.9
	2003	51.5	54.1	63.5	53.0	52.0	72.1	48.7
Change in SOC (Mg ha^{-1})		4.4	5.1	9.4	5.1	4.9	13.8	3.8

Source: Purkayastha et al., 2008.
Note: SOC = Soil organic carbon; FYM applied at 10 t ha^{-1}

Duxbury (2008) suggests that soil has finite capacity to sequester organic-C. Carbon sequestration level is determined by silt + clay content and aggregate percentage. Tillage breaks aggregates and exposes organic-C to biological decomposition. Soils tend to reach equilibrium with regard to SOC. The SOC saturation and equilibrium levels attained may vary depending on a variety of reasons related to soil, organic-C source, cropping systems practiced, cropping duration, and environmental parameters. There is an upper limit for equilibrium SOC level of mineral soil utilized to cultivate maize. Chung et al. (2008) state that SOC saturation is dependent on soil fractions, tillage practices, fertilizer-N supply, and cropping systems. The stabilization of SOC within macro or micro-aggregate or silt or clay fractions seems crucial. Basically, the apparent C saturation of some fractions could be limited.

Soil Carbon Emissions and Loss from Maize Agroecosystem

The agricultural soils emit sizeable quantities of CO_2. It is an important component of global C cycle and it has direct bearing on global climatic changes. Soil and crop respiration returns a part of C sequestered into atmosphere through C emissions (CO_2, CH_4). It is said that CO_2 emissions due to soil respiration accounts for about 25% of global C exchange between terrestrial and atmospheric sources of C (Ding et al., 2007). Soil and crop respiration mediated C exchange may account for 75 Pg C. It is argued that sometimes even a small change in intensity of cropping and SOM dynamics may have important consequences on CO_2 status of atmosphere in the vicinity. Hence, understanding various factors that affect quantity and pattern of CO_2 emissions is important. Maize cropping zones definitely mediate a large portion of CO_2 emissions. The rates of CO_2 emission may however vary depending on the location of cropping and intensity of cropping. Soil management methods, cropping sequences, fertilizer inputs, irrigation, and a range of environmental parameters may all influence CO_2 emissions from maize fields. Ding et al. (2007) mention that it is not easy to measure small variations in CO_2 emission within a maize field. In fact, minor alterations in soil properties and vegetation intensity may influence CO_2 emission from a maize field. Yet, CO_2 emissions from maize cropping zones have been studied and data pertaining to factors that influence it have been accumulated. The CO_2 flux is highly dependent on geographic location, seasonal variations, crop genotype, agronomic procedures adopted, and a range of environmental factors. For example, in the Henan region of China, which is prone to cold climate, CO_2 fluxes in fields cropped with maize increased with plant age. It reached a maximum at 111–220 days after planting that is late elongation stage and then sharply declined at day 131, when maize crops changed from nutritive growth stage to reproductive stage.

Ding et al. (2007) found that during growth season of maize crop, there was a significant positive relationship between soil temperature and CO_2 flux. A simple exponential model explained about 36% of seasonal variation of CO_2 flux. The dependence of CO_2 emission on soil temperature could be explained using a couple of different models and equations. Linear or sinusoidal regressions could be adopted to predict CO_2 emissions. Arrhenius equation is also used because it offers a uniform relationship between temperature and soil respiration. Bauchmann (2000) found an exponential relationship between soil CO_2 fluxes and temperature. Ding et al. (2007) found that in maize cropping zone of North China; soil CO_2 emission could be explained using again an exponential equation. About 18–25% of seasonal variation in CO_2 fluxes could be explained using exponential equation. The relationship between soil CO_2 emission and temperature could be explained with greater accuracy using mathematical equations, if period of observation was restricted to elongation to harvest stage.

Nitrogen fertilization has an impact on seasonal CO_2 fluxes from maize fields. Ding et al. (2007) found that in maize fields around Henan in China, mean seasonal CO_2 emission ranged from 294 to 539 g C m^{-2}, as N inputs were increased from 0 to 250 kg N ha^{-1}. Fertilizer-N inputs caused a 10.5% reduction in soil CO_2 fluxes but did not increase above ground biomass significantly.

Soil moisture firstly affects microbial C cycling. Soil moisture affects microbial respiration, multiplication, and biochemical transformations. Coarse textured soils with relatively low soil moisture holding capacity support reduced levels of soil microbial respiration and CO_2 emission. Fine textured soils with high SOM content obviously generate higher levels of CO_2 due to enhanced microbial activity. Sey et al. (2008) found that in the Canadian plains, soil CO_2 content in pores and emissions were greater in fields under conventional tillage than those under NT system. The peak CO_2 emission occurred at 50th day after planting. The CO_2 in the profile was controlled more by soil moisture than temperature or agronomic practices.

Duxbury (2005) states that in the Corn Belt of USA shift from conventional tillage to zero-tillage has altered C balances advantageously. In case of maize, the net benefit fluctuates around 330 kg ha^{-1} yr^{-1}. A certain fraction of C sequestered under NT system could be offset by extra CH_4 emissions. Following is an example that delineates influence of tillage on C emissions and sequestration trends during maize production in USA:

Green House Gas Source (+) or Sink (–)	Conventional Tillage kg ha^{-1} yr^{-1}	No-tillage kg ha^{-1} yr^{-1}
Soil C Sequestration	0	–301
Carbon Dioxide Emissions		
Ag Inputs	+143	+198
Machinery	+71	+23
Net C flux	+223	–110
Relative C flux	0	–333
CH_4 Emissions	0	+9

Source: Duxbury, 2005.

CROP RESIDUES AND NUTRIENT RECYCLING

Maize farmers utilize a very wide range of crop residues during crop production. The quantity of crop residue used may also vary widely. Stover that includes stem, leaf, twigs, and cobs/panicles/fruits are used as sources of organic-C and nutrients. In some cases, root material is also used as crop residue. Otherwise, generally roots are left to decompose below ground. The crop species that provide residues is important. More importantly, biochemical composition of the crop residue has a major influence on the extent of decomposition and release of C and other nutrients (Johnson et al., 2007; Machinet, 2006). Of course, irrespective of biochemical nature of crop residue, environmental parameters such as temperature, humidity, and soil characteristics, all of them affect rate of crop residue decomposition and release of nutrients. Farmers generally give due importance to amount of mineral nutrients and C present in different sources (Table 5.2). It gives them an idea about the nutrients that could be recycled and possibility for economizing on use of inorganic fertilizers. There are indeed too many studies and reports about nutrients derived from crop residues, ways to improve nutrient recycling and grain yield response of maize. A few reports that deal with different geographical regions, wherever maize is a major crop have only been depicted in the following paragraphs.

Experiments in Illinois have shown that continuous corn responds to crop residue recycling with enhanced grain yield. For example, full removal of crop residue (without recycling) gave 13.1 t ha^{-1}. Partial removal or no residue removal resulted in 5–13% increase in yield over control (Coulter and Nafzier, 2008). Further, recycling crop residue decreased economical optimum N requirement of maize. Firstly, quantity of crop residue recycled influences SOC and nutrient turnover in the ecosystem. At the same time, there are several other parameters of crop residue and soil environment that affect nutrient recycling (Plate 5.1).

Plate 5.1 Maize on Southern fringes of US Corn Belt—Kansas, USA.
Source: Dr. Kraig Roozeboome, Kansas State University, Manhattan, Kansas, USA.
Note: Incorporation of residues from previous crop is a common practice. Incorporation of maize residues improves SOC content, nutrients, and soil quality in general. It enhances soil microbial activity and nutrient transformations. It also reduces soil erosion, percolation of moisture, and dissolved nutrients to lower horizons.

Johnson et al. (2007) studied the decomposition rates of different plant materials like leaf, stem, root organs, and cobs/pod from different plant species. Crop residues derived from alfalfa, corn, cuphea, soybean, and switch grasses were examined for rates of decomposition and nutrient release. The characteristics of crop residues such as C:N ratio, succulence, lignin content, size of residue particles, moisture content, and plant portion all of them affected the rates of decomposition. The quality of crop residue was mostly dictated by the C:N ratio. The C:N ratio is among most useful indicators while predicting decomposition. Generally, high lignin contents retard decomposition processes in soil. Researchers have used several types of computer-based simulations and models to predict the extent of nutrients derived from different crop residues, their impact on various soil biochemical reactions, and net advantages the maize crop may derive. We should note that in addition to quality or composition of crop residue, the microbial load and species that dominate the soil profile also affect rates of nutrient release. The microbial component has a major impact on nutrient dynamics in the soil ecosystem. Models such as CENTURY, CQUESTER, and DAISY are most frequently adopted to estimate the effects of crop residue recycling (Johnson et al., 2007, Machinet et al., 2006; Mueller et al., 1998). There are also sub routines on SOM and crop residue that could be attached to more commonly used crop growth models.

Table 5.2. Nutrient contents of crop residue, organic manures, and green manures utilized during maize production.

Source	Nutrient Content (%)		
	N	P	K
Paddy Straw	0.36	0.08	0.12
Rice Hulls	0.3–0.5	0.2–0.5	0.3–0.6
Sorghum Straw	0.41	0.23	2.1
Pearl Millet Straw	0.65	0.75	0.4
Cassia auriculata	0.98	0.12	1.8
Careya arborea	1.67	0.40	2.3
Terminalia chebula	1.46	0.35	1.2
Terminalia tomentosa	1.39	0.40	1.2
Crotolaria spp.	2.89	0.29	0.7
Tephrosia spp.	3.73	0.28	1.8
Water Hyacinth	2.04	0.37	3.4
Azolla spp.	3.68	0.20	0.2
Cowpea	0.71	0.15	0.22
Dhaincha	0.62	0.41	0.81
Guar	0.34	0.51	0.68
Horse Gram	0.33	0.25	0.18
Mungbean	0.72	0.18	0.15
Black Gram	0.85	0.18	0.12
Sun Hemp	0.75	0.12	0.62
Cattle Dung	0.3–0.4	0.1–0.2	0.1–0.3
Sheep Dung	0.5	0.4	0.3

Table 5.2. *(Continued)*

Source	Nutrient Content (%)		
	N	P	K
Rural Compost	0.5	0.8	0.8–1.0
Farm Yard Manure	0.4–1.5	0.3–0.9	0.3–1.2
Vermi-compost	1.6	5.0	1.7

Source: Krishiworld, 2002.

Field trials in the Corn Belt of USA have shown that decomposition of maize crop residue could be hastened by using inorganic N primers. There are potentially clear advantages to farmers if they could hasten decomposition of crop residue. The pre-plant soil nitrate-N status, N supplied to soil, and C:N ratio of the crop residue are most immediate factors that affect nutrient release and recycling (Bundy, 2001). We should note that C:N ratio of residue from maize or that from any other crop species could alter as decomposition proceeds.

Field studies in the Piedmont region of Northeast USA have shown that, application of crop residue is essential, in order to improve soil nutrients and quality. Residue recycling enhances C sequestration in the soil. Interaction of tillage systems and crop residue incorporation affected maize productivity. If crop residue was removed out of the field, silage (forage) production decreased from 18.8 t ha^{-1} to 17.5 t ha^{-1}. In a long run, crop residue recycling seems essential and impact on maize productivity gets pronounced as time elapses (Cassel and Wager, 1996).

During recent past, maize ecosystem has been experiencing a minor and perhaps yet imperceptible change in its genotype composition. The BT maize endowed with genes for resistance to certain lepidopterous pests is being sown in many areas of the Corn Belt. Field verifications in Northeast Missouri suggest that generally, no differences occurred with regard to leaf or stem tissue composition of BT and Non-BT isoline. In addition, upon incorporation of residues there was no discernable difference in decomposition rates or N mineralization pattern. However, roots from non-BT isoline decomposed faster than BT isoline. Mungai et al. (2005) state that, over all, there were no differences between residues derived from BT or non-BT isoline with regard to crop residue quality nor its decomposition pattern and N dynamics in the field.

In the Northern plains region, legume crop residues serve as major source of N to succeeding maize crop. Recycling legume stover also gains in importance under low input systems. The rates of organic matter decomposition and N mineralization, as well as type of legume residue may all influence the extent of nutrient benefits that maize crop derives. For example, decomposition constant for alfalfa (*Medicago sativa*) stover was 0.283, that for roots was 0.083; for maize stover 0.00047, and for roots 0.0014 day^{-1} (Pare et al., 2000). Alfalfa stover/roots decomposed rapidly releasing N into soil compared to maize stover.

Recycling maize straw is beneficial in terms of C sequestration in soil, improvement of soil quality, microbial activity, nutrient availability, root growth, and maize productivity. Addition of maize residue along with N delayed peak mineralization. It reduced microbial growth rates and easily available C. However, in general, maize residue input enhanced microbial population and activity. The rates of nutrient transformations were also enhanced (Blagodatskaya et al., 2008). Maize farmers in Andalusia of Spain apply crop residues, cotton gin compost, and inorganic fertilizers to enhance productivity. Usually 20–40 t composted cotton residues, 400 kg N, 80 kg P, and 120 K ha^{-1} is applied. Soil analysis indicates that major nutrient concentrations in soil increase perceptibly. Nutrient leaching is generally high in plots given more of inorganic fertilizers compared to plots receiving cotton gin compost (Tejada and Gonzalez, 2006). Over all, productivity of maize provided with cotton gin compost and inorganic fertilizers was greater than control plots.

Crop residue recycling is an important farming systems procedure in the entire Savanna zone of West Africa wherever cereal cropping is practiced. The savanna soils are generally poor in soil fertility. Therefore, crop residues are cycled in order to restore soil nutrient status. Organic residue also thwarts soil erosion, percolation of water, and dissolved nutrients to lower horizons. Periodic supply of crop residues and/or other sources of organic manures are a necessity on sandy acid soils found in the West African Savanna. Maize fields are often provided with residues just prior to sowing. Crop residues add organic-C, N, and a few other nutrients. Woperies et al. (2005) opine that a certain amount of organic-C build up in soil is needed, in order to derive better benefits from inorganic fertilizer inputs. In Northern Togo, they found that fields treated with residue repeatedly accumulated more SOC (13.4 g C kg^{-1}soil), compared to outfields not provided with organic residues (6.3 g C kg^{-1}soil). This build up of organic-C in the soil induced better NPK recovery from soil. For example, N recovery from organic residue treated soil was 0.41 kg N kg^{-1} fertilizer-N supply compared to only 0.33 kg N kg^{-1} fertilizer-N for untreated soils. Such induction of recovery was also discernable with other nutrients like P and K. Hence, they concluded that supply of crop residue firstly improved organic-C, microbial activity, and soil quality. In addition, it had positive influence on major nutrient recovery by maize or other cereals.

Often, farmers in northern Nigeria burn crop residues prior to incorporation with a hoe. Azeez et al. (2007) compared various mineral nutrient supply schedules with burnt ash from crop residue. They point out that burning the fallow vegetation that contained *Tridax, Imperrata*, and *Chromolaena* plants released sufficient levels of nutrients. It improved nitrate-N, available-P, exchangeable-K, exchangeable-sulfur (S), and calcium (Ca) content in the soil. Influence of addition of burnt ash was easily discernable within 30 days after planting. In most plots, increase of soil nutrients was traced right until maize harvest.

Maize and pigeonpea residue are important sources of N to most of the cropping systems adopted in East Africa. The type of crop residue, its source, timing, and method of application are some of the factors that need attention. The extent of N released into soils depends on soil type, its texture and inherent fertility. Nitrogen immobilization is also an important factor that affects nutrient dynamics. Sakala et al. (2000) suggest that application of maize alone or maize plus pigeonpea residue prolonged immobilization. Mixing different proportions of maize and pigeonpea residues did not alter immobilization pattern. However, when a mixture of young succulent leaves and twigs of pigeonpea were added along with small quantity of mineral N and maize residues, it improved amount of N released into maize fields. It has been pointed out that decomposition of maize residues was actually limited by N availability. Clearly, addition of crop residue plus small amount of mineral-N (primer) to soil has advantages. Mupangwa et al. (2007) have found that mulch based on maize residue is effective irrespective of tillage systems commonly adopted by maize farmers in Zimbabwe. Application of mulch improved moisture content in the soil profile. The maize stover/grain yield was improved in plots provided with mulch.

The contribution of C derived from legume residue to cereals like maize grown in rotation is very important, especially in areas adopting low input technology. In the semi-arid plains of India, it is said that share of legume sole crop or as intercrop with maize has declined marginally. It may have influence on residue recycling and N contribution to succeeding wheat. The influence of crop residue recycling on SOC and soil-N were studied during a maize/legume–wheat rotation by Sharma and Buhera (2008). They found that recycling crop residues improved soil-N by 11.5–38.5 kg N ha^{-1} under intercropped (maize/green gram or maize/cowpea) condition and by 17.5–83.5 kg N ha^{-1} if sole legume was rotated with maize. On an average, N economy derived from legume residue ranged from 40 to 56 kg N ha^{-1}. It was concluded that interjecting a legume such as cowpea or green gram and recycling the residue was highly beneficial to maize–wheat rotation systems followed in the Gangetic plains of India.

Incubation studies under green house conditions suggest that source and quality of crop residue affects N released into soil. Among the crop residues evaluated, Nair et al. (1984) found that maize straw released low amounts of NO_3-N compared with those derived from soybean or green gram. Whatever be the source of crop residue, it required 30–45 days to release NO_3-N into soil-N pool. Further, it was found that response of wheat grown in rotation with maize in the Gangetic plains, responded with higher grain yield, if it was provided with soybean crop residue instead of maize residue.

Maize–groundnut rotation practiced in Malaysia involves substantial quantity of crop residue recycling. According to Mubarak et al. (2002), groundnut haulms decomposed rapidly. The dry matter decomposition constant for groundnut haulm was 0.158% week^{-1}. Decomposition rate for groundnut was faster than maize crop residue which had a constant of 0.099% week^{-1}. The pattern of nutrient release from maize or groundnut residue was as follows: K>P=N=Mg>C

In Northern China maize cropping is intensive. Farmers usually apply a combination of crop residue, manures, and inorganic nutrients. In addition, factors like soil moisture, irrigation, precipitation pattern also influences nutrient release and their utilization by maize crop. The N recovery from residues ranged between 28 and 54% depending on residue and soil moisture status (Wang et al., 2007).

GREEN MANURES AND NUTRIENT DYNAMICS

Soils that support maize agroecosystem, especially in the tropics, are highly weathered and leached. Much of the organic fraction in the top layers is often lost due to erosion or emissions. In many regions, it is the organic fraction of soil that imparts greater cation exchange capacity (CEC). Therefore, loss of top layer invariably results in soil deterioration and reduction in nutrient holding capacity. Adequate organic matter in the upper horizon is essential. In order to improve nutrient buffering and soil quality, farmers tend to supply a variety of organic residues and green manures. Slowly decomposing organic residues may have an impact on SOM content and nutrient buffering capacity, but it takes effect after sufficiently long period. Addition of fresh, succulent, and nutrient rich green manure could make decomposition faster and benefits could be perceived relatively quickly. Green manures add organic-C and several of the essential mineral nutrients upon mineralization. Leguminous green manures are excellent sources of N in addition to organic-C.

In the dry savannas of Brazil and Argentina, SOM and P availability are key factors to stability of crop productivity. Farmers adopt several variations of tillage and cropping systems with interjecting green manure crops. Improvement of SOC via incorporation of green manure crop and lessening loss of SOC by avoiding repeated tillage are major aims. Green manures such as sun hemp or pearl millet are also commonly grown in Brazilian Cerrados. A mixture of cereal plus legume often provides higher advantage in terms of N nutrition of the maize that succeeds the green manure (Adriano et al., 2006). On an average, about 173 kg N ha^{-1} contained in the above ground parts of sun hemp and 89 kg N ha^{-1} in pearl millet were incorporated for the succeeding maize to utilize. Major advantages from a sun hemp plus pearl millet green manure were improvement in C:N ratio of soil and N fertility. The total N budgets of the cropping systems were improved by 10–15%. About 65% of N derived from biological N fixation was channeled into maize. Basamba et al. (2006) experimented with several types of crop rotations on acid Oxisols common to South America. Generally, NT systems improved SOC, N, and P compared to repeated tillage. The maize-green manure-maize rotation improved soil quality by enhancing SOC, N, and P. Incorporation of soybean-based green manure increased maize grain yield by 1.5–2.2 t ha^{-1}compared to other agropastoral systems or uncultivated open savannas adopted by South American farmers. Nitrogen derived from symbiotic N fixation and incorporation of succulent soybean stover was important, since it released nutrients rapidly into soil. Basamba et al. (2006) suggest that such maize-soybean-green manure-maize systems may be of great value in mixed farming conditions; where in nutrient dynamics within the farm is entirely integrated.

In Southern African maize farming zones, soil deterioration has been rampant due to continuous farming. Soil nutrient depletion, erosion of surface soil, and nutrients as well as loss of SOC has been the main reason for soil deterioration. Sakala et al. (2009) report that despite loss of soil fertility and impaired nutrient dynamics, in certain pockets of maize belt, farmers have been reaping modest levels of grain/forage. It has been attributed to leguminous green manure species that are able to recycle large quantities of SOC and fix atmospheric-N. In addition, tendency of farmers to supply nutrient through both inorganic fertilizers and green manures has been beneficial in inducing higher productivity within the maize belt. Timing of harvest, in other words collection of pruning from green manure species, succulence of leaves and twigs, C:N ratio of the green manure, and environment may all have an effect on maize productivity. Above all, C and N contents of the green manure species seems crucial. Following are green manure species commonly utilized by maize farmers in Malawi along with their C and N contents:

Legume	C	N	Lignin	C:N	Lignin:N
	------------ % ---------------			--------- ratio ----------	
Mucuna	0.2–0.4	5.5	5.5–16.8	9.8–30.8	1.3–8.3
Crotolaria	0.1–0.3	5.3	3.8–9.8	8.0–32.1	1.0–6.3
Lablab	0.2–0.4	4.1	2.6–11.5	7.4–29.1	0.4–9.8

Source: Sakala et al., 2009.

The average response of maize to green manure application ranged from 2.7 to 2.9 t ha^{-1} for *Macuna*; it was 2.8–3.0 t ha^{-1} for Crotalaria and 2.7–2.9 t ha^{-1} for Lablab within the maize belt of Malawi (Sakala et al., 2009). It has been suggested that since N content of all three popular green manure species is above 4%, it allows fairly rapid release of N and perhaps other nutrients to the maize upon incorporation. Both, slow and rapid release of nutrients seems to have its own advantage during farming. Timing of green incorporation could be adjusted to suit the requirements. Rapid release of N and its accumulation could make it vulnerable to leaching, if it is not absorbed by maize crop synchronously and in matching quantities. In addition to advantages derived in terms of soil nutrients, green manures planted as intercrop or strip crop may help in improving water use efficiency (WUE). According to Chirwa et al. (2007), green manure crops like gliricidia and pigeonpea improves WUE of the system in addition to nutrient supply.

According to Mtambanengwe and Mapfumo (2006), improving nutrient holding capacity and soil quality *per se* is an important aspect within the maize farming zones of Zimbabwe. The sandy soils with <100 g clay kg^{-1} is prone to loss of nutrients via rapid respiration, erosion, and leaching. Green manures and several types of organic amendments are in vogue among the maize farmers. They all aim at setting the nutrient dynamics more congenial to maize growth and productivity. Field trails at several locations in Zimbabwe have shown that green manures like *Calliandria* and *Crotalaria* are relatively more effective in improving organic-C, N, and other nutrients. At about 3 weeks after incorporation into soil, *Crotalaria juncea*, and *Calliandria calothyrsus* released N equivalent to 13 and 24% of total N supplied to the crop. We may note that rapidly released N was at the same time vulnerable to loss via leaching. Soil-N estimates indicated that NO$_3$-N moved quite rapidly to zones much below maize roots. Mtambanengwe and Mapfumo (2006) have reported that among the field experiments in various locations of Zimbabwe, best response was obtained when *Crotalaria* was applied at 4 t ha^{-1}, although it was still prone to leaching loss. In locations such as Makoholi, Mtambanengwe et al. (2006) found enormous response of maize to green manure. It was equivalent 9 folds over control. In some locations, application of residues or saw dust induced N immobilization at rates much greater than green manures. They suggest composting or application of organic amendments much earlier before planting. Fresh leaf material from green manure trees composted for a while could be useful in improving nutrient status of sandy soils. Composting improves C:N ratio of the manure.

Kimetu et al. (2002) have pointed out that decline in soil fertility is a major cause for low productivity of maize in Kenya and adjoining areas in East Africa. In order to restore soil fertility, especially SOM content, farmers tend to use leaves/twigs from a range of green manure shrubs and trees. Commonly used green manures in Kenya are *Tithonia, Senna, and Calliandra*. Maize biomass yield due to green manure incorporation differed based on green manure source. The woody perennials are a major source of nutrient rich leafy green manures. In succulent form, they are comparable to green manures derived from annuals. Yet, synchronizing green manures inputs in order to induce rapid decomposition of leafy manures is essential. Such green manures release large quantities of N and other minerals upon decomposition. Yet, to maximize the efficiency, it is generally recommended that green manure be mixed with small quantity of inorganic N sources such as urea. The organic matter decomposition rates and N release was better with *Tithonia* compared with *Senna* or *Calliandra*. In most farms, green manure plus inorganic fertilizer-N has provided best N release into soil and its recovery by maize crop. Following is an example:

Green Manure	Control	*Tithonia* + Urea	*Tithonia*	Urea	SED
N Supply (kg N ha^{-1})	0	60	60	60	
N Recovery	93.3	114.3	97.6	131.9	±16.4

Source: Kimetu et al., 2002.

Synchronizing availability of N from green manure with crop stage that involves rapid N recovery avoids undue loss of soil-N derived from the green manure. The maize biomass yield was 3.3, 3.6, and 3.9 t ha^{-1} with *Calliandria, Tithonia*, and *Senna* respectively. In most farms, application of biomass from the above green manure species plus 60 kg N ha^{-1} as urea provided maize grain yield advantage. Zingore et al. (2003) have also expressed that, in many locations within the maize belt of Zimbabwe, despite promotion of green manure prunings as a means of C and N supply to soil, lack of synchronization between mineralization pattern and rates affects crop productivity adversely. High quality succulent prunings from *Tithonia* released almost 70% N and 30% C in 84 days after application to soil. In comparison, low quality pruning from *Flemingia* sp released only 25% N and 5% C during the same period. Mixing prunings from different tree species, and application of at least 5 t prunings ha^{-1} resulted in higher grain yield.

A year long field trial was conducted at Maseno in Kenya using three different green manure species namely *Leucana leucocephala, Sesbania sesban*, and *Cajanus cajan*. The leaf droppings and twigs from these green manure crops were added to soil every 2 months during a year. Total green manure supply ranged from 4 to 8 t biomass ha^{-1}. Soil fertility changes were monitored by chemical analysis and by planting maize and bean. Results suggested that Leucana and Sesbania improved maize grain yield by 76% over a control. The residual effects of green manure supply lasted for 3 consecutive years (Onim et al., 1990).

In the tropics of Southeast Asia, green manures are frequently applied to soils that support maize production. Some of the common green manure species in Southeast

Asia are *Leucana, Sesbania, Gliricidia, Macuna,* and *Crotalaria.* Hairiah and Van Noordwijk (1989) have reported that incorporation of green manures prepared from 3-month old *Macuna pruriens* or *Crotalaria juncea* into alluvial soils of Java in Indonesia improved maize grain yield by 2 times over control plots. Organic-C and N contents in soil were improved leading to better crop growth and yield formation. Heide and Hairiah (1989) have reported that in Indonesia, N input due to incorporation of *Crotalaria juncea* and *Macuna utilis* is high at 198 kg N and 71 kg N ha^{-1} respectively. Green manures prepared from *Macuna* decomposed faster and induced higher maize yield. The extra N recovery due to green manure addition was 121 kg N ha^{-1} for *Macuna* and 147 kg N ha^{-1} for *Crotalaria juncea.* It has been suggested that rapidly decomposing green manures like *Macuna* might show effects on soil organic-C, N, and other minerals quickly. In case of slowly decomposing residues such as *Crotolaria*, the effect may show up after a lapse but get expressed for longer duration. Sangakakara et al. (2004) report that incorporation of green manures such as *Crotalaria juncea* and *Tithonia diversifolia* has immense benefits to maize crop that follows them. A field left to fallow and not grown with green manure served as control. During a 3-year period, incorporation of green manures improved soil physical properties, organic-C, and inorganic nutrients such as N, P, and K. Application of green manures improved rooting and below ground C contents. The root length, spread in soil and nutrient contents of roots improved perceptibly due to green manure inputs.

Cover Crops Affect Nutrient Dynamics

Cover crops are common to many of the maize cropping zones of the world. Such cover crops serve the farmer usefully in terms of fertilizer-N, residue recycling, and weed control. There are indeed several reports that depict N inputs and nutrients recycled by leguminous cover crops. Let us consider an example. Zotarelli et al. (2009) compared contribution of crops like rye (*Secale cereal*) and vetch (*Vicia villosa*). It is said that rye is an efficient scavenger of soil-N. Planting rye and vetch mix may improve N recovery and its availability to maize. A mixture of cover crops sometimes lessens N leaching loss. Basically, the sweet corn growth, N accumulation and yield were increased due to cover crops. The extent of N removed by vetch as a crop may vary. There are reports that vetch removes up to 257 kg N ha^{-1}. Much of it, about 70–90%, derived from biological N fixation. We may note that in warm climatic conditions N contained in the vetch is easily decomposable. It has a C:N ratio of 8:1 to 15:1. On an average, vetch crop may add 50–150 kg N ha^{-1} that could be utilized by maize. In the above example, contribution to the sweet corn ranged between 35 and 75 kg N ha^{-1}. Corn yield improved to 2.2 t ha^{-1} due to leguminous cover crop. In addition, cover crops reduced weed biomass by 2–36% compared with control fields.

In the Canadian prairies, cover crops grown following cereals are known to sequester appreciable amounts of soil-N, that otherwise gets vulnerable to loss via leaching and emissions. Reports from Agriculture-Canada indicate that oats, annual rye grass, oilseed radish, peas, and red clover are some of the most useful cover crops. Each of these cover crops may absorb up to 80–95 kg N ha^{-1} into their above ground herbage. Average N increment due to manure application is about 25 kg N ha^{-1} for the legume species and 40–50 kg N ha^{-1} for the cereals. Reasonably, about 40–60 kg N ha^{-1} is sequestered into stover of the cover crop (Stewart, 2005). It could be recycled. Cover crops literally lessen loss of N and C via emissions. They also reduce weed build up in the fields compared to an unplanted fallow.

Maize productivity in some parts of West Africa has declined and reached below 1.0 t ha^{-1}. Such disappointing production trends have been attributed to lack of proper management of N dynamics in the fields. Irregular nutrient recycling practices have impaired availability of other mineral nutrients too. In the semi-arid West Africa and tropical regions around Ivory and Gold Coasts, farmers have generally practiced cultivation of cereals such as maize and several types of grain legumes like cowpea, soybean, groundnut, and pigeonpea. Such a system improved both maize yield and soil fertility (Sogbedji et al., 2006). During recent years, cover crops that are non-food legumes such as *Macuna* or Lablab have also been preferred. Interjection of crop sequence with such non-food legumes means reduction in grain harvests. In this case, it lessens grain legume production. However, use of non-legume cover crops may offset such a reduction by enhancing biomass and grain yield of maize. Reports from International Fertilizer Development Center (IFDC) researchers have shown that use of short duration cover crops like sesbania, leucana, and pigeonpea all improved soil fertility, especially soil-N and maize grain yield. Sogbedji et al. (2006) compared effects of 2-year cycles of cover crops like *Macuna* (*Macuna pruriens*) and pigeonpea (*Cajanus cajan*) with maize. Cultivation of *Macuna* or pigoenpea as cover crop reduced N and P requirements by 37 and 32% respectively during the succeeding season. Analysis of N budgets for 2-year cycle indicated a net gain of >400 kg N ha^{-1} under maize-cover crop systems. The soil NO$_3$-N improved by 57% in the continuous maize fields, but only 39% in maize-*macuna* fields. Fields under cover crop experienced low amount of N loss (<20%). Interestingly, *macuna* or pigeonpea cover crops replenished substantial amount of soil-P, when stover was recycled. It is said that in countries like Togo, Benin, or Ghana where farmers experience resource constraints, cover crop systems are most useful in restoring and improving soil fertility. They also provide biomass, grain and economic advantages to farmers.

Let us consider yet another example. It pertains to Tamale in Northern Ghana, where in, farmers grow non-grain legumes such as *Macuna*, *Crotalaria*, or *Calapogonium* as cover crops. Reports by Fosu et al. (2004) suggests that Cover crops generated 5–15 t biomass ha^{-1} equivalent to 115–306 kg N ha^{-1} that could be recycled effectively during maize phase of the rotation. Amendment with cover crop residue improved maize grain yield by 2 t ha^{-1} compared to fields under fallow without cover crops. The mineralization rate of cover crop residue is an important factor that affects nutrient release into soil and its availability to maize crop that succeeds. Among the cover crops, usually sun hemp residues decomposed rapidly and released sizeable amounts of nutrients. The nutrient recovery by succeeding maize seems most crucial with regard to extent of nutrient advantages. Fosu et al. (2004) found that cover crops were as effective as application of inorganic fertilizers (N, P, K,) (Figure 5.1).

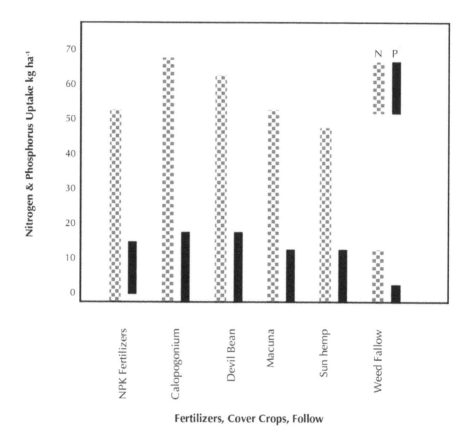

Figure 5.1. Comparison of nitrogen (N) and phosphorus (P) uptake by maize grown with fertilizers or after cover crop or fallow at a farm near Tamale in the Savanna zone of Northern Ghana.
Source: Figure drawn based on Fosu et al., 2004.
Note: N, P, and K fertilizers were applied at 60:17:33 kg ha⁻¹. Cover crops were recycled entirely as above ground dry matter.

Animal Manures

Habtesellassie et al. (2006) examined the influence of dairy-waste on soil microbial process, nutrient release, and productivity of silage maize. They compared application of ammonium sulfate, dairy-waste compost, or dairy waste-liquid as N sources. As a consequence of dairy-waste application, soils contained higher levels of SOC, N, nitrate, available-P, and available-K. Soil organic-C pool almost doubled in 5 years. Nitrate accumulated at soil depth of 60–90 cm. Soil microbial activity was enhanced due to dairy-waste inputs. Nitrifiers were 3 folds more in soils of dairy-waste treated plots than control ones. The silage yield from maize fields treated with dairy-waste was higher than control. It was suggested that to get maximum benefits from dairy-waste application, it should be timed appropriately, so that nutrient release matches the crop's needs.

Hountin et al. (1997) reported that application of liquid pig manure (LPM) results in accumulation of nutrients in soil. Long-term supply of LPM for 14 years has shown that C, N, P, concentration increased significantly in the upper layers of soil 20–40 cm depth. The extent of nutrient accumulation and improvement in maize crop production depended on LPM supply.

Cattle manure is an important organic source that recycles both C and mineral nutrients in integrated farms. In the European plains, cattle manure plus inorganic fertilizers are major source of nutrients supplied to maize farms. Cattle slurry and crop residues effectively recycle C and N. It avoids undue loss via emission and leaching. In the maize farming belt of Italy, about 2% SOC is lost to atmosphere as CO_2 emission. Further, amount of nutrients retained in soil depends on the type of organic manure applied and its timing. For example, if FYM is applied 46% C and 44% N is retained, with cattle slurry only 26% C and 11% N is retained and if crop residues and roots are recycled only 28% C and 10% N is retained in soil. Field tests have shown that application of a combination of cattle slurry, crop residue, and inorganic sources is superior compared to urea alone in terms of forage and grain yield (Bertora et al., 2009). Maize is predominantly grown for silage in Italy. Fields are usually supplied with various cattle slurry preparations or FYM. In order to quantify amount of C, N, and P derived from different animal manures, especially cattle slurry, Grignani et al. (2007) compared two levels each of cattle slurry and FYM with an untreated control. They reported that maize crop responded significantly to inorganic fertilizers, FYM, and animal manure. Nitrogen supplied by organic manures ranged from 215 to 385 kg ha^{-1}. Soil N, P, and other nutrients were rapidly depleted in plots not provided with organic manures. However, when FYM was used, soil C and N surplus was 18% and 45%. The soil C and N build up from cattle slurry was 9% of C and 25% N. Cattle slurry was efficient in terms of nutrient cycling within the integrated farm. At the same time it improved soil fertility and maize forage production.

Cattle slurry is frequently used during maize farming in Netherlands. The N fertilizer value of cattle slurry seems to vary based on time and method of its application, inherent soil fertility status and crop genotype. According to Schroeder et al. (2007), the N equivalence of cattle slurry is greater when it is used for the first time on the maize field. Subsequent applications may not provide similar effect on maize growth and yield.

Poultry litter is an important source of nutrients to silage maize grown in Southern Europe. It is supposedly a cheaper alternative to costly inorganic fertilizers. Poultry residue recycles large quantities of nutrients into maize fields. Juan et al. (2007) have reported that nutrient balance, N recovery, and N-use efficiency improved in fields treated with poultry litter. After 3 years of poultry litter inputs, there was a slight increase of nitrate in the lower horizon of soil. Poultry litter supply increased silage maize production by 1.2–2.8 t ha^{-1}.

In West Africa, organic matter inputs are a major source of nutrients to maize crop. On the sandy soils, organic amendments are crucial factors that regulate nutrient availability, root growth, biomass, and grain production trends. In fact, subsistence farming trends are common in most areas. Therefore, organic amendments are essential in order to sustain soil fertility. Recycling crop residue and animal manure restores soil fertility, nutrient dynamics, and crop productivity. Field trials in humid West Africa have shown that application of leaves and animal wastes recycled sizeable quantities of C and N (120 kg N ha^{-1}) into maize fields. Leaves decomposed rapidly to release C and N into soil. Application of inorganic fertilizers along with leaves or other organic amendments improved mineralization rates. The N uptake improved by 31% and dry matter by 10% over control (Ikpe et al., 2003).

Several types of manures derived from animal sources are utilized by maize farmers. In most of them, 50–75% of total N is bound and occurs as organic N (R-NH$_4$). Such manures have to be mineralized via biochemical process to release N contained in them. The remaining fraction of NH$_4$ is vulnerable for loss via volatilization. The degree of immobilization of nutrients contained in animal manures depends on its C:N ratio. Addition organic sources high in C:N ratio induces immobilization. Addition of animal manures also induces build up of soil microbes and their activity. Among the animal manures adopted by Kenyan farmers, cattle manures showed most rapid decomposition and release of nutrients into soil. Mtambanengwe et al. (2006) found that cattle manure was medium in quality with regard to C:N ratio. They suggest that high rates of animal manure application or mixtures of animal manures and inorganic fertilizers improved nutrient release into maize fields. It overcame the loss of N due to rapid immobilization.

Industrial By-products, Wastes, and Effluents

Maize farmers in US Corn Belt utilize a wide range of industrial by-products and wastes to prop their fields with extra nutrients. A report by Walkowski (2003) suggests that recycled wallboard is available in plenty in the Corn Belt States of USA. It contains sizeable amounts of nutrients and is especially richer in secondary nutrients—Ca, S, and Mg. Gypsum and sulfur compounds are recommended routinely in many locations. Hence, it could be useful to add decomposed wallboards in powdered or granulated form to crop fields. On an average, Gypsum-Wallboards contain 160360 ppm Ca, 133821 ppm S, and 8475 ppm Mg in addition to small amounts of micronutrients like Zn (12 ppm), B (43 ppm), Mn (44 ppm), Fe (885 ppm), Cu (7 ppm), and Cl (194 ppm). In Wisconsin, about 100 kg Ca could be supplied through 450 kg powdered wallboard material. Application of 16 t ha^{-1} powdered wallboard improved soil S status. According to Walkowski (2003), soil test values <50 ppm Mg are considered deficient. Such a soil requires supplementation with Mg either as mineral compounds or via residues and wastes. Chemical analysis suggests that crushed wallboard applied at 2–4.5 t ha^{-1} on sandy Mollisols suffices to improve Mg content. There is a strong need to evaluate a large array of such recyclable material rich in secondary nutrients. Such residues could be used in combination with crop residues richer in major nutrients. There is also need to understand the rates of decomposition of such recyclable residues and nutrients that become available to maize crop.

Rice mill waste (RMW) serves as a useful organic manure during maize cultivation in Sub Saharan Africa. In the Savanna region of Nigeria, farmers tend to harvest very low maize grain yield, mainly due to soil degradation and loss of soil fertility. In order to restore SOM and soil quality, farmers have to supply fairly large amounts of organic manures. Further, if farmers intend to intensify maize production they have to invariably add large quantities of organic manure along with inorganic fertilizers. According to Schultz et al. (2003), RMW applied to field prior to maize planting improves crop growth and productivity. The fertilizer equivalence of RMW may vary depending on source, time, and method of application and extent of decomposition possible. On a farm with degraded Alfisol, about 55% N, 40% C, and 90% P contained in RMW had been released for the maize crop to absorb. Application of RMW improved grain production. For example, on Alfisols, if control plots yielded 0.55 t grain ha^{-1}, RMW (unburnt) supplied at 10 t ha^{-1} improved it by 147% over control. In contrast, application of burnt RMW did not improve maize grain yield appreciably. A cumulative supply of unburnt RMW of 30 t ha^{-1} improved SOC content significantly from 0.77% initially to 1.3% at the end of 3 years. It was concluded that application of 10–15 t RMW ha^{-1} improved both maize grain/stover yield and improved SOM status. Incidentally, improvement of soil C sequestration is very important in Sub Saharan Africa.

Yeast distillery waste and press mud are common industrial by-products that add organic-C and other nutrients into the maize agroecosystem. Both of these by-products contain 1–1.5% N and can partially substitute N requirements of maize crop. Following is the list that compares distillery wastes and press mud with FYM:

Characteristics	pH	EC	Org-C	N	C:N	P	K	Fe	Cu	Zn	Mn
		dSm	%	%	ratio	%	%	----------- ppm ------------			
Distillery Yeast Sludge	7.4	15.8	40.7	1.4	28.1	0.21	2.14	148	38	134	76
Press Mud	6.4	3.12	39.4	1.1	35.2	1.40	0.95	35	45	35	25
Farm Yard Manure	7.1	0.82	48.8	0.6	84.1	0.18	0.40	20	34	12	28

Source: Rajeshwari et al., 2007.

Biofert is an industrial by-product derived during fermentation of lysine and other amino acids. It is characterized as a by-product that contains 6% N and several other ingredients that include amino acids, vitamins, enzymes, micronutrients, etc. (Pepo, 2001). *Biofert* has been tested for its influence on maize productivity in different soil types, locations, and seasons. Field evaluations have shown that *Biofert* affects soil-N dynamics. It contributes large amounts of N based on dosages adopted. The agronomic efficiency is almost similar to N fertilizers commonly used by maize farmers. Interaction between irrigation frequency, in other words soil moisture status and *Biofert* was crucial. Nitrogen released into available pool and its utilization by maize was dependent on soil moisture status at the time application of *Biofert*. On Chernozems, application of 120–190 kg N ha^{-1} using *Biofert* was optimum and it produced 11.0 t grain ha^{-1}.

Organic Formulations

Maize farmers use several types of organic preparations. Some of them contain a combination of organic wastes, microbes, and earthworms. They are sometimes termed as vermin-composts. Let us consider an example. "Revital Maize" is an organic formulation based on vermicasts. It is said that such formulations play a vital role in releasing nutrient slowly to soil. Application of vermin-composts has resulted in enhanced maize grain and silage productivity. A typical organic formulation made of worm casts and soil microbes may contain following microbial ingredients:

Total Bacterial Mass (mg g^{-1}):	249
Total Fungal Mass (mg g^{-1}):	238
Flagellates (number g^{-1}):	11772
Amoeba (number g^{-1}):	94235
Ciliates (number g^{-1}):	2836

Source: http://www.revitalfertilisers.co.nz/files/11852%20Maize.indd2005-6.pdf

The extent of nutrients supplied by such organic formulations may vary based on a range of factors related to source, composition, soil fertility status, etc. In case of "Revital Maize", application of 3 t ha^{-1} releases 50 kg N, 49 kg P, 26 kg K, 7 kg S, 12 kg Mg, and 171 kg Ca. In general, application of such organic formulations improves soil nutrient contents and quality. It enhances root growth and activity. It mainly improves nutrient cycling in the field. Field trials in USA have shown that application of such vermin-based organic formulation improves maize grain yield by 17–21% and forage by 10–26%.

ORGANIC MULCHES AND NUTRIENT DYNAMICS

Maize farmers tend to recycle a portion of nutrients by incorporating crop residues from previous season. They may also replenish soil with minerals by burning the residues *in situ* in the field. Another important procedure they follow is to use crop residue and several other farm or industrial by-products as mulches (Mtambanenegwe et al., 2007; Table 5.3). Maize farms in intensive farming zones of "US Corn Belt" and other locations often adopt organic mulching. In addition to release of nutrients, organic mulches are useful in recycling nutrients *in situ*, conserving soil moisture and regulating temperature regime in the surface layers of soil. Farmers all over the world leave or apply maize residues in the field because of following advantages:

1. Mulches protect soil from rains, sun and wind;
2. Mulches reduce soil erosion. It actually protects top soil from being eroded rapidly;
3. Mulches reduce runoff and conserve water that would otherwise be lost as surface flow;
4. Mulches reduce weed growth. They also suppress weed germination and growth;
5. Mulches reduce evaporation of soil moisture;
6. Decomposing organic matter and crop residues used as mulch, generally improve soil structure and aggregates;
7. Mulches enhance water infiltration and protect soil from sealing and crusting;
8. Mulching regulates soil temperature. It protects seeds and seedlings from frost damage and low temperature-induced seed dormancy. Mulching also protects

maize seeds/roots from high temperature, especially in areas prone to high temperature conditions, for example in Savannas of Africa or Dry lands of South India;

9. Mulching improves soil microbial activity;

10. Mulching improves rooting and root activity; and

11. Mulching aids in recycling nutrients with in the maize fields. It allows nutrients held in residues to percolate to top 40–60 cm soil layer.

Table 5.3. Response of maize crop to different types of mulches applied on Alfisols found in Nigeria.

Mulches	Grain Yield t ha⁻¹
Legume Husks	4.4
Soybean Tops	4.2
Cassava Stems	3.8
Pigeonpea Tops; Rice Husks; Saw Dust	3.7
Hemp (*Eupatorium*); Guinea Grass	3.6
Andropogon Straw; Pigeonpea Stem; Rice Straw	3.5
Mixed Twigs	3.4
Napier Grass; Kikuyu Gras Straw; Maize Stover or Cobs;	3.3
Oil Palm Leaves	3.2
Straw; Fine Gravel	3.1
Bare Soil (Control): Black Plastic Sheets	3.0
: Translucent Plastic	2.7
CD at 5%	0.8

Source: Okigbo, 1980.

Let us consider a few studies on mulches practiced in the Brazilian Cerrados. In this zone natural savannas have been progressively converted into large expanse of maize or soybean monocrops or maize/soybean intercropping systems. At present, about 30% of natural savanna vegetation seems to have been converted into regular crop land. The crop land is prone to loss of SOC due to repeated tillage. As a useful alternative, direct seeded mulch-based cropping (DMC) is practiced in most parts of maize belt in the Cerrados of Brazil (Corbeels et al., 2004). According to Copel et al. (2004) mulching techniques practiced in Brazil revolve around four basic agronomic procedures. These are—absence of soil tillage; direct seeding of maize; maintenance of mulch of maize residues; and adoption suitable maize–maize-based crop sequences. The retention of maize or other crop residues in the field is the key procedure that restricts runoff, soil erosion, and nutrient loss (Plate 5.2 and 5.3). Actually, mulch-based cropping systems have gained in popularity during past 15 years. It is aimed at reducing soil degradation and loss of SOC from the ecosystem. During early 1990s, DMC occupied only 1.0 m ha in Brazil but increased to over 15 m ha in 2001 (Evers and Agostini, 2001; Copel et al., 2004). No doubt, there is still large portion of Brazilian maize belt where repeated tillage induces loss of SOC as CO_2 via soil microbial respiration and tillage. Direct seeding and mulches improve C sequestration in soil. It is estimated that in the Cerrado region itself, DMC induces C sequestration ranging from 3.2 to 3.3 Tg/yr^{-1}. Field evaluation by Corbeels et al. (2004) has clearly shown that mulches improve soil C and N content perceptibly. Long-term accumulation of SOC and soil-N under DMC system, avoids loss of nutrients via erosion, enhances soil-C, improves soil quality, and consequently maize grain productivity. In a matter of 12 years under DMC, soil-C increased from 43 Mg ha^{-1} to 52 M g ha^{-1} and soil-N increased from 38 Mg ha^{-1} to 44 Mg ha^{-1}. It is believed that better rooting and rapid decomposition of senescing roots under DMC contributes to SOC. There was also a 40% reduction in SOC loss due to NT compared to repeated tillage systems. Over all, the C sequestration rate due to DMC was estimated at 0.83 Tg ha^{-1}yr^{-1}. The contribution of SOC by the region that adopts DMC in Cerrados of Brazil is calculated at 3.3 Tg yr^{-1} and it is equivalent to 12.2 Tg y^{-1} CO_2.

Plate 5.2. Maize crop under no-tillage system with mulch in the Pampas of Argentina.
Source: Dr. Eduardo Mulin, Facultad de Agronomia, Universidad de Buenos Aires, Argentina.

Plate 5.3. Maize seedlings growing in a field with residues from previous crop (wheat) retained on soil surface.
Source: International Maize and Wheat Center-CIMMYT, Mexico.
Note: Wheat residue is retained and spread in between rows to serve as mulch and supply organic-C and nutrients to maize crop that follows.

There are indeed innumerable reports regarding improved nutrient and water retention in upper layers of soil and consequent beneficial effects of mulches on crop productivity. Following is an example from maize fields in Brazil:

Effect of mulches on Maize–Cowpea rotation practiced on Oxisol in Eastern Amazonia of Brazilian

	Grain Yield (t ha⁻¹)	
Mulches	**Maize**	⟶ **Cowpea**
Rice Husks	4398	1498
Maize Cobs + Husks	4863	2101
Bare Soil	3539	1169
C.D. at 5%	987	281

Source: Schoningh and Alkamper, 1984.

Field trials and simulations indicate that effect of mulch application or retention of crop residues in field is more pronounced when crops are under low input systems. According to Copel et al. (2004), mulch effects could be site-specific to a certain extent. Trials at La Tinaja in Mexico, a semi-arid site, has shown that surface residue limits water loss as runoff and increases grain yield by at least 2 folds reaching above 1.0 Mg ha⁻¹. At the other site situated in humid tropics of Planaltina, in Brazil, they found that crop residues did reduce water loss via evaporation and surface runoff, but much of this advantage was offset by perceptible loss of water through drainage and percolation. Deep drainage induced nutrient leaching. In fact, in humid Planaltina, a large pool of nutrient rich organic mulch placed on surface was prone to nutrient leaching. Therefore, nutrient recycling measures were needed to overcome loss of nutrients to subsurface or to distant locations via drainage. Clearly, the grain yield increases were marginal due to nutrient leaching. Therefore, surface application of maize crop residue as mulch is common in many parts of South America.

Now let us consider examples from Asia and elsewhere in Fareast. Mulches are known to reduce soil erosion and nutrient loss both in the hilly and Tarai regions of Nepal. The maize crop sown in *kharif* suffers most during the months May to September when rains are at peak intensity. Mulching and reduced tillage are excellent measures that retard soil and nutrient loss from the maize agroecosystem. Mulching reduces soil erosion by 50–66% compared with control. For example, soil loss during May was 245 kg ha^{-1} in fields under farmer's practices and 93 kg ha^{-1} in those treated with crop residue-based mulch (Atreya et al., 2005). As a consequence of mulching, about 7–8 kg N, 10–12 kg P, 20–22 kg K, and 140–150 kg organic matter could be retained in top soil without getting eroded. The intensity of mulching has direct bearing on extent of soil erosion control. Generally, application of 5 t ha^{-1} mulch reduces soil erosion by 52–60% compared with control. Maize–wheat rotations are getting popular in Northern India. In the dry lands, soil moisture recedes rather quickly after the monsoon and it limits seed germination and early seedling growth during post-rainy season. Mulching seems to be useful in conserving soil moisture and dissolved nutrients. Field trials at Ludhiana in Punjab indicate that mulching *kharif* maize or fallow fields improve soil moisture status. Mulching with green twigs improved maize productivity. Wheat that succeeds yielded about 3–3.5 q ha^{-1} higher than control plots (Prihar et al., 1979).

SUMMARY

Maize agroecosystem extends into regions with soils that vary widely regarding organic-C content and quality. Maize cropping systems in sandy soils of West Africa thrive on soil with very low SOC (0.1–0.3% C). Majority of maize belts are supported by soils with moderate levels of SOC ranging from 0.5 to 2.5%. Maize belt thrives luxuriantly and produces over 10 t grain ha^{-1} on soil rich in SOC (3–5% SOC). Maize crop also thrives on peaty soils rich in organic-C (e.g., Podzols of Northern Russia). Maize cropping directly influences SOC status. Several factors related to crop, soil, and environment affect extent of C lost or sequestered into agroecosystem. Repeated deep tillage induces loss of SOC and results in soil deterioration. Therefore, farmers practice NT or reduced tillage system in order to avoid rapid loss of SOC. In addition to NT, procedures like mulching, stubble incorporation, crop residue recycling, cover crops, intercropping, application of green manures, and adoption of suitable cropping systems may all help in sequestering C within the maize belts. There is need to understand root exudation patterns and total C shifts from crop phase to soil through. It allows us to manipulate C dynamics and soil quality.

Maize is a highly preferred crop both due to its nutritious forage and its usefulness during biofuel production. This aspect may wean away a large portion of crop residue from the fields. It really reduces *in situ* SOC recycling. Therefore, farmers may have to carefully partition the amount of residues recycled into soil that used as animal feed and biofuel production. Maize genotypes with profuse rooting and better root:shoot ratio may be a method to improve SOC sequestration. Currently our knowledge contribution of maize roots to C sequestration is feeble. We should note that large amount of C is sequestered into below ground portion of maize agroecosystem via crop roots and their decay. Cultivation of profusely rooting genotypes may compensate for loss of crop residue to biofuel production. Maize genotype that selectively produces larger amount of stem and leaves could also be preferred in order to satisfy different purposes. Over all, it is clear that SOC content should be enhanced to improve soil quality.

KEYWORDS

- **Biofuel**
- **C:N ratio**
- **Nitrogen fertilization**
- **Organic matter recycling**
- **Soil organic carbon**
- **Soil organic matter**

Chapter 6

Maize Roots, Rhizosphere, Soil Microbes, and Nutrient Dynamics

---

INTRODUCTION

Maize root growth and spread has immense influence on nutrient dynamics and productivity of the agroecosystem. Firstly, soil nutrients and water are garnered almost exclusively through roots. Roots sequester a large fraction of C and other nutrients that would otherwise be lost from the ecosystem. To a certain extent, roots mediate the transfer and re-allocation of nutrients in a maize agroecosystem. At any given point of time, roots hold sizeable quantity of C and minerals that could be re-translocated to above-ground parts and for grain formation. Root growth and its architecture do influence maize crop productivity. In fact, simulations and historical data clearly suggest that architecture of root system and water/nutrient capture has direct effect on crop growth and grain formation (Hammer et al., 2009). Further, we should note that agricultural soils are heterogenous with regard to several aspects like moisture, aeration, temperature, pH, osmotic balance, redox potentials, soil microbial flora, organic, and inorganic nutrients. In nature, maize roots negotiate these factors as they grow, absorb nutrients, and translocate them to above-ground portion of maize plants. The above factors may operate at varying intensities in the rhizosphere and surrounding soils. In the natural soil, nutrient dynamics at the interface between soil and root is indeed crucial. We have to note that almost every small unit of soil nutrient has to transit via rhizosphere, root hairs, or mycorrhizae. On a wider scale, rhizosphere, root hairs, and mycorrhizae together coordinate and facilitate the transport of large amounts of nutrients, in both directions between above- and below-ground portion of maize agroecosystem. There is a need to investigate and compile details about maize roots; their growth pattern and nutrient absorption trends; nutrient accumulation and re-translocation patterns; rhizosphere effects; and soil microbial activity on nutrient dynamics and productivity of maize belts. Further, we should realize that maize genotypes exhibit wide variation with regard to root morphogenetics and physiological functions. Therefore, every change of maize genotype may potentially impart definite alteration in rooting, nutrient accumulation in roots, recovery into above-ground phase, and finally nutrient dynamics *per se* within the agroecosystem.

MAIZE ROOTS, THEIR GROWTH, AND NUTRIENT DYNAMICS

In agricultural fields, maize root growth and activity are indeed influenced by innumerable interacting factors, each with varying degree of intensity, at various stages of the crop and for different lengths of period. The composite effects on root growth is important considering that much of nutrient dynamics, especially translocation aspects in soil phase and transit up to stem is highly regulated by crop roots, their formation, senescence, and activity. Let us consider a few examples pertaining to different factors regulating maize root growth.

Soil Physico-chemical Properties and Maize Root Growth

Root growth is actually influenced by soil texture and proportions of silt, clay, and sand. Soils with greater fraction of clay and silt possess higher water holding capacity. In such cases, soil compaction is an important factor that influences maize root distribution in the profile. The rooting depth is particularly affected by soil compaction. Soil factors such as bulk density (BD) that increases soil compaction impedes root growth. The greater BD under no-tillage (NT) tends to retard root growth in the upper 10–25 cm layer. Soil temperature in root zone is an important factor that affects root elongation, thickening, and nutrient absorption trends. Engels (2005) has found that maize root growth, nutrient demand, and shoot growth too was reduced if root zone temperatures were held low at 12–16°C. The concentrations of P and K in the root remained constantly low if root zone temperature is 12–16°C.

Maize root growth specifically its rate of elongation, nutrient absorption, and translocation rates are key factors that determine nutrient dynamics that ensues in the rhizosphere and immediate surroundings. According to Bloom et al. (2005), maize root growth or elongation is influenced by a range of phenomena that operate at soil-root interface. In the heterogenous soil conditions, tropisms such as gravitropism, hydrotropism, thigmotropism, and chemotropism may all influence rooting pattern appropriately. Nutrient distribution and soil acidity or pH gradient may be crucial in inducing chemotactic response and regulating root growth rates. In most of the maize farming belts, fertilizer-N inputs are practiced to induce crop growth. As such, inorganic and organic-N in soil are depleted at faster rate compared to other nutrients. Bloom et al. (2005) state that rhizosphere pH changes as roots absorb and assimilate inorganic-N. The assimilation of NH_4-N induces acidification; whereas depletion of NO_3-N induces alkalization of rhizosphere. Such changes in rhizopshere pH may have immediate impact on maize root growth. Alternatively, alterations in NH_4-N and NO_3-N may itself have direct influence on root elongation. Evaluation by Bloom et al. (2005) has revealed that exogenous inorganic-N input has greater influence on root proliferation.

Tillage and Maize Root Growth

Tillage is a primary land management practice that directly affects maize root growth and nutrient uptake. Currently, major trend world wide is to opt for NT systems, at least for a few years before resorting to deep tillage. It seems maize grown under NT systems usually produces thicker roots than those under conventional tillage (CT). (Holanda et al. 1998; Pereira de Mello and Mielniczuk, 1999; Qin et al., 2005). Thicker roots obviously means greater amount of C sequestration in roots per unit length. The root length density (RLD) in maize field could be either greater or similar in both conventional and NT systems (Qin et al., 2005; Table 6.1). It has been argued that maize crop grown under NT systems consistently produced greater quantity of roots in the upper layers of soil at 0–10 cm depth. It resulted in higher RLD in the upper layers. As a consequence, maize under NT generally explored and absorbed greater amounts of nutrients and water from upper horizons of soil (Pereira de Mello and Mielniczuk,

1999; Qin et al., 2005). The influence of tillage systems, especially those of NT or CT were not consistent at deeper layers of soil.

Table 6.1. Root length density and thickness as influenced by tillage system

Tillage System	Root Length Density		Root Thickness	
	(cm cm^{-3})		(mm diameter)	
	Location 1	Location 2	Location 1	Location 2
No-tillage	4.08	2.94	0.31	0.31
Conventional Tillage	5.30	3.02	0.29	0.30

Source: Qin et al., 2005.

Chassot et al. (2001) have reported that in Swiss Midlands, maize crops grown under CT possessed thinner roots at higher RLD compared to NT systems. Qin et al. (2005) have found that RLD was less, but mean diameter of roots was larger with maize grown under NT compared to those under CT.

Maize root growth is slowed under cool climates and in chilled soils. The NT systems generally render upper layer soils a bit cooler and this may impede the otherwise rapid root growth. Sometimes, poor growth of maize roots is also due to higher BD and compacted soils that are under NT. Soil compaction may induce production of lateral roots. However, Qin et al. (2005) have pointed out that greater number of macropores under NT may actually enhance maize root growth and compensate for retardation caused by higher BD.

Nakamoto (1997) states that biopores are tubular macropores burrowed by soil fauna or left by roots after they decay. Such macropores are known to provide congenial soil physical conditions for rapid and deep growth of roots. It is a kind of biological drilling of soil to allow better penetration of roots into subsoil. It definitely enhances exploration of subsoil for nutrients. However, it may not be entirely a substitute for deep plowing. Reports about biopores and their influence on root growth and nutrient dynamics in soil are feeble. A prior knowledge about root distribution in relation to nutrient recovery aspects by maize root system seems necessary, in order to compare the effects of biopores. Nakamoto (1997) have examined the effect of artificial vertical macropores on maize root growth. Presence of large number of macropores induces rapid root growth. Roots seem to cluster in areas where macropores are in greater number. Obviously, nutrient removal is also greater in regions with more macropores.

Soil Fertility and Nutrient Supply Affects Root Growth

Maize roots tend to grow rapidly into zones richer in nutrients and water (Plate 6.1 and 6.2). The influence of major nutrients inherently available in soil or that applied as inorganic/organic manures have been reported extensively. Generally, nitrogen input enhances root length and surface area of absorption, but root mass per unit length may itself be diminished (Costa et al., 2002). This leads to large number of thin lengthy roots. Using split root technique, Schortmeyer et al. (1993) have shown that ratio of

$NH_4^+:NO_3$ and soil pH has an impact on maize root growth, dry mater accumulation, root surface area, and fine root density.

Phosphorus supply to soil induces production of large number of thicker and lengthy roots (Chassot and Richner, 2002). However, there are suggestions that P supply need not always result in enhanced root length. Interaction of major fertilizers, specifically N × P and their influence on root growth is worth investigating. Maize roots are longer and thinner in soils rich in P and NH_4^+ ions. In the maize fields, roots were longer, thinner, and profuse in regions close to bands where fertilizers were applied. It seems root induction was pronounced nearer fertilizer bands but the effect was not seen on entire root system. This has been attributed to asymmetrical location of nutrients on one side of rows, in plots banded with fertilizers (Chassot et al., 2001). The amount of fertilizers banded or applied to soil may also affect the extent of root growth. Qin et al. (2005) have shown that banding fertilizers resulted in greater RLD and mean diameter at anthesis.

Soil Organic Amendments Improve Root Growth

Maize farmers use several types of organic supplements in order to enhance soil quality, stabilize soil fertility, and improve productivity. They also use several types of fermented organic products sometimes referred to as biodynamic preparations (Goldstein and Barber, 2001). For example, organic manure treated with cellulose degrading bacteria, composts, field sprays of cow manure and silica, cow manure and herbs, etc. Field trials in Wisconsin have clearly shown that treatment of maize crop with such organic amendments improves root growth by several folds. A few of the organic preparations had hormone-like effect on root proliferation. The extent of root growth stimulation due to organic supplements seems important. Such root growth stimulation improves nutrient recovery from soil. Maize grain yield improved from 5.58 t ha⁻¹ under conventional system to 7.15 t ha⁻¹, if biodynamic preparations were applied.

Soil Moisture and Maize Root Distribution

Nakamoto et al. (1992) has pointed out that maize farmers generally impart greater attention to depth of root system and its consequences on water and nutrient extraction from soil. However, knowledge about 3-dimensional distribution of maize root system seems more important (Plate 6.2). A comparison of different cereals like proso millet, pearl millet, barnyard millet, maize and wheat has shown that crops like pearl millet that are tolerant to drought may produce roots profusely to explore soil better. Among the cereal species tested by Nakamoto et al. (1992), RLD was greatest with foxtail millet followed by maize, proso millet, adlay, and barnyard millet in that order. Maize and foxtail millet produced greater amounts of roots even in deeper layer of soil (>50 cm), compared to barnyard millet that tended to produce shallow but laterally well spread out root systems. The root length and distribution also varied based on the degree of susceptibility of cereals to water logging. There is no doubt that root distribution has direct impact on nutrient recovery and C sequestration. We should note that root distribution is influenced both by environmental parameters and genetic nature of maize genotype. For example, crop species such as fox tail millet, maize, or proso millet differed with regard to depth and width of root spread under similar conditions. Maize genotypes differ markedly with regard to several traits that affect root growth and spatial distribution. Currently, there are techniques that allow nondestructive study/ identification of genotypes with root characteristics that confer adaptation to various environments (Van Beem et al., 1998).

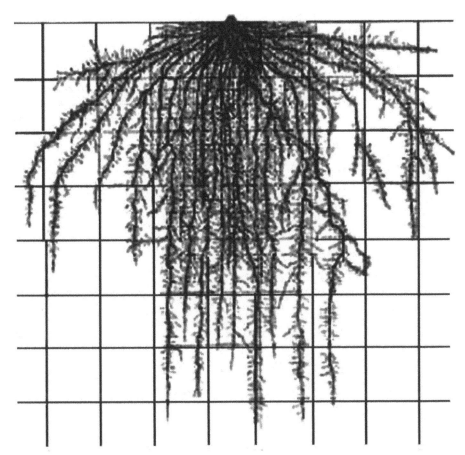

Plate 6.1 Depiction of a typical root system of mature maize plant.
Source: http://www.soilandhealth.org/01aglibrary/010139fieldcroproots/fig85.jpg
Note: Fibrous roots appear in whorl and emanate from the base of the stem (hypocotyle region). The fibrous or adventitious roots produce root hairs profusely.

It is interesting to note that maize roots garner water from different depths in soil at different stages of the crop. Of course it is highly dependent on root and soil moisture distribution. It is said that during a growing season, top 30 cm of soil provides about 40% of water requirement of maize crop. Next, 30–60 cm depth provides for 30% of water requirement of maize crop and 60–90 cm depth provides about 20% of water requirement. Deeper layers beyond 90 cm may provide the remaining 10% of water requirement. We should note that soil moisture extraction pattern is intricately linked with nutrient absorption since almost all of nutrients are extracted by roots in dissolved state. In other words, root distribution and water depletion pattern govern below-ground nutrient dynamics to a great extent.

Plate 6.2 Distribution of maize roots in a soil profile.
Source: Dr. Carina R. Alvarez, Universidad Buenos Aires, Argentina.
Note: Maize root system spreads 1–2 m in radius and depth in the soil profile. Above example depicts maize roots in the profile of a Vertic Argiudoll utilized to produce maize in the Rolling Pampas of Argentina. Fibrous roots are well spread. They explore large quantities of water and nutrients. They also harbor large number of microbial species. Roots support asymbiotic N-fixers in the rhizosphere. Maize roots also form symbiotic association with Arbuscular Mycorrhizas that help in P absorption.

Maize crop grown in dry land areas often encounters intermittent drought stress. A short duration of dry spell is common in many regions of the world, where maize is cultivated. The rapid adaptation of roots to changes in soil moisture distribution is crucial in such conditions. The root architecture and activity needs to swiftly change course in tune with soil moisture availability. In this process, nutrient dynamics is intricately connected to rooting pattern and moisture absorption. Sharp and Davies (1985) reported that soil drying affects moisture and root distribution. Maize roots, generally penetrated up to 60 cm depth in soil profile, if the crop is watered. The greatest density of roots gets localized at 20–40 cm depth. Withholding water altered the rooting pattern. Generally, root production in the upper layers gets reduced due to moisture depletion. However, roots penetrated deeper to explore moisture and nutrient. Therefore, after a lapse of time, rooting gets induced at greater depths of soil profile. Rooting density too increased at deeper layers in fields experiencing water depletion. It is obvious that intermittent droughts firstly affect soil moisture, then rooting pattern and as a consequence nutrient depletion patterns too are affected. In a different study at Hohenhiem in Germany, Engels et al. (2007) found that maize roots do respond quickly and promptly to dry and wetting cycles that are common in drier regions. They found that maize provided with optimum irrigation generally produces large amounts of roots in the top soil (0–40 cm depth) and an increase in rooting in deeper horizons is usually preceded with reduction or partial cessation of root spread in the top soil. Drying the top soil usually induces rapid spread of roots in the subsurface layers. Root growth and density in upper layers recovers rapidly upon rewetting of top soil. It is interesting to note that nutrient dynamics is rather intricately linked to drying and wetting cycles of soil. Drying of top soil reduces P in the above-ground portion of maize crop. However, concentrations of N, K, and Ca were not affected significantly due to wetting and drying cycles.

In addition to roots, drying and wetting cycles may also affect the rhizosphere and adhering soil particles. Soil particles that adhere to roots and aggregates have an important role in maintaining rhizosphere microbial flora, fine roots, and nutrient availability to root hairs (Czarnes et al., 2000). The adherence of soil to maize roots and the aggregate strength is dependent on soil textural traits such as clay, silt, and sand fractions. It has been suggested that soil aggregates and nature of soil adhering to root may actually affect nutrient diffusion and availability at the root-soil interface. Watt et al. (2004) have reported that mucilages elaborated at the rhizosphere, both by maize and microbial flora are important in achieving strong binding of soil particles in an aggregate.

Maize/cotton intercropping systems are common to many regions. They are especially popular in dry land areas of India, Great Plains, Pampas, etc. Invariably, both maize and cotton experience drought spells equally during a season. Comparative studies have indicated that maize root mass increases rapidly between 25 and 65 days. Under prolonged drought stress, root mass of maize was higher in the top soil (0–40 cm depth) during first 65 days compared to cotton. Maize produced 1610 kg ha^{-1} root mass but cotton accumulated only 450 kg ha^{-1} (Devereux et al., 2008). During a maize–cotton rotation, greater maize root mass at the top soil may result in increased macropores which aid water infiltration. This phenomenon may help the deeper root system of cotton to efficiently absorb water and nutrients held in the subsurface.

Maize is an irrigated crop in many parts of the world. Farmers adopt a wide range of irrigation methods. The intensity and frequency of irrigation too vary enormously depending on geographic conditions and crop yield goals. Most commonly, farmers adopt sheet irrigation on flat beds, furrow irrigation if sown on ridges and furrows, sprinkler systems and drip systems. Sometimes nutrients are also channeled using irrigation systems and this system is known as fertigation. Coelho and Or (1998) point out that detailed information regarding root distribution and water uptake patterns is crucial while judging water and nutrient recovery by maize grown under any of the irrigation systems. In case of drip irrigation, surface and subsurface rooting pattern and soil moisture distribution play vital roles in determining the nutrient and water absorption. Comparisons made on silt loams provided with surface and subsurface irrigation has shown that difference with regard to rooting and water absorption pattern are feeble (Coelho and Or, 1998). However, generally, irrigation techniques adopted do affect water absorption and nutrient dynamics in maize fields.

Oliviera et al. (1998) have stated that water distribution and root growth could be affected depending on furrow or surface irrigation systems. Roots were concentrated in the top soil and lateral spread of roots systems were pronounced under furrow irrigation. It was suggested that deep plowing could induce rapid penetration of soil by roots, so that water from subsurface regions too could be garnered efficiently.

Modeling has allowed us some understanding about maize root growth and its influence on water and nitrate recovery from soil. On Mollisols found in the Corn Belt, rooting pattern, density, nitrate leaching, and soil moisture distribution play a role in water and NO_3 uptake (Benjamin et al., 1996). Rooting patterns may induce differential NO_3 depletion trends. It seems bulk of soil explored is more important with regard to water and NO_3 absorption.

During practical farming, especially in tropics of Africa and Asia, maize is often intercropped with trees. *Gliricidia*, *Sesbenia*, and *Grevillea* are examples of tree species intercropped. Odhiambo et al. (2001) found that rooting pattern at different depths of soils of both intercrops was affected. In case of tree species, roots penetrated deeper than those from maize. Yet, root length densities of both trees and maize localized consistently at 20 cm region, because soil profile was recharged repeatedly with water at this depth. Roots concentrated where water availability was greatest. Further, roots of trees flourished most during early phase of crop season, but root density of both trees and maize was almost similar by the time season ended. Regarding root growth and activity, it was conclude that in a maize/tree intercrop, there is temporal separation but spatial separation is least, if any. We should note that maize root growth and activity is also influenced by several other factors such as root exudates, phytoalexins, intercrop competition, etc.

ROOT EXUDATIONS

Crop plants discharge an array of organic and inorganic compounds into rhizosphere through roots. Such root exudates bring about chemical changes in the rhizosphere.

They affect microbial population and their activity. The nutrient availability, turnover rates, and transformations are significantly influenced by the pattern of root exudations. Most commonly, rhizo-depositions include, lysates liberated by the autolysis of sloughed-off cells, exudates released passively via diffusion or actively as secretions (Neumann and Romheld, 2000). In case of most cereals, including maize, 30–60% of photosynthetically fixed C is translocated to the roots and a sizeable portion (70%) of this C is released into rhizosphere or soil environment. The extent of root exudation is generally affected by factors like light intensity, temperature, nutritional status of plants, impedance to rooting, soil characteristics, and microbial activity. Root exudates often include sugars, amino acids, flavin compounds, and propanoids (Walker et al., 2003). Most of these compounds are utilized efficiently as C source by rhizosphere microbes. A few of the compounds may directly influence the availability of nutrients to roots (e.g., organic acids). Several others may actually have inhibitory effect on soil microbes and in turn on nutrient dynamics in the rhizosphere. Some of the early reports on maize root exudates suggest that exudations are greatest at seedling stage. Among amino acids excreted, glutamic acid accounted for 60% followed by alanine. The exudation pattern of these two amino acids fluctuated during the course of crop growth. Stachyose was a major soluble sugar exuded by maize roots. Root exudates derived just prior to heading stage contained other sugars like glucose and fructose. Lactic acid was the most dominant organic acid in the root exudates. Matsumato et al. (1979) believe that such exudates have major role in supporting soil microbes, especially N-fixing microbes.

Typically, a maize root could be demarcated into root tip, elongation zone, maturation zone, and matured zone. The root tip has a cap and meristematic region. The root exudation from these regions of root varies depending on several factors related to plant growth stage, soil physico-chemical conditions, nutrient status of rhizosphere, microbial load and activity, etc. According to Walker et al. (2003), spatial pattern of exudation from the maize root system is one of the least understood aspects. Root exudation from maize root segment is not uniform throughout the axis. There are reports that root tips or regions slightly above it may release more of organic-C bearing compounds compared with other regions on the root. For example, release of phytosiderophores in response to dearth for Fe seems localized more in the apical region of root. Further, knowledge about spatial and physical localization of exudation regions on the root will help us judge plant-microbe relations and nutrient dynamics within the micro environment of rhizosphere better.

According to Sauer et al. (2006), microbiological activity and biochemical processes close to roots are immensely influenced by C and energy source. There is a continuous release of C bearing compounds through root exudates. The rhizodeposits are found at higher concentration nearer the root (rhizosphere) and decreases as we move away from root surface. The soluble compounds in the rhizodeposits generally diffuse faster and may be detected at farther distances away from roots. The rhizosphere extension and microbial activity supported by rhizodeposits is dependent on the soluble constituents. Studying the nutrient profiles of rhizosphere and adjoining soils may be helpful. It may allow us to predict microbial activity and nutrient dynamics that ensues in the root zone. It may also help us in understanding the nutrient recovery mechanisms better.

Kuzyakov et al. (2004) have reported that average CO_2 efflux in soil with in *Zea mays* field was about 22 g C m^{-2} day^{-1}. About 78% of total C efflux was attributable to roots and 2% to soil respiration. Maize plant transferred 4 g C m^{-2} to below-ground portion of the crop in 26 days of growth. A ^{14}C-based analysis of the root exudation pattern revealed that C released was retained at maximum concentration within 1–3 mm away from surface. Root exudations were in feeble quantities at 6–10 cm distance from root surface and found only in traces beyond 10 cm from root surface. The transit of root exudates from root surface to rhizosphere then on to bulk soil is highly depended on substrate, soil traits, moisture, and nature of root exudations. Some of these factors may play an important role in determining the nutrient dynamics in the rhizoplane and rhizosphere.

The quantity of root exudates liberated is an important factor that determines the extent of C compounds transferred to soil phase of the ecosystem. It is a good indictor for the extent of organic-C transfer from above- to below-ground portion of maize crop. For example, Liljeroth et al. (1994) found that 26–34% of net assimilated ^{14}C could be traced in rhizosphere of maize roots. In case of wheat, about 48–56% of assimilated C got translocated and exuded into rhizopshere. Experimentally, we have to quantify the amount of CO_2 produced in the rhizopshere. Respiration by both roots and microbial component of the soil releases CO_2 (see Neubert et al., 2000). Root exudates and energy contained in them is also utilized for N_2O production by soil microbes. In a 4-week old maize plant, Haller and Stolp (1985) found that, root exudation amounted to 7% of total photosynthates. Further, calculations indicate that 25% of organic matter flowing into the root system is actually excreated as root exudates rich in C containing compounds. Extrapolated to a field or agroecosystem it amounts to enormous quantity of organic-C shift from above- to below-ground portion. This amount of C and energy transfer indeed is utilized to sustain large number of soil microbes and their activity. In case, genotypes differ with regard to root exudates and pattern of exudates significantly, it leads to proportionate changes in C sequestered in soil and that available for microbial activity and nutrient transformations. Our ability to regulate root exudates and C transfer to soil seems to decide both crop yield and C sequestration in soil. Maize genotypes need to be screened, characterized, and utilized appropriately in different environments.

The quantity of root exudates librated in the rhizosphere is affected by microbial inoculations. For example, Laheurte and Berthelin (2006) found that inoculation of maize plants with phosphate solubilizing bacteria (*Enterobacter agglomerrans*) reduced exudation of organic compounds. As such, the exudation of C compounds by maize plant in the given experimental conditions was low at 3–6.4% of the total plant biomass. Plant growth promotion observed was attributable to elaboration of plant growth hormones by microbial strains inoculated.

There are some interesting facts regarding role of exudations of roots by maize in imparting tolerance to low-P availability in soil. We know that maize genotypes differ with regard to pattern and quantity of root exudations. Investigation using a P-efficient line-181 and inefficient line-197 has shown that firstly, P-efficient line-181 had larger amounts of lateral roots than P-inefficient line 197. Maize genotype 181 when grown under low-P conditions markedly reduced pH of the soil solution/rhizosphere; increased ATPase activity on root surfaces; but exudation of organic acids was low in 181 (Liu et al., 2004). It was concluded that root exudation pattern and enzyme profile on the root surface of a genotype was responsible for low-P tolerance to a certain extent. Reports from Spain suggest that organic acids exuded from maize roots do affect P availability to roots. For example, in the presence of citrate in the vicinity, maize roots garnered relatively larger quantity of P. However, effect of root exudations, especially organic acids on P solubilization was dependent on soil type. Phosphorus recovery by maize roots was better in acidic soil with high P sorption capacity like Haplic podzol. However, on Eutric Cambisol, root exudations did not affect P recovery from soil. In other words, extent of exudations, their composition, and influence on P dynamics were dependent on soil type.

MAIZE RHIZOSPHERE

There are indeed several reviews/books published periodically during past 5 decades, on rhizosphere and its physico-chemical characteristics; microbial composition, nutrient transformations in the vicinity of roots and finally its role in crop nutrition and productivity (Drinkwater and Snapp, 2006; Hinsinger et al., 2005; Pinton et al., 2007). According to McCully (2005) rhizosphere is a key functional unit that regulates plant root/soil and microbial interactions in the field. It has impact on nutrient dynamics at the crop-soil interface. Drinkwater and Snapp (2006) have highlighted the role of rhizosphere in crop nutrition, productivity plus in maintaining ecosystematic functions. In the present context, it is useful to highlight the role of maize rhizosphere in low input subsistence farming belts and equally so in the all important intensive maize cropping zones of North America, Europe, and China. Rhizosphere and nutrient transformations that occur are of great relevance to crop growth and productivity of maize cropping zones.

Maize roots support an assortment of soil microbes on their surface. They are called rhizoplane flora. Rhizosphere is defined as a region 2 mm thick region surrounding the root cylinder. It harbors, again, a variety of soil microbes that mediate large number nutrient transformations. Rhizosphere microbial load is usually higher than that traced in regions away from roots. This is generally attributed to biochemical substances exuded by roots. Many of the components in the root exudates are rich in organic-C and energy bearing substances.

Roots are major sources of SOC. They add large quantities of SOC and minerals upon decomposition. In addition, sizeable quantities of exudates are also released during active cropping season. Carbon compounds derived from maize roots is a major source of energy and C for soil microbes and their activity. Gregorich et al. (2000) have reported that during maize cropping, source of soil microbial C clearly shifts to maize roots. Investigation on continuous maize has shown that 55% of soil microbial biomass-C (MBC) was derived from maize roots. Quantification of SOC derived from maize root and that stored in soil microbes is important since it has great relevance to C dynamics at the rhizosphere and bulk soil. Liang et al. (2002) found that SOC, water soluble organic-C (WSOC), and MBC change rapidly to that derived from *Zea mays* (C_4) compared to those from C_3 plant species. During the course of a maize crop grown on Hapludolls, C_4 derived C in root zone ranged from 0 to 12.3% of total SOC, 0–30.7% of WSOC and 0–52% of MBC. Let us consider another example involving both C and N in root zone soil. Qian et al. (1997) quantified ^{13}C abundance of microbial biomass derived from *Zea mays* (C_4). In addition to C, they evaluated contribution of *Zea mays* root derived ^{15}N compounds to C and N dynamics in the root zone. Actually root released C accounted for 12% (210 kg ha[-1]) of measured C fixed by 4-week old maize seedlings. It decreased to 5% of C fixed by maize at maturity. Of the C_4-C, 18% was found in microbial biomass that was in active turn over. Regarding ^{15}N, about 16–23% was found in microbial biomass, 64% in inorganic-^{15}N, and 15% in non-biomass organic-N. The above reports indicate that major share of root zone C and microbial C is derived from maize crop and that such transition occurs in short span of time. Rhizosphere chemistry and transformations related to SOC derived from maize needs careful and detailed study. It may have an important role to play in regulating C and N dynamics in rhizosphere and root zone soil. There are clear reports that depending on C:N ratio and root exudation pattern, major C related and N related transformations such as mineralization, nitrification, de-nitrification, and emissions get regulated (Qian et al., 1997). Yoshitomi and Shann (2001) have shown that mineralization of C in the rhizosphere and roots are immensely influenced by root exudation pattern and consequent microbial activity.

Knowledge about nutrient dynamics at the interface between soil and rhizosphere of maize root is essential. There are reports on influence of so called "rhizosphere effect" on accumulation of major nutrients in the root zone soil. It seems that NH_4-N and NO_3 accumulated in the rhizosphere more than in the soil away from it. This phenomenon was influenced by the amount chemical fertilizer-N supply. Soil P, on the other hand got depleted from rhizospheres, if planting density was kept high. Soil available K accumulated in the rhizopshere soil and it was affected by inherent K fertility of soil. Considering the pattern of nutrient absorption, relatively high amounts of N and P were translocated through rhizosphere followed by K.

Fertilizer-N supply to maize fields may acidify the soil. The capacity of fertilizer-N to acidify soil is regulated by the quantity of N input and source. Basically, fertilizer-N undergoes hydrolysis and nitrification in soil. It results in release of H[+] ions. Simultaneously, NH_4-N is absorbed by maize roots. These processes result in acidification of rhizosphere. Absorption of NO_3-N, on the hand may slightly increase pH and alkalinize the rhizosphere. Over all, supply of fertilizer-N rapidly enhances acidity in the rhizosphere region (Rodriguez et al., 2008).

John et al. (2005) stated that knowledge about spatial and temporal variability of nutrients in the rhizosphere solution is crucial. However, for a long time, it was not easy to sample rhizosphere selectively and specially at different points of the root system. Recently, technique to sample rhizosphere solution *in situ* has been standardized. Periodic samples drawn from a mini-rhizotron based experiment revealed that nutrient concentrations in the rhizosphere are highly dynamic. For example, John et al.

(2004) found that during a 10-day period, rhizosphere solution P and K concentrations decreased by 24% and 8% respectively. Xu et al. (1995) have suggested that phosphate distribution in root zone is dependent on its rate of depletion in rhizosphere and surrounding area. A gradient may develop soon based on P diffusion rate in soil. Phosphorus accumulation zone may get created at 0.5 mm distance from root in about 15 days. While for Ca and Mg, gradients did not develop in the maize rhizosphere. Actually, mineral nutrient concentrations in the rhizosphere were highly dependent on crop species. For example, under similar soil conditions, maize accumulated 2–5 folds more K than other crop species such as switch grass. Characterization of nutrient dynamics at the rhizosphere may be useful while deciding on nutrient supply.

Micronutrient availability and uptake is also influenced by physico-chemical changes that occur in the rhizosphere. There are several reports suggesting that bioavailability of Fe, Zn, and Cu is influenced by root exudations and physico-chemical conditions that prevail in the rhizosphere (Cattani et al., 2006; Tao et al., 2003). According to Tao et al. (2003), during cultivation of maize, decrease in redox potential, increase in pH, dissolved organic carbon (DOC), and microbial activity may all result in altered uptake of micronutrients such as Cu. Such root induced alterations in physico-chemical properties could cause changes in Cu speciation. It is interesting to note that maize plants induce changes in the rhizosphere that preferentially enhances micronutrient bio-availability. Reports by Merckx et al. (2006) suggest that organic compounds released by maize roots into rhizosphere are capable of complexing different micronutrients. Comparison of soil extracts from rhizosphere and bulk soil showed that exudates in the rhizosphere helped in complexing more of micronutrients such as Co, Zn, and Mn. It is interesting to note that Cu bioavailability and its mobility is greater in rhizosphere compared with bulk soil. The Cu concentration in bulk soil was generally low at <0.06–0.02%. However, rhizosphere Cu was 3–6 folds higher than in bulk soil. At the same time, maize rhizosphere contained higher levels of DOC, at least 3 fold higher than that noticed in bulk soil. Such higher DOC content actually helps in complexing Cu and reduction of Cu uptake at toxic concentrations (Cattani et al., 2006).

Rooting pattern, rhizosphere effects, and nutrient absorption in an intercrop are worth understanding. Maize–peanut intercropping is a common practice in regions prone to dry land conditions. Such an intercropping helps in build up of soil-N through peanut-bradyrhizobial symbiosis. Roots and even rhizosphere may physically overlap frequently depending on spacing and rooting pattern of intercrops. Inal et al. (2007) found that intercropping helped in improving P, K, and Zn uptake. Improved nutrition was attributed to rhizosphere effect that includes microbial and chemical processes. For example, acid phosphates activity in the maize rhizosphere was much higher than in soil away from roots. There was 2 fold increase in phytosiderophores that helped in Fe nutrition of maize/peanut intercrops. In a different study, Inal and Aydin (2008) found that in a maize/peanut intercrop, peanut facilitated P nutrition of maize. On the other hand, maize improved K, Fe, Zn, and Mn absorption by peanuts. Again, rhizosphere interactions and root exudates from the intercrops were suspected to aid such positive interactions.

Let us consider an example from maize belt in China. Here, maize is frequently intercropped with legumes like faba beans or field beans. Song et al. (2007) have found that in several combinations that include maize, intercrops improved microbiological properties of soil. Examination of microbial community by denaturing the gradient gel electrophoresis showed that both diversity and population were affected. Intercrops resulted in significantly higher microbial biomass in the rhizospheres compared with those of sole crops. Intercropping enhanced organic-C in the rhizospheres and improved nutrient availability, especially N, P, and K. Long et al. (2007) point out that such inter-mingling and overlapping of roots, rhizosphere, and root exudates may have a prefer-ential effect on nutrient availability. Certain nutrients could be made more available to maize, by the root exudates and rhizosphere microflora of surrounding faba beans and *vice versa* could also be true. Antagonistic effect too should be anticipated depending on crop species that is cultivated along with maize. Firstly, Long et al. (2007) report that intercropping improved above-ground portion of maize crop. It could be partly at-tributable to below-ground interaction with faba bean rhizosphere. They used selective permeable membranes to separate out the effects of roots, rhizopshere, and exudate from intercrops. They identified several organic acids, sugars, and amino acids in the diffusates. They concluded that rhizosphere based facilitation of P nutrition was im-portant. Acidification of rhizosphere by microbes improved P nutrition. It resulted in a decrease of pH in the rhizosphere with a consequent increase in soluble-P. Carboxylates from root exudates could have chelated Ca, Al, and Fe. It leads to better P availability. Actually insoluble P gets mobilized. Further, they argue that higher phosphatase activ-ity in the rhizosphere might have decomposed organic-P into inorganic-P and improved available-P pool in soil. Over all, intermingling of rhizospheres of intercrops and ben-eficial effects, if any seems crucial in terms of nutrient dynamics in the below-ground portion of maize crop. It may have significant effect on productivity.

Recently, researchers in Germany have accumulated some interesting and useful information regarding effect of BT maize on rhizosphere microflora. As such fields with BT maize may affect a few non-target insects. Apart from it, BT maize, as it grows may affect by excreting BT toxin into soil and/or rhizosphere. It is difficult to trace BT toxin found in soil quickly. Suitable assays need to be standardized. We should note that BT toxin is also prone to soil degradation. In soil, decomposition of BT toxin (a protein) is accomplished rather rapidly. Micro-organisms found in the rhizosphere of BT maize may be different compared to conventional non-GM hybrids. It is said small amounts of BT toxin may persist in the rhizosphere and bulk soil for more than a crop season (FARC, 2004). Overall, it is felt that expression of BT toxin in plant roots lead to small/negligible structural changes in soil microbial community. This conclusion holds for rhizosphere microflora of BT maize. Therefore, its conse-quences on nutrient transformations, utilization, and accumulation pattern may be nil or imperceptible.

SOIL MICROBES AND NUTRIENT DYNAMICS

The rhizosphere and root zone soil from a maize crop supports a wide range of mi-crobes. The microbial population found in this region of the ecosystem differs with regard to species diversity, quantity, and activity. The microbial activity in the rhizosphere and root zone soils has immense direct influence on the nutrient dynamics in the immediate vicinity plus the entire crop. There are indeed several methods to assess microbial diversity, their activity, and relevance to maize crop nutrition.

Interactions between roots and microbes at the rhizosphere seems important considering that much of the nutrients that maize crop acquires actually moves through this small zone of 2.0 mm thickness surrounding the root. Rhizosphere harbors variety of microbes and the population is kept highly dynamic based on physico-chemical changes that occur due to soil heterogeneity and root exudates. Microbial population in the rhizosphere often responds to changes in the exudation patterns. Schonwitz and Ziegler (2007) reported that rhizosphere microflora of maize often consisted of short rods, actinomycetes, and few species of fungi. Rhizosphere microflora covered about 4% of micro-layer near the root hairs and 20% on older portion of root. The exudations from roots consisted of vitamins and sugars. The amount of exudations is generally enough to stimulate rhizosphere microbes. However, stimulation depends on specific requirements for nutritional factors. The rhizosphere microbial population, specifically diversity, population and activity seems crucial with regard to nutrient dynamics at the interface.

Soil microbes collectively termed "Asymbiotic N fixers" are harbored on rhizoplane, in the rhizosphere and bulk soil. The N dynamics in the root zone is immensely influenced by such asymbiotic N fixers. They convert atmospheric N into organic-N compounds, utilizing organic exudates from roots and SOM as energy sources. Most cereals and grasses including maize support asymbiotic N fixation in their rhizosphere. The extent of N fixation by soil microbes depends on several factors related to soil, crop genotypes, environmental parameters, irrigation, fertilizer schedules, cropping systems, etc. Asymbiotic N fixation as a scientific topic has been reviewed periodically by different authors. Earliest reports by Dommergues et al. (1972) clearly suggested that soil factors like moisture, pH temperature, texture, and fertility affected nitrogenase activity by asymbiotic N fixers. Nitrogenase activity depended on energy rich organic sources in soil and exudations from maize roots. Hence, peaks in nitrogenase activity in the rhizosphere often coincided with root exudation pattern. Therefore, asymbiotic N fixation is partly dependent on the photosynthetic activity of the crop. Previous studies indicate that depending on inoculum, microbial isolates used, crop genotype, agronomic procedures, and environment, N contribution by asymbiotic N fixers, like *Azotobacter*, *Azospirillum*, or there species may range from 20 to 40 kg N ha^{-1} season^{-1}. It is a useful contribution to the N dynamics, considering that maize belts are vast and if extrapolated it amounts to large addition of N into the cropping belts. Currently, several types of microbial preparations containing improved bacterial isolates and mixtures are available for application to maize fields. Nitrogen contribution via asymbiotic microbes is of great value to N dynamics that ensues in subsistence farming belts. Resource poor farmers may sometimes entirely depend on N contributed via asymbiotic N fixers or through legume intercrops.

Arbuscular Mycorrhizal Fungi
Tillage intensity influences soil microbial population and their activity significantly. The extent to which certain microbial species are stimulated in preference to others

needs to be known. The selective microbial process attached with accentuated species is also important. For example, if Arbuscular mycorrhizal fungi (AMF) are stimulated we may expect better P transfer rates into root system. Regarding tillage, Evans and Miller (1988) have shown that soil disturbance, especially after a stretch of zero-tillage period induces loss of soil structure and in turn affects both root growth and AM fungal propagation. Absorption of P by AMF was reduced if soil was disturbed. The AM fungal colonization was higher in maize roots from plots under no-till systems compared to conservation or CT. It is interesting to note that several non-mycorrhizal fungi were also traced in the rhizopshere and rhizoplane (e.g., *Olpidium, Polymyxa, Guamenomyces, Phycomyces, Penicillium, Alternaria*, etc.) and they too might have affected nutrient transformations and availability to maize roots. Tillage affects AMF propagules and their activity measured in terms of P uptake by maize roots (Garcia et al., 2007; Mozafar et al., 2000). Wortmann et al. (2008) state that a long stretch of no-till improves several characteristics of the soil, but much of the improvement is confined to top 5 cm depth. Microbial population and activity are among important characters that no-till affects. Field evaluation with maize in Eastern Nebraska has shown that no-till systems do affect microbial population in the 0–5 m depth. The microbial population decreased by as much as 50% at 5–20 cm depth, and it reduced further at greater depths of 20–30 cm. The microbial biomass was actually assessed using fatty acid methyl ester profiles as biomarkers of soil microbial groups. The AMF were characterized using biomarker C16:1(c11). It is interesting to note that AMF that are aerobic were found in greater quantities, 22% higher in no-till plots compared to conventional till plots.

The cropping trend in a field has immense influence on microbial population, its dynamics and activity. There are indeed several reports about influence of cropping systems on AM fungal survival and perpetuation in soil. In case of Corn Belt, one of the reports from Minnesota states that spores of AMF such as *Glomus aggregatum, G. leptotichum*, and *G. occultum* were abundant in fields with corn-based cropping systems (Johnson et al., 2006). A study of cropping history of various fields has suggested that those with corn usually supported higher levels of spores in soil and colonization in roots compared to those with soybean. Soil pH did not significantly affect AMF in the root zone. The root exudates and rooting pattern might have induced AMF in soil. We should note that maize benefits from enhanced population of AMF in soil. Root exudates from maize roots, AMF, and other soil microbes together add to soil C.

Maize farmers utilize several types of fertilizer-P sources. They may be organic or inorganic P sources. On acid soils, it is possible to use phosphate rock (PR) sources. Such PR sources could be either powdered and used directly or partially acidulated in order to improve P availability to maize roots. Supply of AMF and PR may be helpful in improving soil P status and absorption by maize. Field trial at Appalachian Center in USA has shown that maize crop produces greater amounts of roots if PR or PR plus AMF are applied. The AM fungal colonization was better if AMF plus PR was applied than in plots given only AMF. Interestingly, in addition to mobilization of P, other nutrients (Ca, Mg, and K) that are generally less available to maize roots in acid soils were absorbed in greater quantities, if AMF were inoculated. The AMF inoculated plants had 2–3 folds greater inflow rates (inflow = μg nutrient m^{-1} root length day^{-1})

of Ca, Mg, Cu, Zn, and B than non-mycorrhizal maize seedlings. Further, it was reported that application of PR reduced concentrations of Fe, Mn, and Al in plant tissue. The lower specific rate of accumulation of Fe, Mn, and Cu was attributed to possible precipitation and metal-organic complexes in soil. Similar reports are available from locations in West Africa, where cereals like maize or millet are grown on sandy soils low in available–P (Plate 6.3). Clearly, supply of AMF along with PR affects P nutrition of maize. It may also influence dynamics of micronutrients in soil.

Plate 6.3 Extraction of arbuscular mycorrhizal fungi from sandy soils.
Above: Washing roots of agro-forestry saplings that are often inter-cultured or strip-cropped with maize or other cereals. *Below left:* A make shift facility to extract arbuscular mycorrhizal spores using wet sieving and decanting technique. *Below right:* arbuscular mycorrhizal fungal spores sieved from sandy soils.
Source: Dr. K.R. Krishna, ICRISAT, Hyderabad, India.
Note: The maize-arbuscular mycorrhizal symbiosis is known to influence P dynamics in low-P sandy soils of West Africa. Arbuscular Mycorrhizas improve P recovery from sandy soils deficient in available-P (<3–5 ppm Olsen's P). Arbuscular Mycorrhizas are also known to improve P uptake from partially acidulated rock phosphates. Incidentally, West Africa is endowed with large deposits of phosphate rocks. Further, AM fungi also mediate nutrient transfer between intercrops, for example trees and maize.

In Central America, maize is an important cereal grown under subsistence farming situations. Often, maize genotypes that suit the local population are preferred over hybrids that may need high nutrient supply. In other words, during past decades, maize genotype and soil microbial community, including AMF must have established well suited pattern of interaction. Hess et al. (2005) point out that whenever maize genotypes are changed or new hybrids are introduced and high P supply occurs to support high grain yield formation by hybrids, the soil microflora in general, and AMF in particular may alter. We should expect changes in both species diversity and population of AMF. In Guatemala, for example, a change from local landraces to maize hybrids affected AM fungal colonization and propagule number. High yielding and improved hybrids like (HB83) was colonized less by AMF. We ought to note that hybrids were supplied with high amounts of fertilizers, especially P to support extra yield. High soil P is detrimental to AM fungal build up and activity. In comparison, a local variety––"Cushpeno" was colonized to a greater extent (72% roots). Over all, the study points out that a change in genotype, say from local to hybrids affects AMF and nutrient dynamics, especially P. In due course, as hybrids are consistently planted, AMF that tolerated higher concentrations of soil P may slowly increase in population.

There are reports regarding occurrence of AMF in the sandy soils of West Africa. The low-P soils are known to harbor efficient AMF (Plate 6.3). Several other reports suggest that AM fungus (*Glomus constrictum*) improved P uptake and accumulation by maize grown on low-P soils of Egypt. Tests have shown that maize benefits from AM inoculation when soils are provided 30–60 mg P kg^{-1} soil, but beyond, it crop response subsided. Maize roots were colonized by AMF rapidly extending into 50–70% of roots, but at high P inputs AMF were suppressed (Omar, 1998). Build up of an assortment of AM fungal species in soil seems useful. AM fungal species that tolerate high P inputs could be identified.

Observations on soils of maize fields from Tanzania have shown that AMF do affect P dynamics. The extent of benefits from AMF in terms of P or other minerals need to be quantified. Yamane and Highuchi (2003) reported that indigenous AMF found in maize fields of Eastern Tanzania were efficient in improving P recovery by the crop. The AMF did not affect N uptake perceptibly. It was suggested that build up of AM fungal population in soil was important under subsistence production conditions.

Soil fertility status is a major factor that influences AM fungal symbiosis and benefits derived from it. In practical field conditions, AMF may often encounter highly fertile soils or those supplied with high dozes of N, P, and K. Subsistence farming zones are exceptions. It is opined that maize genotypes that require high amounts of fertilizers may respond feebly to AMF. Further, Pitakdanatan et al. (2007) state that maize cultivars with lower nutrient efficiencies responded to AM inoculation.

Ortas et al. (2001) have found that on P and Zn deficient calcareous soil of Central Anatolia in Turkey, inoculation with mycorrhizal species such as *Glomus mosseae* and *G. etunicatum* improved uptake of P and Zn by maize. The dependence of maize crop on mycorrhizas was greater if soil-P status was below threshold.

On Andosols of Fareast, AM fungal association with maize seems to affect P dynamics. The improvement of P recovery by maize roots was influenced by soil moisture status. It is said that optimum soil moisture conditions improved P uptake to an extent that it sometimes masked influence of AMF (Karasawa et al., 2000).

Arbuscular Mycorrhiza influences P nutrition of maize plants. They may aid better P recovery even in soils with constraints like adverse pH, soil moisture conditions, or even toxic factors. For example, Andrade and Da Silviera (2008) have found that P nutrition of maize is held at normal levels in soils with Cd stress. The P uptake and utilization pattern was normal in mycorrhiza inoculated maize fields compared to those not provided with mycorrhizas. Mycorrhizal maize overcame Cd stress to a certain extent and resisted alteration to P dynamics.

Liasu and Shosany (2007) made a detailed analysis of microbial population in the rhizosphere and mycorrhizosphere of maize grown on sandy soils that occur in Nigeria. Bacterial species such as *Streptococcus pyogenes*, *Bacillus subtilis*, *Psuedomonas aureginosa*, and *Micrococcus* species were dominant in the maize rhizosphere. Fungal species like *Fusarium chlamydosporum*, *Rhodosporum* were also common. Rhizobacteria such as *Rhizobium japonicum*, *R. leguminosarum*, and *R. melilotti* were traced frequently. Some of the rhizobacterial species were inhibitory to strains of *Aspergillus flavus* and few other fungi. Antibiotic effects were easily traceable. We should note that microbial population, especially those of AMF and N-fixing bacteria alter. They may have direct bearing on the nutrient dynamics in the rhizosphere. Soil-N status and P absorption by maize roots would be influenced. Populations of plant growth promoting rhizobacteria (PGPR) bacteria may also affect nutrient availability to maize roots. During recent years, microbial diversity in the rhizosphere has been rapidly assessed using molecular marker and restriction fragment length polymorphism analysis (RFLP). Such molecular typing is available for a range of rhizosphere bacterial species, AMF, N-fixing bacteria, and plant growth rhizobacteria (Chang et al., 2007; Krishna, 2005; Martin Laurent et al., 2006; Oliviera et al., 2009; Payne et al., 2006). These techniques may not allow us to quantify microbial effects, but occurrence and fluctuations in appearance of various microbes in the maize rhizosphere can be accurately gauged. Marcel Gomes et al. (2003) have shown that in addition to identification, molecular techniques are helpful in assessing dynamics of fungal communities in maize rhizosphere and bulk soil. To a certain extent rhizosphere effect on fungal population and diversity could also be assessed.

Tillage affects several types of soil borne fungi that reside in the rhizosphere and non-rhizosphere soil but close to root zone. There are indeed several reports on soil fungi traced in maize rhizosphere. In the Argentinean maize belt, Nesci et al. (2006) have made an evaluation of different tillage systems such as no-till, reduced-till, and conventional till on various soil fungi that colonize soil under a maize crop. The NT without grazing had highest fungal density at 5.7×10^3 g^{-1} soil. Major genera of soil fungi noted were *Aspergillus, Fusarium, Penicillium, Trichoderma, Cladosporium*, and *Alternaria*. Several species of each genus were detected in the rhizosphere of maize. Clearly, tillage operations do affect soil microbiota and it may have an impact on soil quality and nutrient transformations in soil.

Plant Growth Promoting Rhizobacteria

The PGPR are beneficial soil microbes that reside on rhizoplane, in the rhizosphere and bulk soil surrounding maize roots. They help the plant in different ways such as: release of sideropheres that improve nutrient availability, induction of rapid root growth by releasing hormones, increase of nutrient uptake, and shoot growth. A few of them may show antifungal properties and reduce disease incidence in root region. In the present context, we are interested in the PGPRs that elaborate siderophores and improve nutrient uptake. Most commonly traced PGPRs in the rhizosphere are *Azotobacter*, *Azospirillum*, *Bacillus*, and *Pseudomonas* species. Many of these species are also active N-fixing microbes in free living state in soil. They also enhance soil-N status. Obviously, PGPRs could be affecting a series of nutrient related functions in soil. There are several examples where PGPRs are known to improve nutrient uptake by maize and other cereals. Let us consider an example. In the dry regions, evaluation of several strains of PGPRs such as *Azospirillum lipoferum*, *A. brasilense*, *Azotobacter* sp., and *Bacillus* sp. has shown that PGPRs improve absorption of N, P, K, Fe, Zn, Mn, and Cu (Biari et al., 2008). The PGPRs are found to improve availability of Fe by elaborating siderophores. According to Paskiewicz and Berthelin (2006), rhizosphere of plants like maize that adopt strategy II liberate siderophores that help in improving Fe availability to roots. The phyto-siderophores can dissolve iron from gothites and other similar sources. Experiments with maize have shown that, on ferralsols, rhizosphere microflora enhances weathering of Fe and Mn hydroxides. It alters availability of Fe, Mn, Ni, Cr, and Co.

Oliviera et al. (2009) made a detailed study of the rhizosphere microbes involved in P solubilizing activity. They assessed over 350 bacterial colonies derived from Oxisols of Brazilian Cerrado region that are low in P content. They classified the microbes based on their ability to solubilize P found in various organic and inorganic sources. Further, they identified and typed microbes using molecular markers. Greatest P solublizing effect was noticed on medium containing tricalcium PO_4. Bacterial strains like B17 and B5 identified as *Bacillus* sp. and *Burkholderia* sp. were most effective among the hundreds of species screened. The above species mobilized 57–68% of calcium PO_4 applied in 10 days period in the rhizosphere soil. Strom et al. (2002) have shown that mobilization of P in the rhizosphere is mediated by elaboration of organic acids. The extent of organic acid exudation has direct impact on release of P in maize rhizosphere. Obviously, soil microbes collectively termed as P solubilizers must be serving maize cropping belts immensely usefully by mobilizing P, that otherwise would have taken longer period to become available to maize roots. The extent of P derived from such activity by rhizosphere microbes needs quantification. They may be of greater significance in low input subsistence farming and reclaiming soils using maize farming.

SUMMARY

The below-ground portion of maize agroecosystem plays a crucial role with regard to nutrient dynamics and productivity. It includes the all important soil profile that harbors nutrients and moisture, also roots that sequester and mediate transfer of large quantities of nutrients from soil to above-ground phase of the agroecosystem. The below-ground soil profile encompasses a wide range of soil microbes that mediate and help in regulation of various nutrient transformations. Many of these physico-chemical and biological transformations are absolutely essential for crop growth and yield formation, as well as maintenance of various ecosystematic functions.

Maize roots play pivotal role in nutrient absorption, sequestration, and transfer across different portions of the ecosystem. Maize roots are affected by soil physico-chemical properties, nutrient concentrations and availability, and moisture. Root ac-tivity is immensely influenced by soil moisture and irrigation schedules. Maize root system sequesters large amounts of C fixed via photosynthesis. It is said that on an average 25% of C fixed by plant may find its way into below-ground roots. Maize root production and activity are affected by various soil management and crop husbandry procedures adopted by farmers. A careful selection of procedures and their timing is essential. Root system, its size, spread, biomass, mineral content, and moisture absorp-tion patterns are also genetically controlled. Typically, root related characteristics of maize cultivar/hybrid that covers the field or landscape is important. Over all, nutrient dynamics is immensely influenced by root biomass and activity. A change in maize genotype sown in any region has its proportional effect on nutrients partitioned in the agroecosystem. Selection of maize genotypes with better root:shoot ratio seems very useful in enhancing C sequestration in the below-ground portion. This aspect gains in importance considering that during recent years a large fraction of above-ground biomass, organic residue, and nutrients are not recycled. Crop residues are utilized to feed animals and produce biofuel (alcohol). Use of specific genotypes may offset loss of C from the ecosystem.

Root exudation is an important below-ground phenomenon that regulates SOC and nutrient dynamics at the rhizoplane, rhizosphere, and entire root zone. It adds to SOC. It causes a quantum shift of organic-C from crop phase to soil. It is said that almost 7–10% of photosynthates are transferred to soil through exudations. If extrapolated to a field or cropping expanse, the root exudation as a phenomenon has immense rel-evance to C and mineral nutrient dynamics. We ought to realize that root exudation pattern and composition are dependent on maize cultivar used by farmers. There could be genotypes that transfer more of C to soil and allow it to be sequestered. As an alter-native, we may be able to find and breed genotypes that exude low quantities of C and nutrients. The photosynthates conserved in the above-ground crop residue may help us in enhancing forage production for animal feed and/or biofuel production.

Root exudations and SOM are important sources of C, minerals, and energy to soil microbes. Soil microbes and their activity literally regulate most nutrient transforma-tions. In addition, soil microbes like Rhizobium and Azospirllum add to soil-N via biological nitrogen fixation phenomenon. Adopting maize–legume cropping systems should be highly useful since it adds to soil-N fertility. Maize rhizosphere and root zone soil also harbors wide range of microbes that mediate nutrient transformations. Repeated inoculation of beneficial microbes and adopting procedures that accentuate soil microbial activity (organic manure inputs, microbial inoculants) seems pertinent.

Over all, it is clear that roots, their activity, root exudations, and soil microbes regulate nutrient dynamics and productivity of the maize belts, to a great extent.

KEYWORDS

- Arbuscular mycorrhizal fungi
- No-tillage
- Rhizosphere
- Root exudation
- Root growth
- Root length density

Chapter 7

Crop Physiology, Genetic Improvement, and Nutrient Dynamics

INTRODUCTION

Genetic nature and expression of physiological traits of maize crop has immense influence on establishment, sustenance, perpetuation, and productivity of maize agro-ecosystem, anywhere in different continents of the world. Historically, morphogenetics and traits relevant to adaptability and yield formation of maize genotypes have immensely affected preferences for it in different agricultural zones. Basically, physiological traits of a genotype has to suit the given environment, in order that maize belt thrives well. In the present context, genetic constitution of maize crop and physiological manifestations relevant to acquisition of soil nutrients and accumulation, light interception and photosynthesis, carbon partitioning patterns and biomass/grain production are important. There is indeed a large pool of genetic variation available with regard to traits that have direct impact on nutrient dynamics and productivity of maize belts. To quote a few examples, maize genotypes that respond to fertilizer supply with high yield are preferred in intensive cropping zones (e.g., Corn Belt of USA). Genotypes that tolerate drought and low N are preferred in subsistence farming zones of Africa. Rooting pattern for example has direct impact on nutrient absorption and C sequestration below ground. Production of foliage influences extent of organic matter that could be recycled or provided to farm animals. During practical agriculture, physiological aspects like rooting depth, nutrient and water recovery rates, growth and biomass accumulation patterns, and grain yield potential have direct influence on nutrient dynamics, as well as maintenance of ecosystematic functions.

In this chapter, relevance of various physiological traits to nutrient dynamics that ensues in the agroecosystem has been discussed. The impact of maize crop genetic improvement on nutrient dynamics and productivity of agroecosystem has also been highlighted.

Growth and Development of Maize Crop

At any point of time during a crop season, growth habit of a maize crop has direct impact on nutrient distribution within the agroecosystem. Basically, we need to understand the influence of crop development pattern on nutrient recovery from soil, its accumulation in various tissues of plants, its removal via grains and recycling trends. Firstly, let us consider the various stages of maize crop growth and development. Corn researchers and producers in the "Corn Belt States of USA" have meticulously studied plant development and classified the growth stages (IASTATE, 2007a). The development of corn plant has been divided into Vegetative (V) and Reproductive (R) stages. There are several subdivisions of vegetative stage such as V1, V2, V3 until Vn, where "n" is the last leaf stage. Seedling emergence is designated as VE and last vegetative stage, tasseling is designated as VT. The reproductive stages such as silking, milking physiological maturity are designated from R1 to R6 stages (see Table 7.1). Nutrient dynamics in a maize field or cropping zone is intricately linked with different stages of crop growth.

Table 7.1. Vegetative and reproductive stages of corn crop

Growth Stage	Remarks
VEGETATIVE STAGES	
VE = Emergence: Coleoptile and seminal roots appear. Mineral nutrient acquisition, if any is feeble. Carbon fixation is also relatively small in quantity. Seedling survives initially on stored C and energy sources. Nutrients in soil is yet to affect the crop perceptibly.	
V1 = First leaf	
V2 = Second Leaf	
V3 = Third leaf: Seminal roots stop growth. Nodal roots are profuse, grow rapidly and put forth root hairs in great density in order to explore soil efficiently and garner as much nutrients. Nutrient accumulation in roots is rapid. Sizeable amount of mineral nutrients are rapidly translocated to shoot system, especially to growing tips and young leaves. Roots are succulent and vulnerable to implements. Therefore, soil should be inter-cultured carefully and fertilizer placement should be carefully done. Nutrients held in shoot and root increases rather rapidly due to elongation of stalks and leaves, also due to rapid root growth. Weed control is important in order to avoid competition for nutrients, water and photosynthetic radiation. At this stage, loss of nutrients and water to weeds could be detrimental to maize crop growth and canopy development.	
V6 = Sixth leaf: Nodal roots are well developed and distributed in the soil. Hence, fertilizer placed is garnered efficiently. Side dressing with N via soil application is preferred. A larger share of plant C and nutrient is held in nodal root system. Seminal roots are senesced (IASTATE, 2007b).	
V9 = Ninth leaf: The stem becomes stouter and taller. Stalk actually elongates through rapid growth of internodes. The time gap between leaves stages begin to shorten. Tassel begins to develop. Ear shoots begin to appear from the internodes except for top 6–8 internodes. Initially, each ear shoot develops fast, eventually most slow down or degenerate leaving only 1 or 2 to develop into harvestable ear. Stems that show up more than one ear shoot are termed prolific. It has direct bearing on nutrient and photosynthate partitioning. Harvest index is dependent on ear formation. The prolificness is dependent on nutrient status of the plant and planting densities.	
Crop growth—mainly roots, leaves and canopy development is rapid. Nutrient and water is absorbed in larger quantities compared to other growth stages. The nutrient dynamics within the plant, field and entire cropping belt is affected through rather rapid changes in nutrients stored in roots, shoots and that held in soil. Farmers tend to top dress, in other words add nutrients into cropping zone to match rapid absorption.	
V12 = Twelfth leaf: Cob development depends on moisture and nutrient flow into the shoot system. At this stage (V12–V17), ovule number and their development into kernels depends on photosynthate and mineral flow. Number of kernels is determined by one week from silking. Hence, nutrient and moisture deficiencies should be avoided during this period. Early maturing hybrids have shorter time during this stage. It leads to smaller ears than hybrids that require longer duration. Obviously, it affects nutrient partitioning between cobs and stover.	
V15 = Fifteenth leaf: The maize crop is approximately 10–12 days away from onset of reproductive stage and consequent changes in nutrients. This stage is important in terms of plant development and grain yield. (IASTATE, 2007C). Regarding ears; tips are visible at the top of leaf sheath. The tip of tassel is also visible at V17 stage. In many locations of tropics (especially dry lands), corn crop is	

vulnerable to water stress and its effects on grain formation. The largest grain yield reduction occurs if crop suffers water scarcity at silking stage. Similarly, nutrient deficiencies at stages V15–V17 can severely reduce cob yield.

V18 = Eighteenth leaf: Brace roots (aerial nodal roots) appear from nodes near the soil surface. They are helpful in exploring and absorbing nutrients and moisture from upper layers of soil. Water deficit during this stage delays cob development. Silking is also affected. It leads to low weight kernels and decrease in grain yield.

VT = Tasseling Tasseling stage begins approximately 2–3 days after silk emergence. The period from VT to R1 can vary depending on the genotype. Nutrient and water scarcity should be avoided at this stage. Otherwise, it retards kernel development.

REPRODUCTIVE STAGES

About six reproductive stages can be identified during development to maturity of cobs in a corn plant. Usually, top of the ear of a prolific plant is considered while judging the reproductive stage.

R1 Silking The R1 or silking stage begins when silks are visible outside the husks of the cob. Fertilization of ovules occurs during this period. The number of ovules that get pollinated and those which develop into mature kernels are determined at this point. Therefore, environmental stress especially, nutrients, water or temperature should be avoided. The corn crop at this stage would have already absorbed most of K from soil to satisfy its needs during kernel development. Nutrient stored in stover (stem plus leaves) is important because it gets re-translocated to support grain formation. However, uptake of N and P is rapid at silking, hence appropriate top dressing is essential. It is important to note that, leaf analysis for N, P, and K at this stage is highly correlated with final expected corn kernel yield. Fertilizer inputs, mainly split doses applied at this stage lead to rapid and excellent responses in terms of cob/grain yield (IASTATE, 2007d).

R2 Blister (10–14 days after silking)

At R2 stage, kernels appear white and are blister shaped. Endosperm and inner fluid in the kernel could be visualized by dissection. Embryo is still developing, but radicle, coleoptile and miniature leaves are all formed. A miniature plant exists within the developing embryo. Cob is close to its full size by the time R2 stage ends. In terms of crop management, this is a grain fill stage. Starch and dry mater accumulation is steady. Such rapid development of kernel continues until R6 stage. In terms of nutrient dynamics, plants absorb relatively larger quantity of nutrients from soil. There is also significant amount of nutrient translocation into seeds, from other parts like leaves and stem (IASTATE, 2007e). Agronomically, split dose of N seems important at this stage.

R3 Milking (18–22 days after silking)

The kernels are yellowish and inner fluid is milky and thick due to starch accumulation. The cob material and silk are brown and dry. The dry matter accumulation in the kernels is rapid. Nutrient re-translocation is significant. In terms of crop management, any dearth for nutrients or water is bound to reduce crop yield.

R4 Dough (24–28 days after silking)

Starch accumulation makes inner contents of kernels pasty. Embryonic leaves are enlarged and 4 in number. The shelled cob appears pink or light red in color. Nutrient and water shortages should be avoided since only half the grain fill is over by this state. The kernel dry weights are around half the full capacity.

R5 Dent (35–42 days after silking)

At R5 stage, kernels are dented and shelled cob is red in color. Kernels begin to dry and moisture content is 50–60%. A small white layer of starch appears after denting across the kernel. In terms of agronomy, if frost damage is anticipated, it is preferable to select hybrids that mature 3 weeks earlier than first detrimental effects of frost begin (IASTATE, 2007f).

R6 Physiological Maturity (55–65 days after silking)

At R6 stage, all kernels in a cob attain full dry weight or maximum dry matter accumulation. The hard starch layer extends much from base to tip and a black abscission layer develops. Kernel growth for the season almost ends by R6 stage. The husks and leaves are dry and not green. The average moisture in kernels dips to 30–35%. Such a grain is not safe for storage. It has to be dried until moisture content dips to 13–15% and then shelled.

Sources: http://www.extension.iastate.edu/hancock/info/Corn+Develop+Stages.htm;
http://www.extension.iastate.edu/hancock/info/Corn+Develop+Stages.V6htm;
http://www.extension.iastate.edu/hancock/info/Corn+Develop+Stages.V9htm;
http://www.extension.iastate.edu/hancock/info/Corn+Develop+Stages.V9htm;
http://www.extension.iastate.edu/hancock/info/Corn+Develop+Stages.V12htm;
http://www.extension.iastate.edu/hancock/info/Corn+Develop+Stages.V15htm;
http://www.extension.iastate.edu/hancock/info/Corn+Develop+R2++Stages.htm;
http://www.extension.iastate.edu/hancock/info/Corn+Develop+R3+-+Milk.htm;
http://www.extension.iastate.edu/hancock/info/Corn+Develop+R5+Dent+Stage.htm;

On arable soils, maize seeds germinate in 5–8 days provided soil moisture and temperature conditions are optimum. Maize root system contains seminal roots, adventitious and prop roots. Maize root system has two distinct phases. First phase is development of seminal or seed root system. The second phase is development of nodal or crown root system. Seminal roots appear from coleoptile in the initial stages of seedlings. Growth of seminal roots retards soon after emergence and is almost nonexistent by V3 stage. The seminal roots continue to persist and function, but its contribution to nutrient and soil moisture recovery by crop is small or relatively negligible. Maize plant has nodal or adventitious (fibrous) and prop roots. Prop roots are useful in supporting the stem. They usually appear from first three nodes closer to soil surface. The prop roots are thick and waxy. The adventitious roots or nodal roots are initiated at emergence stage. Nodal roots begin to elongate at V1 stage. These roots develop and grow rapidly between V1 and R3 stages. The root growth is negligible after R3 stage since senescence sets in. The nodal root system functions as the major conduit for water and nutrients by V6 stage. Indeed a large portion of soil nutrients held in upper layers of soil (0–2m depth) is absorbed and translocated to above ground portion of maize ecosystem through nodal roots. Nodal roots are larger and thick. Hence, they sequester greater quantity of C that could be held in soil. Senesced nodal roots add C and mineral nutrients to soil. Maize crop may encounter very cold soil temperatures that restrict nutrient absorption and movement into shoot system. In order to hasten root growth and nutrient uptake at early growth stages, it is advisable to place a small amount of fertilizers closer to seedlings so that seminal roots absorb them relatively efficiently (IASTATE, 2007b).

Maize plant develops a profusely branched root system with secondary and fine (tertiary) roots. Under optimal soil moisture and nutrient availability, each maize plant may produce roots that stretch up to 1,500 m in length. The fibrous roots may spread into 1.5 m radius and may reach a depth of 2 m in the soil profile. Rooting is denser and confines to surface layers. Fine roots are active in upper layers of soil up to 60 cm depth. Nearly 80% of active roots are localized in the upper 0.8–1.0 m of soil profile. Maize crop extracts almost all of its water and dissolved nutrients from 1.0 to 1.7 m depth in soil profile. At this stage, we should note that root growth rate, its biomass accumulation pattern, regeneration/senescence rates all influence carbon and mineral dynamics in the maize ecosystem rather perceptibly. At any given time, root: Shoot ratios, rate of translocation of mineral nutrients and accumulation pattern literally decide the amount of nutrients held in the above and below ground portion of maize belt. There is strong need to quantify extent of nutrients held in the roots and shoots of different composites and hybrids and perhaps group them. Such information allows us to judge effect of a particular group of genotypes on nutrient dynamics in the ecosystem better, at any stage of the crop.

The number of leaves formed on a mature plant varies depending on genotype. Generally, a maize plant may develop 8–20 leaves on a single vertical stem. Leaves are arranged alternately on opposite sides of the stem. Typically, maize leaf has a sheath, ligules, auricles, and blade with parallel venation. The leaf blade is long, narrow, undulating, and tapers at the tip. Leaf is mostly glabrous. Leaf is the main photosynthetic organ. In maize, like any other cereal species, foliage has immense influence on C-fixation, biomass and grain formation, kernel weight and final yield. Egharevba et al. (1976) have shown that defoliation reduces dry matter accumulation rates. Complete defoliation reduces grain yield by 82% compared to control. Partial defoliation reduces grain yield by 2–37% compared to control. Most importantly, defoliation affects grain fill and kernel weight decreases by 13–52% compared to control. Clearly, leaves and translocation of photosynthates plus minerals are key to grain formation and final yield.

Generally, stem height varies from 1.6 m to 3.0 m. Stem is solid, cylindrical with nodes and internodes. The number of internodes varies with genotype. Internodes that occur near soil surface may not expand appreciably. Lateral shoot bearing the main ear develops from the bud on 8th inter node above the soil surface. The five or six buds directly below the cob give rise to rudimentary lateral shoots, but one or two of them may occasionally develop into ears.

Male and female flowers are borne on the same plant, but as separate inflorescence. Male flowers are borne on the tassel and female flowers on the ear. The maize ear or the female inflorescence appears on lateral branches. Bracts enclose the ear. The silk of the flowers remain receptive to pollen for approximately 3 weeks, but after 10th day. Maize is primarily a cross-pollinated species. Cross pollination contributes to morphogenetic variability and adaptability.

Maize cultivars exhibit wide range of variation with regard to physiological traits, especially growth habit, duration to flower and maturity. Short duration cultivars may ripen in 60–70 days, but certain long duration genotypes require up to 300 days in order to attain physiological maturity. Harvest index (HI), total biomass accumulation pattern, and grain yield formation also varies depending on a range of physiological traits. At a given point of time in the crop season, several physiological traits directly influence distribution of biomass (carbon) and mineral nutrients in the above and below-ground portion of the agroecosystem. Productivity of maize is physiologically associated with interception of photosynthetically active radiation. The average radiation use efficiency before silking (4.14 g MJ^{-1}) and after silking (2.45 g MJ^{-1})

is correlated with biomass formation, HI, and grain yield (Otegui et al., 1995). In temperate regions, carbon fixation and maize cob yield is more stringently linked to amount of radiation received, especially around silking period. Sowing time needs to be adjusted in order to utilize photosynthetic radiation efficiently.

The number of cobs formed per plant, its size, number of kernels, and related traits like test weight, and nutrient accumulation in endosperm may all influence nutrient dynamics in the agroecosystem, commensurately. Maize kernel or grain consists of an endosperm, a pericarp, and tip cap. The endosperm stores carbohydrates. For example, a dent or flint corn kernel contains 84% carbohydrates, 10.9% protein, 4.5% fat, and 1.3 % minerals. The embryo harbors parts that germinate to provide the next generation seedling. Recently, Halvorson and Johnson (2009) have compiled some interesting data on corn cob and its characteristics relevant to farmers who produce it as cellulosic feedstock. Field investigations in Colorado and Northern Texas have shown that cob yield increases with nutrient supply. The cob:stover ratio is crucial and it fluctuates between 0.14 and 0.25.

Nutrient Uptake Pattern of Maize Crop

Nutrient recovery by maize seedlings is small. Further, fertilizer-based nutrient supply is also relatively small during early seedling stage of the maize crop. Yet, it may be beneficial to have higher concentration of major nutrients in the root zone. It is said that growth of leaf, ear, and other parts depends to a certain extent on nutrient supply during early seedling stage. In other words, decisions regarding nutrient supply at seedling stage may have significant impact on nutrient dynamics that ensue immediately, plus even at later stages of the crop, because it affects crop growth and physiology. In many of the temperate countries maize may encounter cold climatic conditions at seedling stage. The seminal roots serve as the main root system during this stage. Fertilizer placement, if any, has to be banded carefully without affecting seminal root growth. At later stages, nutrient requirement of maize crop is much higher. Roots effectively absorb nutrients from moist soil. Fertilizer supplements (top-dressing) is effectively absorbed by fibrous roots. A soil test sufficiently high in N, P, and K is required for optimum nutrient uptake and accumulation. Due care is needed during supply of fertilizer-N. Timing, dosage, and method of application should be carefully chosen, so that plants effectively garner most of N. Potential loss of N via leaching, seepage, and volatilization should be avoided.

Table 7.2. Dry matter and major nutrient accumulation pattern at various stages of maize crop.

Growth Stage/	Seedling	Rapid Vegetative	Silking	Grain Fill	Maturity	Total
Days	(1–25)	(26–50)	(51–75)	(76–100)	(101–125)	(1–125)
Dry matter (C fixation)						
Quantity (kg ha⁻¹)	524	3597	6369	6745	1499	18734
Percentage of Total	8	36	34	19	3	100
Nitrogen						
Quantity (kg ha⁻¹)	19	84	75	48	14	240

Table 7.2. Dry

Growth Stage/	Seedling	Rapid Vegetative	Silking	Grain Fill	Maturity	Total
Percentage of Total	8	35	31	20	6	100
Phosphorus						
Quantity (kg ha^{-1})	2	12	16	11	3	45
Percentage of Total	5	27	36	25	7	100
Potassium						
Quantity (kg ha^{-1})	18	88	62	28	4	200
Percentage of Total	9	44	31	14	2	100

Source: Johnston and Dowbenko, 2009.

During a crop season, maize crop recovers a mere 8–10% of major nutrients during first 3 weeks and only 2–7% at crop maturity. Most of the nutrient absorption amounting 86–88% of total nutrient recovery, occurs between 26 and 100th day that is during vegetative growth, silking, and grain fill stages (Table 7.2). The dry matter accumulation too occurs most rapidly during these above three stages. If extrapolated to agroecosystem, it means major quantity of soil nutrients, inherent in soil or that applied as fertilizer is rapidly translocated to above-ground portion during 75 days. Nutrient supply schedules should be carefully tailored to match the peak nutrient recovery by maize crop/genotype. Further, genetic traits that affect nutrient absorption and accumulation during the vegetative, silking and grain fill stages may have significant impact on nutrient dynamics during crop season. Genetic selection for higher nutrient uptake and efficiency should target traits that are most relevant to these three stages of the crop-namely vegetative, silking, and grain fill. Since, dry matter accumulation is high during 26th–100th day, genes active and relevant to nutrient-use efficiency during this period should be accentuated. This way, it may improve nutrient use and dry matter production efficiency significantly. Obviously, whenever, a new genotype with improvised genes for nutrient uptake and use efficiency is introduced, it does bring about alteration to nutrient dynamics proportionately.

Let us consider a few examples. In the corn belt of USA, average productivity of a sole maize crop is 10.2 t ha^{-1}and annual production is 332 m t. In order to produce 10.2 t grain ha^{-1}, a sole crop absorbs approximately 240 kg N, 45–50 kg P and 200 kg K ha^{-1} (see Table 7.2). Since 86% of N, P, and K absorption occurs in 75 days, during the three physiological stages namely vegetative growth, silking and grain fill, it means 205 kg N, 39 kg P, and 178 kg K is rapidly absorbed by maize roots and translocated to stem and grains above-ground. We can easily estimate the quantity and rate (per day) of nutrient translocation in a large agroecosystem. See chapter 1 for annual grain yield in different countries. In case of India, greater share of maize production occurs in Southern Indian states. The productivity is 2.1 t ha^{-1}. It means 35.2 kg N, 8 kg P, and 34 kg K is translocated between 26th and 100th day of the crop. Similarly, we can easily calculate the exact amount of nutrients translocated at different stages of maize crop for as many cropping zones, for example Pampas in Argentina, Cerrados in Brazil, Northeast China, etc. First, there is advantage in estimating nutrient transfers in a field or in entire ecosystem. Based on it, we may rearrange nutrient disbursement schedules within a single field or a cropping patch or even an entire agroecosystem. We can also take to precautionary changes in nutrient supply, if and when farmers change the maize genotypes. The extent of area covered and expected grain yield for a genotype should dictate the nutrient supply in a field or agroecosystem *per se*. Over all, we should note that physiological stage, genetic potential, nutrient needs of a geno- type can have far reaching effects on nutrient dynamics within a cropping ecosystem. It may also affect decisions regarding fertilizer transport and application schedules in a given region. It is generally advisable to have gross idea about consequences of chang- ing a maize genotype in a given area with regard to nutrient needs, actual quantity removed via forage and grain. Further, each physiological stage has its unique impact on nutrient dynamics in the ecosystem.

It is interesting to note that during 2008, to produce 332 m t grains, Corn Belt of USA and adjoining areas would have experienced a shift of 0.64 m t N from soil to above ground crop in a matter of first 25 days of crop (seedling stage). It is relatively a small percentage of total N that a maize agroecosystem sucks up from soil phase. Next, in the following 75 days, about 6.8 m t N would have moved from soil to shoot system. During crop maturity period, a small share of 0.48 m t N should have got translocated from soil to above ground portion. Depending on genotype, its root:shoot ratio and HI, large portions of N would have been held in roots, stover, and grains. Plausibly, physiological manifestations of a genotype that dominates the area will have greater influence on the large scale nutrient dynamics that ensues in the agroecosystem. Any change in the composition of maize genotypes planted could be noted. Then, decisions regarding nutrient supply could be accordingly matched. Currently, there are several models and simulations dealing with crop and soil nutrient dynamics. Computer based simulations can help in arriving at better judgment and recommendations on fertilizer movement and disbursement in a given area.

Harvest Index and Nutrient Translocation Index

Regarding relative importance of HI and total biomass accumulation to grain yield, Kawano (2004) states that "potential for higher biomass production" as a genetic trait is of greater relevance in low yielding regimes. The HI is of greater importance in high fertility and high yielding situations. However, in some cereals, biomass production is equally important both in high and low yield regimes. The HI and nutrient transloca- tion index both are important physiological parameters that have great influence on portioning of biomass, C, N, P, K, and other mineral nutrients within the crop phase of the ecosystem. Historically, maize grain yield increase has been partly attributable to enhanced HI of improved composites and hybrids compared with landraces/local cultivated. The HI affects C and mineral nutrient dynamics in the ecosystem at several points. For example, HI affects nutrients removed via grains. It affects nutrients held in stover which is recyclable *in situ*. It also affects nutrient removed via forage. There are innumerable reports about influence of HI on maize grain productivity. Let us con- sider few examples. Evaluation in India has shown that increase in HI from 0.3 to 0.42 has contributed to enhanced grain harvest possible with recently released genotypes compared to local varieties (see Ikisan, 2008; Kaul et al., 2008). Studies in Ethiopia has shown that among the several composites and hybrids cultivated during 1970s to 2000, HI had a major influence on nutrients recovered/recycled, biomass and grain harvested. The mean HI increased from 0.31 (Bako Composite) to 0.45 (BH450). It indicates the extent of variation in photosynthate partitioning into stover and grain. The grain yield increased from 4.3 t ha^{-1} to 7.2 t ha^{-1} as a consequence of changes in HI (Worku and Zelleke, 2007).

Nitrogen and Phenology of Maize
Nitrogen stress affects phenology and growth rate of maize. Nitrogen stress delays appearance of leaves. At the end of silking stage leaf number may still not differ between N stressed and non N-stressed fields. Low N supply and N-stress are often associated with delay in silking and increase in anthesis-silking interval (Singh and Wilkins, 2001). For example, days to silking increased from 78 to 108 days due to N deficiency. Nitrogen stress has opposite effect on grain fill stage. It shortens the length of this phase. Actually, a combined effect of dearth for N and shortening of grain development reduces productivity. For example, grain-fill stage reduces from 58 to 51 days due to N stress and grains per ear reduce remarkably from 555 to 80–100 per ear, if N stress is severe. According to Singh and Wilkins (2001), N and water are two crucial agronomic factors that control phenology and productivity of maize. We should also note that, if N stress occurs, it does affect nutrient dynamics *per se* in the ecosystem. The biomass and nutrient accumulation pattern in the below and above-ground portion and nutrient recycling is also influenced proportionately.

Planting Dates, Nutrient Dynamics, and Productivity
Planting date is a crucial agronomic decision that farmers have to make. It is usually matched with genotype, its duration, season, expected harvest dates, and economic benefits. In most of the developed countries, planting dates suggested are based on careful evaluation of genotypes, precipitation pattern and predictable yield. Planting dates are also altered to escape from cold spell (Racz et al., 2003), drought, and disease. Actually, planting date may affect a series of genotype x environment interactions. In the present context, planting date may have direct consequences on nutrient dynamics and productivity.

In the North American Corn Belt, maize is planted early to exploit precipitation and soil nutrients more efficiently. For example, in Kansas, dry land corn is planted a trifle early to match with duration to maturity, peak precipitation period and nutrient needs of the crop (Norwood, 2001). In many of the Corn states there was a significant relationship between planting date with nutrient recovery and grain formation. Management of planting dates improved grain yield by 19–23% in Nebraska, South Dakota, Minnesota, Iowa, Wisconsin, and Michigan. Actual grain increase ranged from 0.06 to 0.14 t ha^{-1} (Kucharick, 2008). It is interesting to note that about 8% yield increase (0.031 t grain ha^{-1}) between 1930 and 1970 in the maize cropping zones of Minnesota was easily attributable to advancing of planting date by 10 days. Further, contribution of earlier planting on grain production ranged from a high of 1.86 t grain ha^{-1} in Iowa to a low of 0.58 t ha^{-1} in Wisconsin.

On many occasions, success of maize crop especially in Sudanian or Savanna zone, depends immensely on planting date. In the Nigerian Savannas, maize farmers tend to plant early. They intend to avoid pest attack and derive better exchequer for their crops. At the same time, early planting helps them to increase crop productivity through better nutrient recovery, and utilization. Early planting avoids dearth from soil moisture at crucial stages such as rapid vegetative growth, silking, and grain fill. We ought to realize that soil nutrients are all channeled in dissolved state and as such soil moisture x nutrient interactions are important. Field evaluation by Kamara et al. (2009) has shown that to reduce drought and nutrient stress, it is preferable to plant the crop by June last week or July first week. Late planting reduces productivity by 21% compared with June plantings.

Planting Density, Nutrient Dynamics, and Productivity

Planting density has direct influence on nutrient recovery, dry matter, and grain yield. Planting density has immediate effect on rooting pattern, extent of soil explored, competition for soil nutrients, and moisture. Planting density also influences above ground parameters such as leaf area index, photosynthetic light interception, nutrient partitioning, and recycling. Planting density influences maize growth and yield when sown as intercrop with legumes such as maize or cowpea. Coulter (2010) has reported that grain productivity was high at a planting density of 87–93 thousand seedling ha^{-1} (Table 7.3).

Table 7.3. Influence of planting density on grain yield and nutrient recovery.

Planting Density	Grain Yield (t ha^{-1})	Nutrient Uptake (kg ha^{-1})		
		N	P	K
76300–98600	15.9	350	79	333
87400–93700	16.1	354	81	339
87400–95700	15.4	339	77	323

Source: Coulter et al., 2010.

Note: Nutrient recovery has been calculated assuming that Maize crop absorbs 22 kg N, 5 kg P, and 21 kg K to produce 1.0 t grains.

Dobermann (2001) has reported that for maize grown in Nebraska, planting density is an important factor. The grain yield improved from 10.8 t ha^{-1} to 15.9 t ha^{-1}, if planting density was increased from 31,000 to 44,000 ha^{-1}. The mineral nutrient recovery is high in order to match the higher biomass and grain formation. The uptake of major nutrients, N, P, K, and secondary nutrient Mg and S improved, if maize was planted closely. Duvick and Cassman (1999) evaluated a series of maize genotypes that were released into Corn Belt since past 70 years between 1930 and 1995, for their response to different planting densities. They examined all the genotypes under uniform soil fertility conditions but at three different plant densities namely, 10,000, 30,000, and 79,000 seedlings ha^{-1}. Genotypes released during 1930 had relatively low grain yield potential, yet those planted at higher density of 30,000 and 79,000 produced over 5.5–6.0 t grain ha^{-1} compared to those sown at only 10,000 seedlings ha^{-1}

(3.6 t grain ha^{-1}). Maize genotypes released during 1990s performed exceedingly well at high plant densities. They yielded up to 11.0 t grain ha^{-1}, but those of 1930s were severely repressed and yielded only 4.0–4.2 t grain ha^{-1}. Planting density had immense influence on rooting, shoot growth, leaf area, nutrient recovery, and grain formation. It seems genotypes developed during early 1930s and 1940s lacked ability to withstand intense competition for nutrients and moisture that ensues at high plant densities. Clearly, planting density affects nutrient dynamics and grain formation, but it is dependent on genotype and its characteristics. For genotypes released during 1990s, enhancement of planting density from 10,000 to 79,000 plants ha^{-1} increased grain yield from 5.5 t ha^{-1} to 11.2 t grain ha^{-1}.

A different evaluation at Ames in Iowa by Sangoi and Salvador (1997) has suggested that aspects such as leaf area, light interception, stover, and grain formation by maize decreases if planting density is increased from 25,000 to 1,00,000 plants ha^{-1}. Nutrient recovery and partitioning were affected negatively as population density increased. The extent of plant competition and reduction in grain yield was dependent on genotype.

Field test at Ottawa in Canada has shown that, planting density affects leaf area, canopy light interception, dry matter accumulation, nutrient recovery and grain production by leafy maize hybrids (Subedi et al. 2006). On an individual plant basis, leafy hybrid had 20–25% more leaf area compared to non-leafy genotype. The leafy hybrids were sensitive to high plant density (>90000 ha^{-1}). High planting density reduced HI. Further analysis indicated that grain yield was not affected significantly, but silage production by leafy genotypes improved linearly with increase in planting density.

In the Mexican highlands, maize is often intercropped with beans and squash. The planting density is varied depending on various factors related to natural resources, soil fertility and farmer preferences. The planting density of maize has an immediate effect on nutrient dynamics, biomass generated, and grain yield. Following is an example from Tabasco in Mexico:

	Mono culture				Poly culture
Planting Density (Pl ha^{-1})	33000	40000	66000	100000	50000
Grain yield (kg ha^{-1})	990	1150	1230	1170	1720
Biomass (kg ha^{-1})	2823	3119	4487	4871	5927
N Recovery (kg ha^{-1})	21.8	24.2	26.4	24.2	37.4

Source: Gliessman, 1998.
Note: Nitrogen recovery was calculated assuming that 22 kg N is required to produce 1.0 t grain. Polyculture = Intercropping of maize, beans, and squash.

Clearly, maize grown under poly culture offered better biomass and grain yield productivity. Nitrogen is a key element that affects productivity. Nitrogen was absorbed better if planting density was 50,000 plant ha^{-1} and in poly culture with beans and squash.

According to Monneveux et al. (2005), increasing plant density is an effective measure that enhances capture of solar radiation, leaf area index, tolerance to N stress and dry matter formation. Grain productivity under high plant density was influenced by anthesis—silking interval and total number of ovules at anthesis.

In New South Wales, evaluation of a series of planting densities ranging from 10,000 to 50,000 plants ha^{-1} indicated that short duration hybrids performed better at 20,000 plants ha^{-1}. Long duration hybrids absorbed greater quantity of nutrients and produced greater amount of grain at 20,000–30,000 plants ha^{-1} (Simons et al., 2008). Planting density affected water use efficiency, HI, dry matter accumulation, cobs, and grain yield ha^{-1}.

In the Pampas, maize farmers strive to obtain higher N-use efficiency (NUE). They adopt several agronomic procedures that improve NUE. No-tillage or conservation tillage, split doses of N fertilizer, appropriate planting procedures are some examples. Planting density, especially narrow spacing seemed to help in greater recovery of fertilizer-N applied. Field test at Balcarce, Buenos Aires showed that narrow rows improved NUE and dry matter/grain formation by 12–15% over normal practices (Barbieri et al., 2008). Improvement in light interception was attributed to better yield under high density planting. Further, N recovery rates increased appreciably with narrow spacing. Some of the effects of narrow spacing on NUE were nullified if N supplied to fields were high. Obviously, interplant competition, if any, would be over come under high N supply. Roots of closely planted maize seedlings may not encounter stiff competition for N, that occurs if there is N stress.

Ogunlela et al. (1988) aimed at understanding the effects on planting density and N supply on micronutrient recovery by maize. An increase in planting density from 25000 to 75000 improved parameters such as dry matter production, grain:stover ratio, ears pl^{-1}, total N and micronutrient recovery. However, micronutrient (Zn, Mn, Cu, and Fe) concentration in the plant tissue did not alter response to N supply or variations in planting density.

Planting methods and density have a major impact on maize grown as sole or intercrop. Normally, several combinations of planting density, row spacing and ratio of intercrops are examined in a given region before deciding on specific procedure. Evaluations at Faisalabad in Pakistan have shown that among various combinations, maize planted at 90 cm row spacing provided maximum land equivalent ratio (Ullah et al., 2007). A sole crop yielded 6.71 t grains ha^{-1}. Interactive effects of maize/soybean with planting density were significant.

Site-specific N management (SSNM) methods have been adopted in many of the maize cropping belts. The variable rate application of N may affect maize crop response to several agronomic procedures. Reports from Nebraska indicated that interaction between SSNM and planting density was not significant. Ping et al. (2008) have argued that lack of interaction effects on N recovery is due to plasticity of yield components and response to planting density. Marginal variations in seed rates did not affect grain yield if maize was grown under SSNM.

Hammer et al. (2009) point out that continuous increase of maize biomass/grain yield through many years in the US Corn Belt has been related to planting density. They examined the mechanisms involved in such increases in productivity. Field experiments and simulations based on a hybrid (Pioneer 3394) indicated that a change in root system architecture and water capture (with dissolved nutrients) had a direct impact on biomass and grain yield formation. Further, a change in the canopy architecture had little effect. However, indirectly, it affected via leaf area retention and partitioning of carbohydrates.

MAIZE GENETICS AND NUTRIENT DYNAMICS

Genetic Variation for Nutrient Uptake and Accumulation

Researchers have indeed identified genetic variation for several traits that are either directly or indirectly connected with nutrient acquisition, partitioning, and yield formation. Nitrogen is supposedly the most important nutrient factor that affects crop growth and productivity. At the same time, most soils are deficient for this major nutrient that is required in relatively higher quantities by maize crop. In this context, tolerance of maize genotypes to low soil-N availability and NUE seems most pertinent. Phosphorus dearth is also fairly wide spread in many of the maize cropping zones. Identification of maize genotypes with efficient P uptake and utilization traits is also important. Potassium levels are satisfactory in many soils types. Yet, K may get depleted due to incessant cropping. Hence, it needs to be replenished. Under such circumstances, identification and use of genotypes that are efficient in K uptake and utilization is important.

Low-nitrogen Tolerance

Maize composites and hybrids vary with regard to their response to N availability in soil. It reflects genetic differences in their relative abilities to absorb inherent soil-N as well as that applied as fertilizer (Costa et al., 2002). Identification and use of maize genotypes capable of optimum growth and yield formation even under low soil-N availability seems important. Paucity for soil N is a rampant problem that limits N nutrition of maize crop. Such low-N soils are wide spread and are encountered across different maize cropping zones. Maize genotypes sown in areas with low soil-N fertility may recover and partition proportionately smaller quantities of N into stover and grains. Generally, it is preferable to grow maize that is tolerant to low-N conditions. Several genetic traits that affect N uptake and use efficiency are involved in bestowing low-N tolerance. Most of the genetic factors that impart low-N tolerance are related to roots, their spread, activities of enzymes related to N acquisition and absorption rates. Low-N tolerance is also affected by amount of biomass and grain formed per unit N absorbed that is N-use efficiency. Again, series of several genetic traits and loci may affect the net NUE achieved by a maize genotype. Let us consider an example. In India, about 60–70% of maize is cultivated under rain fed and dry land conditions. The fertilizer N inputs are restricted owing to climatic risks. Therefore, cultivation of genotypes with low-N tolerance is important. Researchers have developed several low-N tolerant maize genotypes using population improvement. For example, Ageti-76, Navjot, Kiran, D-765 yield optimum amounts of stover and grain even under low N conditions. The average productivity is under 1200 kg grain ha^{-1} if 60–80 kg N ha^{-1} is supplied. Similarly, there are few low-N tolerant early maturing hybrids (double top crosses) like D-741, Pop-31, Pop-49, Mahi Kanchan. These hybrids recover relatively higher amounts of N despite N paucity. They explore soil better and yield 1200–1500 kg grain ha^{-1}. According to Joshi et al. (2004) full season genotypes (e.g., IC1768) withstand low-N stress better than short duration genotypes. We should note that several of these genotypes that tolerate low-N perform efficiently even in optimum and high N fertility conditions. Such low N tolerant genotypes have a strong impact on nutrient dynamics within the maize belts. Firstly, they produce higher amounts of biomass and grains per unit of N consumed. Crop residue recycled is comparatively higher. Most importantly, larger fraction of soil-N is translocated from soil to above ground portion of the maize ecosystem. Nitrogen recycling too is enhanced, if appropriately larger fraction of stover is incorporated into fields.

The tropical condition prevalent in Nigeria is highly congenial for maize production. However, maize grain yield is low in many areas mainly due to paucity of soil-N (Kamara et al., 2005). Incessant cropping and rampant soil degradation lead to N deficiency. At the same time, we know that maize genotypes vary widely for their ability to tolerate low soil-N and still yield optimum levels of grains/forage. Field evaluation by Kamara et al. (2005) suggests that genotypes previously selected for tolerance to low soil-N performed well with optimum grain yield, if 30 kg N ha^{-1} was supplied. Repeated selection for tolerance to low soil-N improved the performance of genotypes.

Low soil-N is a problem frequently encountered in many of the Southern African countries. Incessant cultivation without appropriate replenishment leads to this situation. Therefore, farmers prefer to cultivate genotypes that are endowed with a certain degree of low soil-N tolerance. A wide range of genetic variability does exist for low soil-N tolerance (Worku et al., 2007). Evaluation at Harare in Zimbabwe indicates that genetic traits relevant to post-flowering N uptake and utilization contribute significantly to low soil-N tolerance. Nitrogen recovery prior to anthesis seems to have effect on ability of maize genotypes to perform under low soil-N. Clearly, planting low soil-N tolerant genotype in the subsistence farming belt of Southern Africa does affect N dynamics *per se*. It allows farmer to operate at low N application. It also induces higher amounts of biomass accumulation and C sequestration compared with genotypes that need higher levels of N in soil.

Low soil-N fertility is common in many Asian countries such as China, Thailand, Philippines, and Vietnam. Low-N fertility affects N recovery. The problem may get accentuated if it occurs along with drought. In order to overcome imbalance in N dynamics, research programs have aimed at examining genetic variation for N uptake and utilization, when the crop is sown in N limiting condition. Low-N tolerant genotypes are known to improve both grain and forage production (Chantachume et al., 1998; Hong, 2004; Logrono and Lothrop, 1998). The N dynamics in a maize belt could be severely or marginally affected depending on the extent to which genotypes tolerant to low-N are planted. The use of low-N tolerant genotype depends on extent to which the problem spreads in different maize cropping zones. Over all, a low-N tolerant genotype planted in Southeast Asian countries will enhance N recovery from soil. It improves fertilizer-NUE if chemical fertilizers are applied. A low-N tolerant genotype increases forage/grain production per unit of N in soil. On a wider scale, such genotypes affect N recycling in the ecosystem.

Genetic Variation for N-use Efficiency

Firstly, NUE is defined as grain yield per unit of N supplied or available in soil. There are two aspects to NUE. The NUE is a product of N uptake efficiency and NUE. The N uptake efficiency pertains to N absorbed or drawn into above ground portion via roots. On the other hand, NUE denotes amount of biomass or grain produced per unit of N absorbed by plant (see Gallais and Hirel, 2004; Krishna, 1998, 2002; Sangoi et al. 2001). It is well known that N is a key element that regulates productivity of maize crop. Further, there are several reports indicating that maize genotypes differ widely with regard to N nutrition, especially low soil-N tolerance and NUE (Akintoye et al. 1999; Banziger et al.1997; Presteri et al. 2003). Genetic variation has been identified for several of the traits that contribute to N uptake and N-use efficiency. More recently, several quantitative trait loci (QTLS) that govern N uptake and assimilation have been identified. Let us consider a few examples of physiological components that affect NUE of maize. Evaluation of maize germplasm lines at Ibadan in Nigeria has shown that principal components that enhance performance of maize genotypes are root penetration and spread, N uptake, N accumulation, and HI (Kamara et al. 2003). Maize genotypes may also differ with regard to N acquisition based on soil-N source. Generally, maize genotypes absorb greater quantities of N found as NO_3-N than from sources that provide NH_4-N (Debreczeni, 1999). Akintoye et al. (1999) have shown that NUE differences among synthetic lines and hybrids were related to both N uptake and utilization. They observed that hybrids were generally more efficient in N utilization. Hybrids possessed greater number of component traits that enhanced NUE compared with synthetic lines. Oikeh et al. (2003) found that maize cultivars differed with regard to N recovery. Genotypes with traits that allow rapid N uptake during grain fill were important. Greater N recovery at grain fill actually avoided undue loss via leaching. Therefore, it improved N-use efficiency. Maize genotypes vary widely with regard to root traits that contribute to higher N uptake. Yet, Beem and Smith (2004) have expressed that identifying maize genotypes with better root traits may only provide partial advantage with regard to N use efficiency. Maize genotypes with ability to partition greater quantities of N to seed seem to perform better in the field. According to Worku et al. (2007), maize genotypes that produced consistently higher grain yield under low soil-N conditions were associated with higher post anthesis N uptake, increased grain harvest per unit N accumulated and improved N HI. Paponov et al. (2005) have shown that N-efficient genotypes exhibit better kernel set even under low N conditions. It is said that low photosynthate availability reduces N efficiency and agronomic yield. Researchers at Rostov University, Russia have found that in addition to genetic variation for a single nutrient (e.g., N or P or K), ratios of major and micro nutrients may affect the biomass and grain formation. For example at given time, N/Zn or N/Cu ratio may affect genotype performance. Identification of genetic variation in maize for performance under different ratios of nutrients may be helpful. They have also reported that among various traits examined, high yielding maize genotypes often partitioned greater quantities of N into seeds. Nitrogen translocation index seemed to play an important role in grain yield potential of a maize genotype. Overall, we should note that each genetic trait which directly or indirectly affects NUE and biomass/grain yield is important. However, immediate relevance and extent to which each genetic trait influences N dynamics in the ecosystem varies depending on local conditions.

Let us consider a report depicting performance of maize hybrids released into Brazilian Cerrados during past 5 decades from 1960s to 2000. Sangoi et al. (2001) estimated grain yield, N recovery pattern and traits that contribute to grain yield for various hybrids grown by Brazilian farmers since 1960s. Irrespective of N input level, hybrids released more recently during 1990s performed better and yielded 6–9.5 t grain ha^{-1} compared with hybrids of 1960s that yielded 3.9–6.8 t ha^{-1}. Clearly, at any level of soil-N status tested, recently released hybrids were more efficient in absorption and use of N. In other words, during past 40 years maize breeders have been consistently improving genetic traits relevant to N uptake, its use-efficiency and higher grain/biomass yield. Obviously, each of the above genotypes as they entered the ecosystem and spread, they had their influence on N dynamics and productivity of maize. According to Sangoi et al. (2001) genetic traits that contributed to yield such as grain ear^{-1}, days to anthesis, ear leaf N content, N content in shoots, 1000 grain weight were all improved. Each of these traits could have proportionately affected N dynamics and productivity of maize in the ecosystem. Over all, during 40 years, N input to Brazilian maize belt has been gradually enhanced from 30–60 kg N ha^{-1} to 200–225 kg N ha^{-1}. As a result, grain yield has improved from just 3.5–4.0 t ha^{-1} to 9.5 t grain ha^{-1}. Genotypes with better N recovery, NUE and grain yield potential have contributed to intensification of Brazilian maize belt.

We should realize that when farmers replace the cultivars/hybrids, it can have profound effect on the N dynamics of the field or cropping zone. Introduction of N efficient genotypes has its immediate effect on amount of biomass/C-fixation per unit N supplied into the cropping zone. The amount of C sequestered into agroecosystem per unit extraneous N added may vastly improve. When extrapolated to large cropping zones it amounts to significant amount of increase in C sequestration. Next, per unit fertilizer N applied, farmers reap higher amounts of grain and stover. Farm animals may derive greater quantities of forage. At the same time, N and C removed via grains may also increase. Over all, introduction of N efficient genotypes may directly alter N and C dynamics in maize belts. As a corollary, to produce a given amount of biomass an N-efficient maize cultivar requires lesser quantity of fertilizer supply. Therefore, it is reasonable to believe that N-efficient maize hybrids or composites that dominate a region may have greater influence on the N dynamics and productivity of the agroecosystem.

Some of the generalizations regarding uptake and utilization efficiency apply equally to several other mineral nutrients including P, K, secondary, and micronutrients (see Kovacevic et al. 2001; Krishna, 2002, 2010; Parentoni and Lopes de Souza, 2008). There is indeed vast amount of knowledge accrued about genetic variation available with regard to acquisition and utilization of these elements. Let us consider an example dealing with genetic variation from P uptake and utilization efficiency. Parentoni and Lopes de Souza, (2008) evaluated a set of 28 tropical maize genotypes on soil with high and low P content. They assessed several physiological traits relevant to P nutrition including P uptake and utilization pattern. Traits like root phosphatase activity has relevance to P accumulation pattern in a given maize genotype. Maize genotypes are known to show wide genetic variation for phosphatase activity (Machado and Furlani, 2004). Root traits like spread, fine roots and mycorrhiza are important aspects that affect P uptake and its dynamics. Phosphorus uptake was highly correlated with surface contact and root surface exposed to soluble P pool in soil. Based on availability of variability for P nutrition and grain yield, it was concluded that maize breeding programs should concentrate on P acquisition efficiency and HI values in order to obtain better P efficiency. Phosphorus utilization efficiency was less effective. Selection for lower levels of P concentration in seeds improved utilization efficiency (see Table 7.4). Clearly, planting P-efficient genotype will mean lowered supply of P to crop. We can avoid undue accumulation of P in soil. Further, extent of loss due to fixation, leaching and weeds could be reduced. Interactive effects of N with other elements in the cropping belt could be affected. For example C:P and N:P ratios in crop tissue and residues may change, the extent of P recycled via stover and that removed through seeds also differ if P-efficient genotype dominates the cropping belt. Overall, a P efficient genotype will affect P dynamics in the ecosystem.

There are several reports about tolerance of maize genotypes to low K availability in soil. Potassium deficiency has far reaching effects on leaf formation, photosynthesis, biomass, and grain formation. Potassium deficiency affects root growth and its activity. However, we ought to know that maize genotypes grown since decades have varied in their ability to negotiate low levels of exchangeable-K in soil (Minjian et al. 2007). Genetic differences in K uptake, K-use efficiency and grain yield formation has been reported. For example, Doberman (2001) reported that a hybrid such as 33G27 accumulated 30–40 kg K ha^{-1} more at both low and high K input levels. It also produced significantly higher biomass and grain yield. Therefore, consequences of variations in K uptake, K-use efficiency, biomass formation, crop residue recycling, and grain productivity needs to be carefully judged while expanding cultivation of a maize genotype. The difference among genotypes was dependent on stage at which K uptake and biomass is estimated. The K recovery rates of maize genotypes may vary enormously at various stages. Therefore, it is wise to adopt tillage, K supply and irrigation schedules based on genotype, its K recovery pattern and K use efficiency.

Table 7.4. Productivity, Phosphorus accumulation and Efficiency of Maize genotypes sown under high and low P environment in Brazilean Cerrados.

Envrionment	Yield		Phosphorus Accumulation		Phosphorus Efficiency	
	Grain	Stover	Grain P	Stover P	P Uptake	P Utilization
	(t ha^{-1})		(%)		kg kg^{-1} Av P	kg kg^{-1} P
High P						
2005	4.39	8.94	0.22	0.07	0.17	284
2006	5.65	7.30	0.29	0.07	0.22	261
Low P						
2005	2.29	4.31	0.19	0.06	0.33	334
2006	2.82	2.99	0.26	0.08	0.32	283

Source: Parentoni and Claudio Lopez De Souza, 2008.

MAIZE GENOTYPES, NUTRIENT DYNAMICS, AND PRODUCTIVITY

Maize genotype is one of the primary factors that engineered alterations in nutrient dynamics and productivity of maize belts all over the world. Maize genotypes have played an important role in inducing intensification of agroecosystems across various locations. Their effects on nutrient supply, recovery, loss, sequestration, and turnover of nutrients have been highly perceptible in some areas and marginal in others. Maize genotypes selected by farmers have also altered pattern of nutrient supply, recovery and turnover at different stages of the crop. During recent years, fertilizer-based nutrient disbursement in maize belts have been appreciably influenced by the genotype a farmer selects to plant. Genetic traits such as total biomass, grain yield, HI, root:shoot ratio, nutrient acquisition pattern, nutrient translocation index and efficiency have all contributed to alteration in nutrient dynamics within the maize belt. For example, the HI that has improved from a mere 2.4 to 4.7 has meant that sizeable amount of nutrients is held in seeds that could be removed from the ecosystem. A high shoot to root ratio may minimize nutrients held in roots and that sequestered in soil phase of the maize belt. Several other traits such as crop duration, forage component, grain yield, seed appearance, cooking quality may affect spread of a genotype in the ecosystem.

Historically, maize belts all over the world have experienced a certain degree of alteration in nutrient dynamics that is easily attributable to progressive changes of genotypes and their specific traits. Simplest of the examples to quote relates to landraces grown in Mexico or Great plains. The landraces yielded marginally at <1.0 t grain ha^{-1} and nutrients supplied were entirely through extraneous or recycled organic sources. This was the situation prior to invention of chemical fertilizers, say in 1800s or early 1900s. However, availability of improved composites and hybrids with greater yield potential and advent of chemical fertilizers induced drastic changes in the maize agroecosystem of North America. Progressively, nutrients impinged into soils increased and resultant removal via grains and forage also increased. Nutrients unused but accumulated in roots and soil also increased. In 7 to 8 decades, since 1930s, nutrients impinged increased from mere 5–10 t FYM to 250–300 kg N, P and K plus 10–15 t FYM ha^{-1}. According to Duvick and Casssman (1999) introduction of various genotypes since 1930 has improved grain yield by 85 kg ha^{-1} per year. Dobermann et al. (2002) have reported that fertilizer inputs have increased at a rate of 1.1 kg ha^{-1} yr^{-1}. Grain yield potential of maize genotypes sown in the Corn Belt has jumped from 5.5 t ha^{-1} to 12 t ha^{-1}. Parallel increase in nutrient supply during the same period was solely to support higher yield potential. In addition to total nutrient supply, as stated above, nutrient scheduling too has changed especially with regard to N. Nitrogen is being channeled in split dozes to match the nutrient demand of the maize genotype. This is vastly a different situation compared with landraces during early 1900s or even composites/hybrids planted during mid 1900s. Currently, a maize genotype sown in the Corn Belt receives N as basal and at least 4–5 split dozes. The N dynamics that ensues is quite different from maize belts where in farmers tend to supply much of the fertilizer-N as basal input. Maize genotypes planted have also engineered enhanced nutrient recovery from soils the maize ecosystem (Figure 7.1).

During past 50 years, nutrient supply into Corn Belt has increased approximately at a rate of 2.5 kg N, 0.8 kg P, and 2.2 kg K ha^{-1} yr^{-1} from 1950 to 2000. Nutrients partitioned into forage, recycled via stubble/stover, removed as forage for animals and as grains for human consumption have all changed enormously compared with yester years when land races or composites with marginal yield potential (3–4 t grain ha^{-1}) were cultivated in the Great Plains. Practically, plant breeders and farmers together have induced intensification of Corn Belt by using corn genotypes with high yield potential. Perhaps, there is still a gap large enough to be achieved by using improved hybrids and better nutrient schedules. At present, we have no idea regarding the extent to which the US Corn Belt could be intensified. However, we should note that unused fertilizer-based nutrient, if any, will build-up in soil and become vulnerable to loss via emissions. The ground water quality may also be affected if genotypes are not efficient in scavenging nutrient from soil. It may induce a certain amount of green house effects, if not checked in time. The situation described above is also encountered in other maize belts such as those in Northeast China, European Plains, Cerrados of Brazil, Pampas of Argentina, and where ever maize cropping is highly intensified using fertilizers (see Table 7.5).

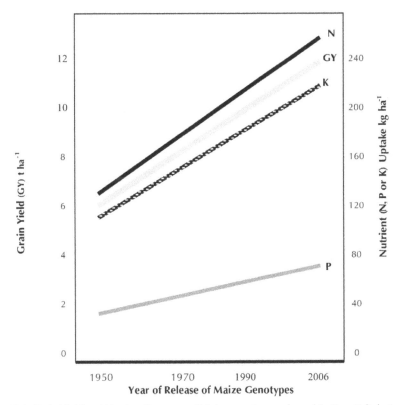

Figure 7.1. Grain Yield and Nutrient Recovery of Maize genotypes released in Corn Belt during past 5 decades.

Note: Introduction of different maize genotypes has affected grain production and nutrient recovery from soil in the ecosystem. Grain yield improved by 6.5 t ha^{-1} from 5–5.5t ha^{-1} in 1950s to 11.0 t ha^{-1} in 2000 partly due to crop improvement. Nutrient supply and recovery increased proportionately in order to support higher grain/forage productivity. In particular, nutrient recovery rates improved remarkably. Nitrogen recovery from soil to above-ground portion improved from 131 kg N ha^{-1} in 1950 to 255 kg Nha^{-1} in 2000. Similarly, in 50 years, P absorption increased from 39 kg to 78 kg P ha^{-1}; and K absorption from 119 to 232 kg k ha^{-1}. Since, over 75% of nutrient uptake by maize occurs between 26 and 100th day, nutrient uptake rate exhibited by genotype during this period is crucial. It regulates nutrient dynamics and productivity to a great extent. In the above graph nutrient uptake has been calculated based on assumption that on an average maize crop recovers 23 kg N, 7 kg P, and 21 kg K to produce one ton grain. Grain yield data of maize genotypes is derived from Duvick and Cassman (1999).

In India, maize is cultivated in the entire South India, parts of Gangetic plains and hills. For a long period since its introduction in medieval times, maize landraces have occupied these regions sparsely compared with major cereals like wheat, rice or sorghum. Until mid 1900s landraces were cultivated with marginal yields. They were confined to dry lands and regions with low soil fertility. The subsistence level nutrient supply was solely achieved via organic manures and crop residue recycling. It supported low levels of grain yield (0.5–0.8 t ha^{-1}) plus 3–4 t forage ha^{-1}. During 1970s, several hybrids and composites were released into Indian peninsula (e.g., Deccan, Ganga, Ganga safed, etc.). These genotypes induced farmers to supply higher levels of nutrient to match their higher grain/forage yield potential (3.5–4 t grain ha^{-1}). Farmers tended to supply 30–40 kg N, 10–20 kg P ha^{-1}plus FYM. During past decade (2000–2010), hybrids with greater yield potential at 5.5–8.0 t grain ha^{-1} have been spreading rapidly into South Indian maize belt (see Kaul et al., 2007; Ikisan, 2008). In order to support higher yield expectations, farmers in this region are advised to apply much higher levels of nutrients (Table 7.5). The South Indian maize belt has got intensified from subsistence farming zone to moderately high yielding, partly due to introduction of improve hybrids. Higher amounts of organic and inorganic manures have induced this intensification. Nutrient supply increased by 50–80 kg ha^{-1} and grain yield jumped by 2.0–2.5 t ha^{-1}. Intensification is also attributable to change over from composites to hybrids with higher grain yield potential. Currently, 90–95% of maize ecosystem in South India is filled up with hybrids (Table 7.6). It is a clear case of maize genotypes inducing rapid changes in nutrient dynamics and productivity.

Table 7.5. Impact of maize genotypes on nutrient supply, yield and nutrient recovery within different maize belts of the world—A gross view and few examples.

Maize Cropping Zone	Nutrient Supply kg ha^{-1}	Yield (t ha^{-1}) Grain	Stover	Nutrient Recovery kg ha^{-1}
Corn Belt of USA				
Landraces (1800s–Early 1900s)	Subsistence, FYM 2–5 t ha^{-1}	1.25	3–4	20 N, 5 P, 18 K
Hybrids (during 2005–2010)	180 N, 60 P, 200 K, 12 t FYM ha^{-1}	11.2	14	235 N, 60 P, 240 K
Middleast Asia (Egypt)				
Landraces (until 1950s)	Subsistence inputs 2–3 t FYM ha^{-1}	0.5–0.6	3–4	15 N, 3–5 P, 18 K

Table 7.5. (Continued)

Maize Cropping Zone	Nutrient Supply	Yield (t ha⁻¹)		Nutrient Recovery
	kg ha⁻¹	Grain	Stover	kg ha⁻¹
Hybrids	120N, 30P, 5 t FYM ha⁻¹	2.2–2.9	7–8	56 N, 18 P, 55 K
East Africa				
Landraces	Subsistence inputs, 2–3 t FYM ha⁻¹	0.5–0.7	3–4	15N, 5 P, 18 K
Hybrids/Composites	25–50N, 25 P, 5.0 t FYM ha⁻¹	3.2–4.0	5–8	70 N, 25 P, 75 K
(Pioneer H3253, KH600-11D, H625)				
South India				
Landraces (until mid-1900s)	Subsistence level, 2–5 t FYM ha⁻¹	0.5–0.8	3–5	18 N, 5 P, 18 K
Composites/Hybrids 1970–1995	40–60 N, 20 P, 40 K; 5 FYM t ha⁻¹	1.5	5–6	30 N, 10 P, 35 K
Hybrids (1990–2010)	80–100N, 40P, 100 K; 10 t FYM ha⁻¹	3–4	6–9	70 N, 25 P, 75 K

Sources: Duvick and Cassaman, 1999, Dobermann et al., 2002; Soliman, 2006; Abendroth and Elmore, 2007, Ikisan, 2008; Mureithi, 2008, Krishna, 2010.

Note: Nutrient recovery by maize genotypes has been computed considering that 22–23 kg N, 5–9 kg P. and 22 kg K is needed to produce 1.0 t grains plus proportionate forage based on HI.

Table 7.6. Influence of maize genotypes on nutrient supply, recovery and productivity in India.

1960-1970s

Composites: Kiran, Renuka, Jawahar, Vijay, Rattan, Vikram.

Hybrids: Ganga, Deccan, Deccan 101, Ganga Safed,

Nutrient Supply: 30 kg N, 20 kg P, 2–5 t FYM ha⁻¹

Nutrient Recovery: 20–25 kg N, 5-7 kg P, 25 kg K ha⁻¹

Grain/Forage Yield: 1.0-1.5 t grain ha⁻¹ plus 3-4 t forage

2005-2010

Composites: Pusa Comosite, Pratap Makka, Jawahar 216, NAC 6002, Narmada Moti

Nutrient Supply: 60-80 kg N, 40 kg P, 80 kg K, 25 kg ZnSO₄ plus 10 t FYM ha⁻¹

Nutrient Recovery: 50-60 kg N, 14-18 kg P, 50 kg K ha⁻¹

Grain /Forage Yield 3-4.0 t grain ha⁻¹ plus 3-8 t forage ha⁻¹

Hybrids: HQPM, Vivek, COH5, HM8, COMH4, DMH2, Pratap

Nutrient Supply: 90-110 kg N, 60 kg P, 80-100 kg K, 25 kg ZnSO₄ plus 10 t FYM ha⁻¹

Nutrient Recovery: 90-120 kg N, 22-30 kg P, 110 kg K ha⁻¹

Grain /Forage Yield 5.5-6.0 t grain ha⁻¹ plus 10-12 t forage ha⁻¹

Increase (ha⁻¹ season⁻¹) in 40 years due to introduction of Improved Composites and Hybrids

Composites

Nutrient Supply: 30 kg N, 20 kg P, 30 kg K ha⁻¹ 3-5 t FYM ha⁻¹

Nutrient Recovery: 30-40 kg N, 8-10 kg P, 30 kg K ha⁻¹,

Grain/Forage: 2.0-3.0 t grain ha⁻¹ plus 4-5 t forage ha⁻¹

Hybrids

Nutrient supply: 60-80 kg N, 40 kg P, 50-70 kg K ha-[1] 5-8 t FYM ha-[1]
Nutrient Recovery: 70-80 kg N, 12-15 kg P, 75 kg K ha-[1]
Grain/Forage Yield: 3.0-40. t grain ha-[1], 7-8 t forage ha-[1]

Sources: Ikisan, 2008, Kaul et al., 2008, Krishna, 2010.

Note: Increase in nutrient supply was induced by introduction of maize genotypes with greater yield potential. Maize agroecosystem, particularly in South India has been intensified using high yielding hybrids and higher nutrient supply. Average productivity has increased by 3–4 t grain ha-[1] in 40 years since 1960s.

Maize genotypes may get preferred by farmers for various reasons other than just grain/forage yield. Agroclimatic conditions, adaptability, human preferences like grain color, and cooking quality also affect preference for a genotype. For example, white grained maize is preferred in tropical and subtropical regions. Yellow grain maize is mostly grown in temperate countries (Dowswell et al., 1996). White maize is most widely grown in Eastern and Southern African countries. Spread of such genotypes may not have immediate change in nutrient supply into the fields, unless such genotypes have higher grain yield potential and specific need for higher nutrient inputs. Similarly, during recent years there are several transgenic maize genotypes being introduced across different maize belts. Most of them relate to herbicide or insect tolerance or grain quality. As such, these may alter neither the grain/forage potential nor the nutrient requirements perceptibly. The alterations, if any, in nutrient dynamics will be solely derived from biomass/grain yield advantage due to pest resistance—for example, in case of BT maize. Let us consider an example in support of the above suggestion. Subedi and Ma (2007) compared parameters related to N dynamics and productivity of a BT hybrid and Non-BT near isoline at Ottawa in Canada. They employed ^{15}N methods to decipher alterations in N dynamics, if any. They found that N accumulation, its partitioning and NUE did not differ between the two genotypes. About 47% of applied N was recovered by both BT and Non-BT hybrids and out of it 70% was partitioned into kernels. There was no difference in pattern of N recovery between Bt and Non hybrid right until silking stage. The extra N found in BT hybrids was attributable more to higher dry matter accumulation

SUMMARY

The influence of maize cropping on nutrient dynamics and ecosystematic functions begins earnestly as seedlings establish. The growth rate and nutrient accumulation pattern has direct impact on carbon status and mineral nutrient partitioning between above and below ground portion of the agroecosystem. Nutrient supply schedules should aim at matching crop's demand at various stages. Certain cultivars may absorb slightly higher amounts of nutrients at early growth stages and accumulate it in different parts, then translocate it to grains. In such cases, nutrient supply has to be regulated accordingly. Knowledge about nutrient uptake pattern, HI, potential biomass/grain yield, rooting patterns and root:shoot ratio is immensely useful, since they affect nutrient distribution within maize fields. Many of these physiological traits have direct effect on grain/forage yield. Among nutrients, physiological aspects of N, its absorption rates and accumulation pattern has received greatest attention. Nitrogen supply has immediate and perceptible effects on growth rate and yield formation. Physiological manifestations of maize are also influenced by planting dates ad planting density. Historically, it is said that N supply, irrigation and planting density have engineered the highly productive maize belt in USA. Obviously, we can aim at replicating this effort in other maize cropping zones.

Maize crop improvement has helped us to improve biomass and grain yield generation. Genetic variation for traits like nutrient uptake rates, nutrient use efficiency, low-N tolerance, drought tolerance and water use efficiency need greater attention with regard to their influence on productivity of the ecosystem.

KEYWORDS

- **Endosperm**
- **Micronutrient**
- **Mineral nutrients**
- **Nodal roots**
- **Silking stage**

Chapter 8

Integrated Nutrient Management

Integrated Nutrient Management (INM) encompasses use of a combination of several resources and agronomic procedures. The INM basically involves supply of nutrients using a variety of sources both organic and inorganic in nature. The INM also involves an assortment of agronomic procedures. It begins with conservation tillage or no tillage. Soil conservation procedures like ridges and furrows, contour planting, mulching, growing green manure, wind breaks, planting agroforestry species in strips (strip cropping), intercropping, and others that aim to reduce soil erosion are included under this concept. Methods that enhance fertilizer efficiency are most important under INM. A few examples are deep placement of fertilizers, splitting fertilizer-N, combination of inorganic and organic manures, use of rock-phosphates plus compost, adopting foliar sprays if nutrient deficiencies are to be corrected rapidly, etc. Methods that lessen loss of fertilizer-N or P are practiced rather stringently. During recent years fertilizer application has gained in accuracy. Intensive soil sampling and periodic tests allow us to map variations in soil fertility. Therefore, fertilizer supply schedules are mostly site specific. Fertilizer recommendations based on large cropping zones is getting replaced by use of Site-specific Nutrient Management (SSNM). Computer based models that predict crop response to fertilizer supply is available. The INM considers yield goals and soil properties carefully while devising nutrient supply schedules to the crop. The INM also involves due consideration to weather and precipitation pattern as well as irrigation.

Almost all of the procedures or their combinations are obviously aimed at conserving soil nutrients, soil quality, and moisture. Primarily, INM aims at maximizing crop productivity and improving agronomic efficiency of fertilizer-based nutrient and organic manures. It also tries to preserve long term productivity of soils and avoids soil deterioration. The INM aims at least disturbance to soil biological processes and ecosystematic functions. There are indeed too many reports on these topics. In this chapter only examples from few maize cropping zones have been included for detailed discussions.

CROP RESPONSE TO NUTRIENT MANAGEMENT PROCEDURES

Historically, for a considerably long stretch of time, maize production was accomplished predominantly by adopting subsistence farming practices. Agronomic procedures aimed at managing soil nutrients were meager. They included *in situ* recycling of nutrients via crop residues and FYM supply. Organic manures were the main stay to sustain soil fertility. Precipitation pattern dictated the crop growth and grain yield response. Irrigation, if any was once again sporadic. Nutrient loss via soil erosion, percolation, and emissions may have reached higher proportions due to lack of appropriate technologies. Farms that adopted soil conservation practices like ridge planting, bunding and mulching may have preserved soil nutrients *in situ*. Yet, in USA and other maize cropping zones of the world, average crop yield during 1800s to 1930–1950s never increased beyond 0.8 to 1.2–1.5 t grain ha⁻¹ plus forage (Egli, 2008). Egli (2008) terms this period since 1800s to 1930 that involved subsistence farming procedures as era of "low input agriculture". Whatever be the forage/grain yield potential of land races or specific cultivars, subsistence farming or even traditional farming procedures practiced by farmers curtailed high grain yield. Literally, there was not much soil nutrients and moisture provided in the root zone for maize genotypes to respond. Subsistence farming approaches often resulted in maize grain yield <1.0 t ha^{-1} and forage based on harvest index. In the absence of nutrient replenishments, soil fertility got depleted and fallows were introduced to refresh and replenish soil fertility. Even today, subsistence farming procedures are adopted in Savanna regions of Africa, Southern plains of India, or elsewhere in Cerrados of Brazil. In these locations, maize forage/grain yield averages are low, ranging form 0.8–1.0 t grain plus 2–3 t forage ha^{-1}.

Beginning in 1950s, "Corn Belt of USA" got intensified due to large scale use of chemical fertilizers, irrigation, and introduction of high yielding composites/hybrids of maize. This period from 1950s till date is termed "high-input era" by Egli (2008). The maize grain yield consistently increased since 1950s from 1.7 t ha^{-1} to 10.5–11 t ha^{-1} in 2008. A clear 5 fold increase in grain yield equivalent to 8.0 t grain ha^{-1}. According to Duvick and Cassman (1999) genetic improvement of maize hybrids and enhanced plant population had a major effect on grain yield increase since 1930s. On average, maize grain yield response improved annually by 52–87 kg grain ha^{-1}. Hybrids contributed a major share (90%) of grain yield increase. Rapid increase in fertilizer-N, P, and K supply too contributed to higher grain yield. Adoption of INM, further enhanced crop response. The INM procedures also aimed at preserving and improving soil quality, fertility, and ecosystematic functions. Some of the best maize crop responses to soil fertility and crop management procedures in the Corn Belt have ranged between 21.2 t grain ha^{-1} and 23.2 t grain ha^{-1} (Duvick and Cassman, 1999). Such rapid increase in grain yield of maize was experienced in other continents at different periods during past 5 decades. For example, introduction of high yielding hybrids into South Indian Plains in 1970s markedly enhanced increased grain yield from 0.7–0.8 t grain ha^{-1} in 1970s to 3.5–4.0 t grain ha^{-1} in 2005. It was also attributed to higher amount of nutrients impinged into the ecosystem. We should note that, yet, in most of the maize farming zones of the world, grain/forage yield potential of genotypes in the given soil and atmospheric conditions is considerably higher than what we actually reap currently (Table 8.1; Figure 8.1; Plate 8.1).

Table 8.1. Yield potential and current yield of maize crop raised in different parts of the world.

Region	Yield Potential (Current Yield) t grain ha^{-1}	
	Low or Moderate Soil Fertility	High Soil Fertility
Fareast and Southeast Asia	5.0 (2.5)	8.0 (3.0)
South Asia	5.0 (1.7)	7.0 (2.6)
West Asia and North Africa	5.0 (3.2)	7.5 (3.2)
Sub-Saharan Africa	5.0 (0.6)	7.0 (2.5)
Latin America and Caribbean	6.0 (1.1)	10.0 (4.0)
Corn Belt of USA[@]	6–7 (5.5)	12 (10–11.2)

Sources: Duvick and Cassman, 1999; Egli, 2008; Ping et al., 2008; Ofori and Kyei-Baffour, 2008.

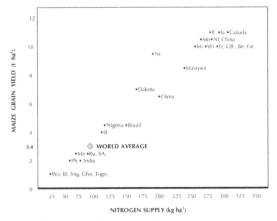

Figure 8.1. Response of maize crop to nitrogen supply.

Note: Nitrogen supply level to Maize cropping zone varies markedly in different countries/regions. The N-use efficiency ranges from 20–22 kg grain kg^{-1} N supplied in high input intensive farming zones to 40–45 kg grain kg^{-1}N in subsistence farming regions. Fertilizer-N use efficiency is generally higher in areas with low N input trends and it lessens as fertilizer-N supply increases. In the above graph, fertilizer-N use efficiency is 40–45 kg grain kg^{-1}N in countries practicing subsistence or low input technology and fertilizer-N supply is around 25–60 kg N ha^{-1}. As fertilizer-N input is increased to 140–180 kg N ha^{-1}, fertilizer-N use efficiency decreases to 30–35 kg grain kg^{-1} N and in some of the highest N input (180–280 kg N ha^{-1}) regions such as US Corn Belt or European plains, fertilizer-N use efficiency decreases further to 20–22 kg grain kg^{-1} N. High grain yield reaching up to 20 t ha^{-1} have also been reported from Corn Belt of USA.

Clearly, yield gaps small or large exist between harvest size achieved by farmers at present and that possible by manipulating nutrient dynamics.

Plate 8.1 Response of maize to fertilizer-n supply being examined at Arlington in Wisconsin in USA–2008.

Source: Carrie Laboski and Todd Andraski, Department of Soil Science, University of Wisconsin, Madison, WI, USA.

Note: Maize cultivation is intensive in the Corn Belt of USA. Fertilizer N and other nutrients are supplied at relatively high levels. Maize responds to fertilizer N supply with 7–10 t grain ha^{-1} plus forage.

Farmers Practices (Traditional Methods) for Maize Production

At present, quite a few farmers in many of the maize cropping belts adopt techniques that were actually evolved from traditional farming procedures that were in vogue for centuries. Organic and inorganic manure supply methods, timing and quantities are often dictated by factors like local environmental conditions, soil fertility status, irrigation facilities, crop genotype, and yield goal. Farmer's practices vary widely across continents and sub regions. Quite often, nutrients supplied adopting farmer's practices may not suffice to reach grain/forage yield potential. Therefore, grain/forage yield under Farmer's practices are usually low. It could be revised to reach higher yield goals. Further, in most cases, only fertilizer N, P, and FYM are applied. Therefore, it creates an imbalance with regard to other nutrients. Incessant cropping may lead to loss of SOC and soil quality. Farmers may apply nutrients considering inherent soil fertility status and soil test values for major nutrients at the time of sowing. Yet, when deciding nutrient supply it overlooks soil fertility variations at individual field level. Nutrient loss from the system could be rampant, since extent of nutrient removals via grain/forage and loss to atmosphere often far exceeds that supplied via fertilizers, manures, and recycling of residues. Nutrient supply may not match rates at which they are depleted from soil. Further, it is not balanced. Ratios of nutrient supplied may not match crop's preferences. Overall, Farmers Practices leads to relatively low turnover rates of nutrients in the field. As a consequence, it results in marginal grain/forage yield. Nutrient recycling too is insufficient to reach higher potential grain yield possible in the given environment. Farmer's Practices are well suited and popular in subsistence farming or low input agricultural cropping regions (e.g., Sub-Saharan Africa, Semi-arid regions of Brazilian Cerrados, Drylands in Southern Africa, Middleast or Southern India). Fertilizer application to maize is almost nil or meager in some regions. It may range from nil to 40 kg N, 10–20 kg P, and 5 t FYM ha^{-1}. This leads to grain yields around 1–1.2 t ha^{-1}. It is much below the grain yield possible under State Recommendation or INM. For example, in South India, maize cultivated adopting Farmer's Traditional procedures yields only 1600 kg grain ha^{-1}, but that provided with nutrients as stipulated by State Agency Practices yields 3150 kg ha^{-1}. Farmer's practice involves application of 60–80 kg N, 20–40 kg P, 30 kg K, and 5–10 t FYM ha^{-1}. State Agency recommendations include 80–120 kg N, 40 kg P, 60 kg K, 10 t FYM plus biofertilizers (see Krishna, 2010). Micronutrient inputs are lacking under Farmer's practice. Therefore, nutrient imbalances that set in, may curtail response to high N or P inputs. Hence, farmers are advised to apply larger dosages of organic manure that can replenish micronutrients to a certain extent.

State Agency Recommendations

Production technologies or package of practices are developed by governmental agencies of different countries for a specific region, soil type, a crop genotype, or a definite cropping pattern. The agronomic procedures are prescribed for a large maize cropping belt or agroclimatic zone. For example, maize farmers in a state such as Iowa in the Corn Belt of USA may be advised to supply 240 kg N, 60 kg P, and 180 kg K ha^{-1} depending on soil type and maize hybrid selected for cultivation. Nitrogen is supplied in 4 splits. Lime may be prescribed for the entire region in order to enhance soil pH. Such recommendations apply to small or vast maize planting regions.

Let us consider yet another example. In South India, maize farmers are asked to apply 80–120 kg Urea, 45–60 kg DAP, 40–45 kg MOP plus 10 kg $ZnSO_4$ kg ha^{-1} in order to achieve a targeted grain yield of 1.7–2.5 t ha^{-1} on the low fertility Alfisol. In the Vertisol region, maize crop may receive much higher fertilizer dosages. An irrigated *rabi* crop is supplied with 120 kg N, 60 kg P, and 40 kg K ha^{-1}. Nitrogen is applied in three split dosages. Micronutrient mixtures are also supplied to achieve balanced nutrition.

In the Pampas of Argentina, state agencies stipulate application of 40–80 kg N, 60 kg P, and 50 kg K ha^{-1}, micronutrients plus FYM at 12 t ha^{-1}. We should note that such State Agencies Recommendations consider background soil fertility, yield goals and economic advantages of farmers in large, yet definite region. Such blanket recommendations overlook site-specific or within location variations in soil fertility or other geographic parameters. Nutrient dynamics is ensured by adopting higher fertilizer inputs rather than accurate distribution based on soil fertility variations and crop demand. Nutrient accumulation in soil is a clear possibility and it could be uneven depending on extent depleted by crop. Accumulated nutrients are prone to loss via leaching, percolation, and emissions. Nutrient supply is stipulated for individual crop, not the entire rotation. Therefore, if considered for entire crop rotation, fertilizer efficiency is still not the best or highest. Crop response could be uneven, considering that soil fertility variation in small locations or a field persists. Most importantly, State Agency Recommendation may not be the economically or even physiologically the optimum in terms of nutrients. Sometimes, fertilizers supplied could be less than sufficient to close up yield gaps considering the potential yield possible in the given locality. Chemical fertilizer supply rate in different countries has varied. It has resulted in differences in maize grain yield per unit land. Following is an example that depicts effect of state agency recommendations on maize productivity:

Country	Chemical Fertilizer Supply			Grain Yield
	N	P_2O_5	K_2O	t ha^{-1}
	----------	(kg ha^{-1})	----------	
USA	150	70	90	8
Argentina	50	25	20	5
China	188	63	25	5
India	60	20	40	2

Sources: Ikisan, 2007; Zhang et al., 2008.

Note: In case of India, fertilizer supply and grain yield data are for rain fed crop in dry lands. Irrigated farms yield higher quantity of grain (4–5 t grain ha^{-1}) and forage.

Best Management Practices

"Best Management Practices (BMPs)" is a term used more commonly in the "Corn Belt of USA" and other locations. The BMP usually envisage high inputs of fertilizers, in balanced proportions along with larger dosages of organic manure. A combination of major nutrients—N, P, K, and micronutrients is supplied based on maize yield goals set for a given region or location in the "Corn Belt". Often, BMPs consider the extent of profit derived by the farmer due to fertilizer supply. Maize crop response to BMP is generally excellent considering that no mineral element deficiency is allowed to persist. Soil moisture deficit is avoided by periodic irrigations.

Adoption of BMP has improved maize grain/forage yield in United States of America. Grain yield has fluctuated between 8 and 10 t ha^{-1}. The BMPs vary based on geographic location, region or nation. For example, BMPs in South Africa involves addition of 80–160 kg N, 40–60 kg P, and 80 kg K ha^{-1}. Micronutrients are supplied based on soil test and yield goals. Maize farmers, especially those holding commercial farms provide their fields with large amounts of FYM ranging from 10 to 20 t ha^{-1}.

In Australia and New Zealand, BMPs aim at maintaining soil quality, minimizing nutrient leaching, improving crop yield and profitability. Any BMP aims firstly in minimizing yield gap, if any. Yield gap here is defined as grain harvest possible if maize crop was grown in excellent soil physical, chemical, and biological conditions minus that actually attained. Some of the BMPs recommended to farmers include analysis of soil for major and micronutrient just prior to sowing; maintenance of high plant population up to 90,000–1,20,000 and fixing fertilizer schedules accordingly based on yield goals. Obviously, to attain high grain/forage yield farmers are asked to apply 275 kg N, 45–60 kg P, and 300 kg K ha^{-1} for maize crop that yields 10–12 t grain ha^{-1}. Organic manure application is mandatory if SOC is < 2.5%. Application of compost or FYM usually suffices to restore SOC content to optimum levels. Fertilizer-N is usually banded to avoid loss via leaching and emissions. Fertilizer-N is supplied in split doses, but almost entire P and K are supplied prior to sowing. Weeding at seedling stage is preferred. Fields are kept weed-free for at least 45–60 days, in order to avoid loss of nutrients to weeds. A "Maize Calculator"—a computer program is used to calculate grain yield possible. It is predominately based on maize crop response to fertilizer-N supply (Reid, 2006).

The BMPs could be revised based on inherent soil fertility status, yield goal, and profitability. Often, BMPs are tailored to suit the entire rotation. This way, it economizes on nutrient supply. Errors that may occur if only one crop in sequence is considered are avoided. Let us consider an example. In the Chinese maize belt, maize-wheat rotations are common. Such rotations perform better under BMP than under Farmer's Practices. The extent of advantages offered by BMPs compared with Farmer's Practice, especially with regard to grain yield, N saving, N recovery, and reduction in N loss from the ecosystems are as follows:

Advantage of Best Management Practice over Farmer's Practice				
Cropping System	Yield Increase (%)	N Saving (%)	N Recovery Increase (%)	Reduction of N Loss (%)
Maize-Wheat in China	41–59	5–10	12–15	43–69

Source: Zhang et al., 2008.

Maximum Yield Technology

Maximum Yield Technology is a nutrient and irrigation supply related concept that is being tested by International Plant Nutrition Institute and other agencies in different parts of Southeast Asia. Yield maximization is a concept that strives to improvise on maize productivity to the extent possible in a given location. A concept suited to areas where maize productivity is currently low or much below average for nation/region in question. For example, potential maize yield in South Indian plains is 5,500 kg grain plus forage, but actual harvest reaped is only 2,800–3,500 kg grain ha^{-1}. It leaves large yield gap that needs to be covered using different agronomic procedures. Maximum Yield Technology actually aims at closing the yield gap between potential and actual harvest reaped in farmer's fields. Basically, maximum yield technology adopts improvements in fertilizer based nutrient supply to maize crop. Aspects like quantity of fertilizer, its timing, methods of application are adjusted to derive maximum photosynthetic C fixation, biomass formation and nutrient partitioning into grains. Rapid rooting with least interplant competition allows better nutrient and water recovery. Maximum Yield Technology is often applied to entire maize-based cropping sequence instead of considering a single season that supports maize production.

At this juncture it is apt to consider various types of yield gaps that we know and their definitions. Total yield gap is the difference between what farmers reap and the potential yield possible in the given location and/or season. However, according to Pasuquin and Witt (2007), yield gaps can be identified at three stages. Yield Gap-1 is explained as the yield gap between potential yield (Yp) and maximum yield (Ymax) attainable. The Yp of crop is defined as theoretical Ymax in a given season determined solely by climate and genotype, since nutrients, water and other factors are sumptuously attended. The Ymax in a given season could be close to Yp, if management methods are excellent and weather conditions are highly congenial. Yield Gap-2 is the difference between Ymax and Attainable Yield (Yat). Attainable yield (Yat) is defined as the yield achieved in farmer's field with BMP including nutrient, water and pest control. Attainable Yield varies with season and location. Yield Gap-3 is the difference between actual yield obtained on farmer's fields (Yac) and attainable yield (Yat). Most farmers strive to manipulate nutrient dynamics and irrigation schedules to minimize yield gap -3, that is, Yat minus Yac. Following is an example that illustrates maize grain yield levels and gaps:

Location	Farmer's Yield (Yac)	Yield Attainable (Yat)	Maximum Attainable Yield (Ymax)	Potential Yield (Yp)
			t ha^{-1}	
Central Java	4.9	5.7	7.3	12-14
Lampung, Indonesia	7.2	9.2	13.7	12-14
An Giang, Vietnam	8.3	8.8	10.3	14-16

Source: Pasuquin and Witt, 2007.

Integrated Nutrient Management (INM) Systems

The INM methods are currently popular in most of the maize cropping zones of the world. The INM procedures aim at supplying all major and micronutrients through as many organic and inorganic sources. The INM aims at supplying nutrients in balanced proportions. It also tries to match nutrient supply with crops demand at various stages.

Therefore, major nutrients like N and K are split into 2–5 portions. Nutrient recycling procedures through residue incorporation is encouraged. Excessive use of inorganic fertilizers may induce rapid loss of nutrients from the soil profile and it a may pollute ground water. Nutrients accumulated in profile, especially N may become vulnerable to emissions. In order to overcome adverse effects of using inorganic fertilizers, INM methods envisage use of both inorganic and organic sources. Method of nutrient supply is also altered. Fertilizers are often placed in spots or banded, not broadcasted, mainly to avoid rapid loss of nutrients. The INM methods stipulate modifications in cropping systems. For example, intercropping to avoid weeds and soil nutrient loss; growing green manure crops to improve SOC content; rotating with legumes, etc. The INM procedures also aim at reducing soil erosion and nutrient loss by adopting contour planting or strip cropping. All of these measures put together allow better control over nutrients in the field and agroecosystems in general. In addition INM also improved maize productivity. To quote an example, in Karnataka state of India, INM stipulates application of 120 kg N, 40 kg P, 80 kg K, 12.5 t FYM ha^{-1} and microbial fertilizers like Azospirillum and Phosphobacterium. Maize grown under INM may yield 3.5 t ha^{-1} but that under Farmer's Practices yields only 1.6 t ha^{-1}. There are also several reports regarding use of combination of inorganic and organic sources to supply N to maize crop. Often 50% of N is supplied using inorganic fertilizers and other half using organic manures like FYM, residues or animal manure. Overall, INM conserves or improves soil quality and fertility but at the same time induces higher grain and forage production. Several variations of INM that suit each location, region or large cropping belt are available in different continents.

Site-specific Nutrient Management (SSNM) or Precision Farming (PF)
The SSNM or precision farming involves detailed analysis or prior knowledge (accrued data) about soil fertility variations within a maize farm. Often, both soil and plants are analyzed to assess within field variations in productivity. Actually, GPS guided methods allow accurate mapping of soil fertility variations. A computer aided model rapidly calculates the nutrients required to be applied based on yield goals. A variable rate nutrient applicator is used to distribute fertilizers. In most cases, fertilizer required to be supplied to fields for a given grain yield level is marginally or sometimes markedly lower than that envisaged under improved practices or BMPs. Obviously, SSNM firstly reduces fertilizer-N supply into fields. It avoids undue accumulation and loss of nutrients (N) via percolation, seepage or volatilization. It also avoids accumulation of nutrients in subsurface layers and/or contamination of ground water. The SSNM procedures also impart uniformity to soil fertility in a maize field. As a consequence, crop growth response is relatively more uniform. Such SSNM or GPS-based assessment of soil fertility variations and disbursement of fertilizer nutrients has been in vogue on many field crops including maize. The SSNM has been adopted in many areas within maize cropping zones of USA, Argentina, East Africa, China and Southeast Asia. The SSNM has been actively evaluated for adoption in many other regions.

There is great interest and enthusiasm to use SSNM methods in the Corn Belt of USA. Farmers tend to adopt variable rate N application methods mainly to improve grain yield, its quality, especially protein content and economize on fertilizer-N inputs (Ping et al. 2008). According to Ferguson et al. (2002), SSNM is yet another technique available to maize farmers to enhance fertilizer-N use efficiency, improve productivity and at the same time lessen use of chemical fertilizers. Nitrogen is one of the most important factors that affects maize productivity in the Corn Belt of USA. Crop demand for N varies spatially in most farms (Miao et al., 2007). Spatial variability in soil-N fertility and crop response is in fact a pre-requisite, if variable-N application methods are to be used. Homogenous high soil fertility fields may not easily offer advantage via SSNM. There is firstly a need to accrue sufficient knowledge about spatial variability in corn yield and grain protein contents. Let us consider an example. Over 15 field locations in Illinois, USA were utilized by Ruffo et al. (2006) to assess influence of variable-N rates on maize grain yield and protein content. The fertilizer-N requirement ranged from 0–300 kg N ha^{-1} to achieve maximum grain yield in the locations tested. In general, grain yield in fields not provided with N was more variable (6.3–13.3 t ha^{-1}) compared to 12–16 t ha^{-1} in those provided with optimal N supply. Grain protein responded positively to variable-N inputs. Grain protein contents varied within fields. The spatial variability of grain starch was less perceptible compared to grain protein (grain-N). Often, starch decreased as fertilizer-N supply was increased to 300 kg N ha^{-1}. Ruffo et al. (2006) state that effects of variable-N rate on oil content was negligible and precision techniques may not offer great advantage. However, in general, variable-N rate methods (SSNM) can improve grain harvests, grain-N, and quality to a certain extent. At the same time, variable-rate-N methods can lessen N inputs into maize belts and still maintain uniformity of soil fertility, especially N.

Miao et al. (2006b) found that supply of N with due consideration to within field variations resulted in increased grain yield. Average protein content of grains and starch was also improved due to variable N supply. The economically optimum N rate was 125 kg N ha^{-1} with a range of 93–195 kg N ha^{-1} based on soil fertility variation. Selection of maize hybrid was also important. In fact, hybrid specific N management was needed in addition to SSNM. According to Miao et al. (2007) application of State Agency Recommended levels of N did not reduce accumulation of soil NO_3-N at subsurface. Fertilizer-N use efficiency was relatively low. Adoption of variable-N rates reduced economically optimum N rates by 69–75 kg N ha^{-1} without affecting maize forage/grain production. A third approach suggested by Hendrickson and Han (2000) envisaged application of 20–50% of recommended N at emergence and additional (or split) N using variable-N technology based on plant-N status and/or soil N fertility status. Ariel photographs of N stress maps and leaf color charts could also be used. There was a significant reduction in N inputs, if N was split and spatial variability in soil and plant N status was considered. Kitchen et al. (2010) have examined the use of canopy reflectance to measure variation in plant N status. This information could then be used to side dress maize fields with a variable-N applicator. For example, sufficiency index calculated based on canopy sensor readings tallied with fields under optimal N supply. Leaf color charts have also been used to demarcate areas that need N top dressings. Fertilizer-N supply could be accordingly adjusted at various spots within the field so that within field variations are removed. Such quick N measurement techniques have great potential to be used in maize farms world over. They are both time and cost effective and less cumbersome to use in fields. Active sensors too could be used to calculate site specific N requirement of maize. In this case, it considers canopy reflectance and economically optimum N rates while stipulating N dosage to each strip or zone.

At Lincoln in Nebraska, Ping et al. (2008) have evaluated the well known BMP with SSNM at different planting densities. On Arguidolls, BMP has proved useful and provided 12–15 t grain ha^{-1} with appreciable gross profits. Firstly, historical data on soil-fertility (NPK) variations in the fields were obtained. Then soil was tested for variations in SOM and NO$_3$-N contents at different spots using a grid in the field. Among the fields tested, one of them showed significant increase in yield and NUE with variable-N supply. However, Ping et al. (2008) suggested that, there is a need to integrate SSNM methods further by including in-season water and nutrient management. Such an integration of water and N supply under SSNM approaches may help us in obtaining yields greater than that derived under high fertility conditions and BMPs.

Maize is an important cereal crop in many of the Southeast Asian countries. Maize agroecosystem extends into 8.6–9.5 m ha depending on precipitation pattern. Let us consider a few examples depicting usefulness of SSNM and comparative advantage over farmer's practice or other improved practices. In Indonesia, Thailand, and Vietnam average national yield is 3–4 t grain ha^{-1}. In Philippines it is a bit lower at 2–2.5 t ha^{-1} (Witt et al., 2006). Farmers adopt several different agronomic procedures depending on resources, inherent soil fertility, pattern weather and most importantly, yield goals and economic advantages. The traditional farmer's practice may envisage application of a single major nutrient or combinations, depending on known soil fertility regime of the location. Farmers may have to apply larger doses of major nutrients plus micronutrients to achieve Y^{max}. Yet, potential grain yield in a region may differ from Y^{max} harvested. Following is an example that depicts response of maize to nutrient supply:

| | Grain Yield Response to Nutrient Supply | | | | Maximum Yield | Potential Yield |
| | t ha^{-1} | | | | t ha^{-1} | t ha^{-1} |
	+N	+P	+K	+NPK	+NPK	+NPK
Vietnam	0.9	0.7	0.6	7.2	8.2	11.3
Indonesia	3.1	1.2	0.9	8.7	11.4	12.0

Source: Witt et al., 2006.

Witt et al. (2006) have reported that highest grain yield of 8.7 t ha^{-1} was achieved in fields supplied with all three major nutrients NPK and maintained under INM procedures. Grain harvests were 19% greater than that achieved under Farmer's Practices. Highest grain yield recorded was 11.4 t ha^{-1}, but physiologically potential grain yield under the given environment was still higher at 12–12.2 t grain ha^{-1}. Clearly, yield gaps are conspicuous between farmer's practice, maximum yield technology, and potential physiological yield. One of the main reasons for yield gap is no doubt improper nutrient dynamics. Aspects like soil fertility variations, SOC content, nutrient ratios, timing, nutrient supply rates, crop demand for nutrients and water at various stages, and distribution need attention. Adoption of SSNM seems to overcome soil fertility variations. It also matches crop's demand for nutrients with availability in time and space at various growth stages of the crop. Witt et al. (2006) opine that maize grain yield could be increased by fine-tuning fertilizer supply using site-specific techniques. Farmers may have to adjust both timing and quantity of nutrients distributed at each spot in the field. Maize forage/grain productivity seems to depend immensely on planting density. Hence, there are clear suggestions that fertilizer supply under SSNM should consider plant density and yield goals set by farmers.

There are several reports that compare farmer's practice (FP) and SSNM with reference to nutrient supply pattern and grain yield harvested. Following is a summary that compares FP with SSNM at Tanchau, An Gaing in Vietnam:

	Farmer's Practice		Site-specific Nutrient Management	
	FP regular	FP - ICM	SSNM - IPD	SSNM - ICM
	75 x 20 cm	75 x 18 cm	50 x 30	75 x 18 cm
Fertilizer Supply (kg ha^{-1})	180 N: 91 P: 71K		200 N: 120 P: 100 K	
Grain yield (t ha^{-1})	9.14	9.42	9.85	9.50

Sources: Khuong et al., 2008; Witt et al., 2006.
Note: FP regular =Farmer's regular practice; ICM = Integrated Crop Management Procedures; IPD = Improved Planting Density (75000 Plants ha^{-1}).

Witt et al. (2006) have remarked that maize grain yield of 8–10 t ha^{-1} is possible using maximum yield technologies. On alluvial soils, improved procedures like use of hybrids and adoption of state recommendations regarding fertilizer and organic manure supply may yield up to 8.58 t grain ha^{-1}. However, under SSNM, yield goals could be revised. Grain yield under SSNM ranges from 9.2 to 10 t ha^{-1} depending on planting densities. The grain yield advantage due to adoption of SSNM ranges from 0.3 to 0.5 t ha^{-1}. In many cases where high planting density was adopted yield advantage due to SSNM was slightly higher at 0.7 t ha^{-1} over FP.

Reports by International Plant Nutrition Institute (IPNI) also suggest that adoption SSNM is beneficial to maize farmers. In Indonesia, grain yield under SSNM was 8.3 t ha^{-1} compared to 7.5 t ha^{-1} under improved methods (Kartaatmadaja, 2007). Binh (2007) found that average grain yield (7.3 t ha^{-1}) from 13 SSNM sites in Vietnam was higher by 12% (0.8 t ha^{-1}) over farmer's practice (6.5 t ha^{-1}). However, full benefit from SSNM was possible only if other yield constraints like planting density were suitably modified. Adoption of INM procedures along with SSNM meant a high yield equivalent to 8.2 t grain ha^{-1}. Hence, adoption of SSNM on a wider scale was suggested. In addition, SSNM procedures generally lead to lower amount of nutrient supply compared with maximum yield technology or even farmer's practice. This is because of identification of soil fertility variations in a specific field and accurate nutrient distribution using variable applicators. To obtain similar grain/forage yield, SSNM procedures often envisage relatively lower supply of nutrients to soil. In a long run, SSNM has definite impact on soil and above ground nutrient dynamics of the maize belt. Reports from Central plains of Thailand state that even with rice, farmers adopting SSNM apply only 63–70% of fertilizer recommendations to obtain same grain yield.

Application of precision farming methods to small farms found in South and Southeast Asia may require certain modifications, in soil testing kits, computer based-decision support systems, and fertilizer application techniques (Attanandana et al.,

2004). Attanandana and Yost (2003) examined the use of hand held computers for calculating NPK fertilizer requirements for maize, based on soil fertility variations *in situ*. They utilized Decision Support Systems of Agrotechnology (DSSAT-CERES) based on CERES model and a simplified Phosphorus Decision Support System to arrive at most economical fertilizer input levels. In Thailand, average maize grain harvest based on farmer's practice is about 3.7 t ha^{-1}, but high productivity at 6 t grain ha^{-1} is possible using site-specific techniques. Following is an example that compares fertilizer inputs and productivity under farmer's practice with those of fields under SSNM systems. The SSNM involves intensive soil testing, Decision Support Systems and SSNM:

Soils	pH	Nutrient Status	Fertilizer Supply (kg ha^{-1})		Grain Yield (t ha^{-1})		
		N-P-K	FP	CERES-Maize	FP	CERES-Maize	Predicted
			N-P-K	N-P-K			
Clayey	7.5	VL-VL-H	25-25-0	94-44-0	2.78	6.06	5.5
Loamy	7.0	VL-VL-H	18-25-0	94-50-0	2.93	4.47	7.0

Source: Attanandana and Yost, 2003.

Note: We may note that fertilizer recommendation under SSNM varies based on yield goals. Fertilizer schedules are also revised due to improvements in decision support system and accurate knowledge about extent of nutrient loss/additions into fields. Fertilizer efficiency and economic benefits too vary. In the above case, in Thailand during 2004, farmers consistently reaped 100–400 US$ more compared to farmers practice.

FP = Farmer's Practice; CERES-Maize = Computer model to predict N requirements; Vl=Very low; H= high.

According to Witt et al. (2007) there are few aspects that need careful attention before finalizing fertilizer input schedule under SSNM. They are: (a) extent of crop response to fertilizer (N, P, K) supply; (b) agronomic efficiency of fertilizer N possible in farmer' fields; (c) total nutrient demand of the crop during the entire crops reason; (d) variability in background soil fertility and variations in crop response to fertilizer inputs; (e) total nutrients removed into harvested products; (f) total attainable yield and Yp possible in the given environment; (g) constraints other than nutrients that may affect expression of crop response. We should also note that window for fertilizer N supply occurs well within the vegetative growth stages from V0 to V12 or Vt and not beyond as reproductive phase begins. Therefore, split N input schedules should occur within this period preferably. Following is an example for estimating nutrient needs and supply under SSNM for maize crop grown in Philippines and Indonesia (see Witt 2007):

Yield Response to N		Low	Medium	High	Very High
Expected Yield increase over Zero-N plot (t ha^{-1})		2-3	4-5	6-7	7-8
Expected Agronomic Efficiency					
(kg grain increase kg^{-1} applied N)		17-25	25-31	30-35	32-36
Growth stage	**Leaf Color**	**Fertilizer N Supply (kg ha^{-1})**			
Pre plant –V0	--	36	48	60	66
Split Application at V6-V8	Yellow Green	48	66	83	92

	Green/Darkgreen	42	56	70	77
	Yellow Green	48	66	83	92
Splitt Application at V10 or later	Green	42	56	70	77
At V14	Green	-	25	35	35
Total Fertilizer Input		120	160	200	220

Source: Witt et al., 2007.

Note: Yellow Green = LCC<4.0; Green = LCC 4.0–4.5; Dark Green LCC = 4.5. Such supply schedules could be easily prepared for other two major nutrients P and K.

MODELING TO IMPROVE NUTRIENT DYNAMICS AND MAIZE PRODUCTION: EXAMPLES

Models, simulations, and computer-based decision support systems perhaps are most important developments during recent years. They promise to influence and alter nutrient dynamics in the maize agroecosystem. In future, much of our decisions that regulate nutrient supply and utilization pattern would be based on crop models.

A wide range of aspects related to soils, agronomic procedures, weather, crop physiology, and yield goals have been modeled using several different computer based models. Each has its own advantage with regard to certain aspects of maize cropping. Crop models and Precision farming systems together are generally designed to improvise fertilizer disbursement and efficiency. Yet, selection of crop model most appropriate to a given crop or cropping system could be important.

Researchers and farm workers in maize cropping zones world wide, have indeed utilized several models and their variations to arrive at accurate predictions and recommendations. Following are few examples:

The NUTMON (NUTrient MONitoring) is an integrated model that helps us in understanding nutrient dynamics within tropical farming systems. The NUTMON is used to assess nutrient requirements of a field, small farm or even a large farming enterprise (see FAO, 2008; Stoorvogel and Smaling, 1990). The NUTMON software and manual has been developed to integrate and assess nutrient inputs, stocks, and flows with due consideration to economic advantages. The NUTMON model has been utilized to study nutrient balances, especially major nutrients N and P in many agroecozones that support maize production. For example, use of NUTMON predicted that annually, nutrient balance in fields supporting maize is -104 kg N, -13.6 kg P, and -82 kg K ha^{-1}. The basic data required to operate NUTMON and assess nutrient balances in a field often includes:

Inputs: Mineral fertilizer, organic manure, leaf litter, mineral nutrient deposition, biological nitrogen fixation.

Outputs: Harvested product (grain and forage), crop residue, leaching and denitrification, soil and nutrient erosion, volatilization.

The NUTMON model is also amenable for nutrient budgeting. According to Surendran et al. (2005), in the semi-arid regions of South Indian plains, soil nutrient mining is a problem that has got accentuated. It leads to soil deterioration and affects optimal nutrient dynamics in the cropping ecosystem. Therefore, to develop a nutrient budget at field or farm level they assessed nutrient mining/loss and enrichment.

They adopted NUTMON at multi-scale and assessed nutrient stocks, their supply and removal from field as well as geographical unit. As shown above, nutrient inputs and outputs were weighed. For example, at a farm near Coimbatore in South India, they found that nutrient management trends were insufficient and not sustainable. Soil nutrient mining was significant. Soil nutrient pool had to be supplied with extra fertilizers to offset negative balance of N and K. Farmers were suggested to adopt INM methods that include several sources in order supply major and micronutrient nutrients in more balanced proportions.

The CERES (Crop Environment Research Synthesis)-maize is a crop model used to assess the effects of various agronomic factors and simulate a variety of aspects related to crop production. For example, growth pattern, forage, and grain production could be modeled. The effect of precipitation pattern, irrigation, crop genotype, nutrient supply, timing and methods of fertilizer supply, and weeding could also be simulated.

Several approaches could be adopted while analyzing spatial variations for nutrients. First approach involves dividing field into grids and each grid cell could be simulated for nutrient supply and crop response. A second approach involves applying the model at different points within a field. The simulated values are then interpolated and surface maps for whole field or a cropping segment is developed. Third approach is based on management zones. For example, aerial assessment of crop growth and vegetation index could be applied to demarcate management zones. Suitable crop models could then applied to study nutrient inputs and their impact on crop productivity within a management zone. Kiniry et al. (1997) evaluated two crop models namely, CERES-maize and ALMANAC in the Corn Belt and other locations of USA. Simulated value compared well with those reported by national agricultural agencies. Predictions were within 5% range of actual grain yield.

Binder et al. (2008) state that maize yield increased steadily in North China since 1980s, but it has stagnated in some parts at 5–6 t grain ha^{-1}. In order to quantify potential grain production levels, CERES-maize model was calibrated and validated. Simulations were carried out for different soil textures considering weather patterns for past 30 years. Average simulated grain yield was 4800 kg ha^{-1} for summer maize and 5700 kg ha^{-1} for spring maize. The grain yield varied depending on growing degree days and a delay in sowing by 30 days affected growth rate, nutrient recovery, and grain production.

Planting at most appropriate date and soil moisture level is important, especially during maize hybrid cultivation. It is said that length of growing season and crop maturity time is crucial to achieve maximum grain/forage yield formation. Crop models are useful in assessing the impact of varying the date of planting on growth period and grain formation. The CERES-maize model has been clubbed with RZWQM that allows us to consider soil moisture availability along with other parameters. This system devised by Saseendran et al. (2005) allows us to decide planting date based on specific hybrids, their maturity period, water and nutrient requirements, forage and grain yield goals. Obviously, a decision made using CERES-maize has far reaching influence on water and nutrients in the soil and crop phase of the agroecosystem. The CERES-maize model has also been used to simulate the effect of drought and moisture stress conditions that occur frequently in the dry lands (Lopez et al., 2008; Xie et al., 2001). Effect of variations in irrigation schedules on maize crop growth rate and grain yield has also been simulated using CERES-maize models.

Regarding N management, Miao et al. (2006a) state that optimal N supply to "Management Zones" based on soil fertility, moisture and crop yield potential or yield goals is feasible. A second approach commonly proposed by fertilizer industries is broad. It considers a large patch of maize cropping belt. It involves certain compromises on soil N variability within small zones. Fields could also be divided into strips and N requirements for each strip could be decided by using the model. Strips could be grown with maize at different soil fertility levels and yield goals. Maize fields could also be divided into "management zones" and optimal N rates could be decided based on historical data regarding soil fertility, precipitation patterns, soil moisture variations, yield goals.

Agricultural soils including those used for maize production emit N_2O. Intensified cropping zones like "Corn Belt of USA" or Northeast China are examples, wherein fertilizer-N supply is high and many a times far exceeds that actually utilized by the crop. They do emit sizeable quantities of N_2O. Nitrous oxide emission reduces fertilizer efficiency and pollutes atmosphere. Delgrosso et al. (2008) adopted DAYCENT, a biogeochemical model that simulates nutrients (N and C) in soil and predicts crop yield under irrigated conditions. According them, DAYCENT predicted the impacts of tillage intensity and fertilizer-N supply on crop yield, soil moisture content and SOC rather accurately. The N_2O emission predicted using DAYCENT matched measured data. The N_2O emission increased with higher fertilizer-N supply. Emissions from no-till fields tended to be lower than fields under conventional tillage.

Currently, researchers and field workers in maize production zones worldwide do use DSSAT (Decision Support Systems for Agro-Technologies) software to assess fertilizer-N requirement of maize crops. It is based on various factors like soil fertility status, agronomic procedures adopted, genotype, irrigation levels, and grain/forage yield goals. For example, in Thailand, DSSAT has been utilized to predict fertilizer-N needed and its time of application (splits) (Sipasueth et al., 2007). Actually, it was aimed at making fertilizer-N inputs more efficient. Crop response to fertilizer-N inputs was accurately forecasted in some soil series but not in others. The main reason was amount of N derived from mineralization, that is, nitrate-N release into root zone. Nitrogen mineralization is also affected by wetting and drying cycles imposed on soils. Nitrogen mineralization of some soil series like Pc and Lb was higher. Hence, N inputs were not effective in improving crop growth and grain yield. The DSSAT has also been used to predict P and K requirements of the crop accurately. Phosphorus Decision Support System (PDSS) software is available for use, while judging effects of P fertilizer supply on maize crop growth and grain yield.

Kucharik et al. (2005) have reported that C and N cycling could be modeled in maize cropping system of Wisconsin using IBIS crop model. A Precision Agriculture version of the IBIS model is also available to simulate various aspects of SOC and soil-N and relate its impact on crop growth.

The APSIM (agricultural production systems simulator) modified to accommodate sub-routines that simulate soil nutrient dynamics are available. They have been used to assess effects of fertilizer-N supply on maize grain yield (Whitebread et al., 2004).

The CERES-maize has been utilized to simulate effects of irrigation on maize productivity. It has also been used in conjunction with spatially variable precision irrigation methods. It is believed that such techniques will reduce quantum of irrigation needed and improve economic advantages (DeJonge and Kaleita, 2006).

Hybrid-maize model is a computer program that simulates growth and yield of maize, under both irrigated (water non-limiting) and moisture limiting conditions (Yang et al., 2004a; 2004b). The current version of the Hybrid-maize model simulates potential crop growth. It allows us to assess site yield potential and variability based on weather data; evaluates effects of changing planting date, maturity period of hybrid, and plant population. It allows us to forecast consequences of changing timing of silking and maturity; it allows us to assess effects of various options with irrigation, its timing and quantum on grain yield. Over all, it helps us in simulating grain/forage yield considering a range factors that operate during maize production (Yang et al., 2006). Hybrid-maize computer model could also be modified and applied to assess the effect of various agronomic procedures on productivity of maize-soybean rotation. Further, Hybrid-maize model could be improvised by coupling it with best soybean crop growth model and CENTURY ecosystems model. It then allows us to assess nutrient dynamics in the maize-soybean cropping belt. Aspects like C sequestration, N inputs/outputs, N-use efficiency and dynamics of other nutrients could be studied.

The RZWQM (Root Zone Water Quality Model)-CERES-maize is a model developed to assess water and nitrogen uptake in maize fields. The CERES-maize model has been coupled with RZWQM. The RZWQM supplies CERES model with daily soil water and nitrogen content in soil profile, soil temperature variations, evapo-transpiration and weather data. The CERES model supplied plant growth variables like root growth rates, its distribution, leaf area index, grain formation, etc. According to Hoogenboom et al. (2006) growth and yield forecasts derived from RZWQM-CERES-maize model was similar to that derived using original CERES-maize. Of course, predictions under hybrid model were based on several more soil moisture and nutrient related parameters.

The CERES-maize or similar models have also been used to assess the effects of various stress factors like uncongenial soil/ambient temperatures, soil moisture dearth, acidity, salinity, and alkalinity. Each of these stress factors may affect maize and/or its intercrop to different extents. Simulations will provide us with a semblance of reductions in forage/grain yield. Decisions regarding correcting stress and improving nutrient recovery by crop productivity could be appropriately arrived (Castrignano et al., 1998).

Many of the models currently available could be adopted to study the nutrient dynamics of an entire crop rotation. Such models usually consider weather pattern, nutrient requirements, fluxes, transformations, losses, as well as crop productivity during entire crop sequence that includes maize. For example, a newly developed crop model called SALSA (Semi-Arid Latin America Simulation model for Agroecosystems) has

been adopted to assess factors like weather, fertilizer supply, organic manure, irrigation on performance of maize-pasture rotation (Peinetti et al., 2008).

Computer based models and simulation have also been used to assess inter-plant
competition in the field for nutrients, water, and photosynthetic radiation (Rossiter and
Riha, 1999). Models like ALMANAC and CERES-maize have been modified to study
various intercropping strategies. Appropriate combinations of crop species in addition
to maize, row ratios, nutrient supply schedules, water supply, and yield goals could be
set based on simulations.

Maize is often intercropped with agroforestry trees in many of the tropical countries. Nutrient dynamics that ensues in a maize-tree (green manure) combination is
unique and needs to be studied and simulated, so that appropriate decisions could
be arrived at. For example, WaNulCAS is tree-crop interaction model that simulates
tree and crop growth using data on nutrient inputs and daily increments in growth/
productivity. Area bordering a tree line is divided into four parallel zones. A tree zone
and three field crop zones occur. Each zone has four soil years demarcated. In all, 16
cells to study the crop-tree interaction. It is useful to study nutrient availability, depletion and uptake trends. Tree-crop competition for nutrients and moisture could also be
predicted accurately (Radersma et al., 2005; Van Noordwijsk et al., 2000).

AGRONOMIC EFFICIENCY OF FERTILIZERS

Maize agroecosystem is impinged with fertilizer-based nutrients, manures, and several
other soil amendments at different rates in order to enhance biomass/grain productivity. Among major nutrients, N is the most important fertilizer-based nutrient supplied.
Nitrogen is often added at higher rates because of greater demand for it by the maize
crop. Further, to achieve balanced nutrition and higher agronomic efficiency of fertilizers, other essential nutrients like P, K, and micronutrients mostly Fe and Zn are also
supplied. Major emphasis is however on improvement of agronomic and physiological efficiency of fertilizer-N during maize production. There are indeed innumerable
reports/reviews on fertilizer efficiency that emanate from maize belts across different
continent/regions. Let us consider only a few examples to explain various methods by
which agronomic efficiency of fertilizer-N applied to maize crop could be enhanced.
Fertilizer N recovery is an important aspect of nutrient dynamics. In case of fertilizer-
N, recovery percentages often fluctuate between 20 and 44% based on several soil,
plant, and environmental factors. For example, in North Central USA and Corn Belt,
fertilizer N recovery was 38 kg N out of 103 kg N ha^{-1} applied. It amounts to 37%
recovery (Roberts, 2008). In comparison, for other crops like rice it is 40% in Asia,
49% for wheat in Gangetic plains, and 18% on maize grown in dry lands of India.
Nitrogen supply is altered based on its agronomic efficiency. The amount of N still left
in subsurface soil, amount leached to lower horizons, lost via emissions all depend on
N recovery rates and agronomic efficiency. In terms of grain/forage yield response,
maize crops supplied with high amounts of N may produce high grain yield. Yet,
nutrient-use efficiency could be low. On the contrary, at low N input, grain yield is low
but nutrient use efficiency could be high (Dibb, 2000). Achieving balanced nutrition in
soil is important. If not, crop response to N could get curtailed due to Leibig's Law of

Minimum. Following is an example that depicts effect of supplying other two major nutrients—P and K along with N:

Crop	Grain Yield (t ha⁻¹)			Agronomic Efficiency (kg grain kg-¹ N)		
	Control	N alone	N + P, K	Control	N alone	N + P, K
Maize	1.67	2.45	3.23	19.5	39.0	39.5

Source: Roberts, 2008.

We should note that soil gets depleted of micronutrients due to incessant cropping. Again, to achieve balanced nutrition and appropriate ratios, we have to supplement soil with micronutrients. It then ensures optimum response to fertilizer-N. Hence, under INM, deficiencies of all nutrients are corrected using wide range of organic and inorganic amendments.

Land preparation is an important factor that affects fertilizer-root contact and nutrient uptake efficiency. Land preparation and fertilizer placement affects rate of nutrient absorption. For example, on Reddish Brown Lateritic soil found in Thailand, Suwanarit et al. (2006) found that maize planted on flat beds and ridges differed in terms of fertilizer recovery. Broadcasting fertilizer in 50 cm bands was more effective than banding below the cereal seeds prior to sowing. Side dressing fertilizer at 4 weeks after sowing maize seeds on ridges was more effective. Interaction of fertilizer with soil moisture seemed to affect nutrient recovery rates and maize growth.

Maize farmers adopt different methods to supply fertilizer-N. Practically, they aim at maximum recovery of N into crop phase of the agroecosystem. Placement of fertilizer-N formulation accurately is important. Most common fertilizer placement methods adopted are broadcasting, 2 x 2 placement, placement below seed, In-furrow placement, surface banding, surface dribble, etc. Each of placement methods adopted to disburse fertilizer-N has its impact on nutrient recovery by maize roots, N loss from the surface soil, seepage, emission and finally N dynamics in the ecosystem. The agronomic efficiency of fertilizer-N placement methods adopted varies with regard to actual grain yield. Following is an example from Florida in USA:

Method of Starter –N Placement	Grain Yield
	t ha⁻¹
Control	11.8
In-furrow	12.1
2 by 2	14.7
Surface Dribble	13.2
2 inches below	12.5
5 inches below	12.4
8 inches below	13.2

Source: NCSU, 2009.

Timing of fertilizer-N, P and other nutrients is also important. A portion of fertilizer-N is often supplied as starter doze and rest in 2–4 splits. This way it increases fertilizer-N efficiency enormously. It avoids undue accumulation of N in the soil profile. It also avoids excessive loss of N via leaching, percolation, seepage and emissions. Splitting enhances N recovery by maize roots. It actually helps farmers to match fertilizer-N supply with crop's demand at various stages of growth. There are indeed several reports regarding splitting fertilizer-N and its impact on maize crop productivity (see Chapter 3). Fertilizer-based P, K, and micronutrients are often supplied, at or prior to planting in single dosage. In rare situations, fertilizer-K could be split into two halves.

As stated above, splitting fertilizer-N has perceptible advantages in terms of nutrient recovery, reduction in loss of fertilizer-N, improvement in agronomic efficiency of fertilizer-N and finally maize crop growth and grain productivity. A step further, it is clear that size of the splits and timing also affects fertilizer-N efficiency. Field trails at Brazilian Cerrados have shown that although total fertilizer-N supplied is same, if size of split application is different, then its effect on maize alters. Following is an example from Brazilian Cerrados, where in amount of fertilizer supplied via splits is varied:

Total Nitrogen Applied	Grain Yield
(Applied at Vegetative + Silking)	(t ha^{-1})
Total N applied = 80 kg N ha^{-1}	
80 + 0	10.4
30 + 50	7.2
Total N applied = 130 kg ha^{-1}	
130 + 0	11.4
80 + 50	11.1
30 + 100	8.0
Total N applied = 180 kg ha^{-1}	
130 + 50	12.1
80 + 100	11.3

Source: Da Silva et al., 2005.

In order to enhance fertilizer-N efficiency, inorganic sources could be mixed in different proportions with organic manures and supplied to maize fields. Such combinations improve N dynamics and recovery by maize roots. Farmers are often advised to use nitrification inhibitors in order to reduce loss of fertilizer-N via emissions. Mixing nitrifying inhibitors at 2% improves N recovery and reduces loss of N via leaching and emissions.

Slow-N release fertilizers could also be used. Let us consider an example. Maize cultivation on newly reclaimed zones in Egypt is prone to soil problems related to erosion, leaching, seepage and runoff. Physical loss of top soil and N applied via fertilizer often reduces fertilizer-use efficiency. Soluble fertilizers containing NO_3-N or NH_4-H are excessively prone to loss via leaching. Therefore, farmers are advised to use slow- release N fertilizers. Slow-N release fertilizers possess a coating of S, or P, or nitrification inhibitors or neem extracts.

Generally, in Egypt, soluble fertilizer-N sources such as NO_3 or NH_4 based fertilizers are preferred to improve maize productivity. As stated earlier, NH_4 and NO_3 that may accumulate in soil profile are highly mobile and move away from root zone. To avoid undue loss of soluble-N as well as to develop an appropriate match between soil-N status and plant's demand for N at various stages of maize growth, farmers use slow-release N fertilizers. Slow-release N fertilizer improves firstly N recovery by maize crop. It leads to better fertilizer-use efficiency. As a consequence, grain and forage production also improves (see Table 8.2).

Table 8.2. A comparison of urea and slow-release N fertilizer in the Egyptian maize belt.

Fertilizer Source	Grain yield	Biomass Yield	N Uptake (kg fed[-1])	
	t fed[-1]	t fed[-1]	Before Anthesis	After Anthesis
Urea 120 kg N fed[-1]	2.93	5.52	60.1	23.1
Slow Release-N 60 kg N fed[-1]	3.34	6.51	66.8	27.7
Slow Release-N 80 kg N fed[-1]	3.98	6.75	83.2	29.8
Slow Release-N 100 kg N fed[-1]	4.23	7.98	99.5	37.4
C.D. 5%	0.41	0.89	10.5	3.7

Source: Bahr et al., 2006.

Motavalli et al. (2008) believe that fertilizer-N use efficiency could be further improved by adopting a combination of fertilizer sources at variable rates based on soil tests and yield goals decided. Literally, it improves on SSNM by introducing 2 or 3 different sources of N that vary with regard to percentage N, rate of N release from fertilizer and efficiency with which it is absorbed by maize roots. In this case, Motavalli et al. (2008) have opted to use slow-N release and conventional urea. A combination of two sources is applied based on soil test or leaf color chart or reflectance measurements. Such variable application of two or more fertilizer-N sources adopting SSNM may be slightly cumbersome, yet useful with regard to nutrient use efficiency. Loss of N via percolation and emissions could be enormously reduced.

Maize genotype has its influence on nutrient dynamics. The dominant genotype grown in larger patches within a cropping belt and its physiological traits relevant to fertilizer use efficiency needs due consideration. Maize genotype efficient in nutrient recovery and use may need lower amount of fertilizer compared to genotypes that are less efficient in uptake and use efficiency. High amounts of fertilizer-N inputs and undue accumulation could be avoided by using N-efficient genotypes. Selection of appropriate genotype that enhances nutrient recovery and grain yield is an important aspect of INM methods.

Over all, agronomic efficiency of fertilizers is an important aspect that has direct and immediate bearing on nutrient dynamics and productivity of maize agroecosystem.

IRRIGATION AND WATER-USE EFFICIENCY AFFECT NUTRIENT DYNAMICS

Knowledge about water budgets, components that contribute to total moisture supply to maize crop and fate of water supply is important. Soil moisture status at a given crop growth stage may have direct bearing on nutrient recovery and utilization pattern. Soil moisture patterns vary enormously with land use pattern and agronomic procedures. For example, Brye et al. (2000) made comparative evaluation of prairies, no-tillage (NT) maize and chisel plowed maize. The prairie ecosystem generally maintained higher soil moisture content in deeper layers of soils when compared with maize crop. Major factors that caused differences between prairies and maize crops were evapo-transpiration, drainage loss, primary productivity, and precipitation interception by residues. For example, total drainage was 198 mm for prairies, 563 mm for no-till maize, and 793 for chisel plow maize. The interception of precipitation was 477 mm for prairies and 681 mm for maize crops. Brye et al. (2000) suggest that combination of higher soil moisture storage and low drainage loss under no-till makes it more feasible for farmers.

Pikul (2005) has reported that WUE plays a vital role during various maize-based crop rotations practiced in the "Corn Belt of USA". The WUE affects N utilization and grain formation. The WUE of continuous corn was lowest (12.5 kg ha^{-1} mm^{-1}) compared with other rotations such as corn-soybean-wheat (20.5 kg ha^{-1} mm^{-1}) or corn-soybean-alfalfa (18.1 kg ha^{-1} mm^{-1}).

In the Brazilian Cerrados much of the maize cropping zone encounters semi-arid environment. Precipitation use efficiency needs to be maintained at high levels. Further, it is said nutrient x WUE interaction plays a vital role in maintaining growth and grain yield of maize-based intercrops. The soils of Cerrado region are acidic in reaction and they need pre-sowing application of lime and the usual levels of N, P, K. Gaiser et al. (2004) found that biomass production was 2.6 times greater in fields supplied with N, P, K plus lime compared with only N, P, K or only lime. They adopted models such as HILLFLOW and EPICSEAR to estimate soil water balance. Evaluations indicated that fields under complete fertilizer plus lime had higher transpirational WUE (+63 to +80%). It was attributed to fertilizer inputs that enhanced dry matter formation. Lime did not have appreciable effect on WUE. However, there are instances where liming has improved WUE of maize /cowpea intercrops.

Maize is cultivated both under irrigated and rain fed conditions in the Argentinean Pampas. Maize grown in dry land often encounters intermittent drought spells or sometimes extended periods of moisture scarcity. Droughts affect both nutrient dynamics and maize biomass/grain productivity (Plate 8.1). The nutrient x water use interactions are crucial to high grain/forage productivity. Lack of soil moisture reduces nutrient recovery. It also limits photosynthetic activity and C fixation *per se* in the agroecosystem.

Accurate knowledge about soil water storage, water balance and nutrient dynamics in the soil profile is important in any agroecosystem. In West Africa, fluctuation in soil moisture is attributable to erratic precipitation patterns and soil profile characteristics. Nigerian maize farmers adopt different agronomic procedures to maximize water storage and use efficiency. Actually, there is greater interest in improving water use x soil nutrient interaction during the crop season. According to Tijani et al. (2008), firstly, precipitation pattern and inherent soil moisture decided the extent of water recovered and WUE of maize crop. Most commonly, maize farmers resort to fallow treatments.

Planted fallows that support Guinea grass (*Panicum maximum*), Kaliko (*Euphorbia heterophylla*), Kudzu (*Pueraria phaseoloides*) or native vegetation are most popular in tropical Nigeria. Native fallows with fertilizer applied to the following maize crop provided best results in terms of dry matter, grain yield and nutrient recovery. Nutrient supplies were crucial aspects since any deficiency could easily reduce WUE perceptibly. Actually, water and nutrient availabilities were found to interact to increase WUE until water becomes limiting, at which point further additions of fertilizer-based nutrients did not have any effect (Tijani et al., 2008). Soil moisture storage was consistently higher under native fallows without nutrients (25%) compared to fallows with *Pueraria phaseoloides* (19%). Water use efficiency (WUE) calculated in terms of grain yield was best with native fallow with subsequent fertilizer inputs (0.018 t ha^{-1}mm^{-1}). It was followed by *Euphorbia* or *Pueraria* planted fallows (0.015 t ha^{-1} mm^{-1}). Native fallows without fertilizer gave least WUE at 0.011 t ha^{-1} mm^{-1}. However, Nigerian maize farmers utilize *Pueraria*-based short fallows to improved WUE and nutrient recovery by subsequent maize crop.

Plate 8.1. The influence of mid-season drought on maize growth, nutrient dynamics and productivity.
Left: Drought has affected crop growth. Note that stem and foliage development has been retarded, canopy is relatively small. Nutrient held in roots and above ground in a drought affected crop is much lower than in a well irrigated crop. Mineral nutrient accumulation pattern and C fixation are usually impaired due to water stress.
Right: An irrigated crop showing luxuriant growth with succulent leafs.
Source: Alberto Quiroga and Alfred Bono, Anguila Agricultural Experiment Station, Anguila, Argentina.

Let us consider a few examples from Indian subcontinent. Maize is grown both for fodder and grains in the Northwest plains. Water resource is an important factor that influences maize fodder production. Reports indicted that maize provided with irrigation at 7.5 cm produced about 20 t fodder and recovered 26 kg N ha^{-1}. However, increase of irrigation frequency and quantity enhanced N uptake to 36 kg N ha^{-1} and fodder productivity increased to 23 t ha^{-1} (Niaz et al., 2007). In most parts of Gangetic plains and peninsular India, maize requires 460–600 mm of water during a crop season. Silking stage is critical. Water deficit at silking affects nutrient recovery and grain formation. Grain yield reduces by 20% if water deficit persists for 2 days during silking stage. Maize is cultivated under relatively drier water regime. Maize responds to water deficits by producing extensive root systems. Roots extract water and nutrients more from deeper layers of soil. An irrigated maize crop may receive 3–5 irrigations during a crop season. Grain yield fluctuates between 1.7 and 2.3 t ha^{-1} and WUE ranges from 6.3 t ha^{-1}m^{-1} to 7.23 t ha^{-1} m^{-1} (Bandyopadhyay and Mallick, 1996). In the South Indian plains, maize is an important alternative cereal to irrigated rice crop. Rice needs over 900–1200 mm per crop season. Maize grown mostly as a dry land crop or with moderately low rates of irrigation can be efficient. Maize needs about 400–550 mm water per season. About three crops of irrigated maize could be cultivated for every rice crop (CRIDA, 2008). The biomass conversion rates and nutrient turnover is also efficient with dry land cereals like maize or sorghum or finger millet.

As stated above, maize is an important cereal in South India. Yet, it is exposed to vagaries of natural factors like variations in precipitation pattern and soil fertility. Fluctuation in rainfall affects nutrient dynamics and productivity of ecosystem rather markedly. Much of the rainfall is received during monsoon from June to September and then winter (October–December). Since almost entire nutrient absorption occurs in dissolved state via soil moisture, rooting pattern affects nutrient recovery rates immensely. A well developed root system is a pre-requisite for efficient utilization of both irrigation/precipitation water and soil nutrients. Selvaraju et al. (2004) pointed out that cropping systems, especially those including cereals like maize have changed enormously based on availability of irrigation. As a consequence, nutrient dynamics has also been affected. Farmers tend to exploit both surface and ground water irrigation rather efficiently. The water balance components of irrigated maize grown at Coimbatore in Tamil Nadu during warm, cold or neutral years are as follows:

	Temperature Regime		
	Warm	Cold	Neutral
Effective Rainfall (mm)	161	196	198
Evapotranspiration (mm)	640	562	615
Gross irrigation Requirement	600	450	500
Irrigation Efficiency (%)	77	83	82

Source: Selvaraju et al., 2004.

Irrigation requirements of maize crop cultivated in South India is similar to most arable or dry land crops. It is relatively small during early seedling stages until 30–35th DAP. Irrigation requirements increase markedly from 35th day until 60th day when it reaches almost 60–70 mm for every 10 days (see Figure 8.2). As stated above, nutrient absorption rates are tightly linked to rate of soil moisture depletion. Maize recovers major nutrients like N, P, K, and micronutrients at higher rates during 35–80th DAP. Irrigation requirement of maize crops declines after 100 days.

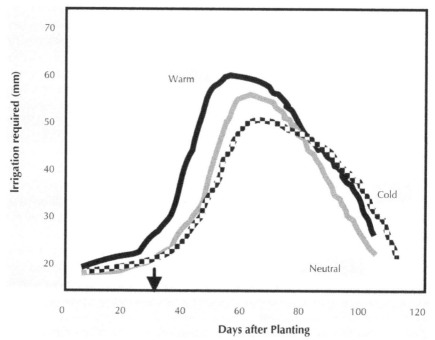

Figure 8.2. Irrigation requirement of Maize crop cultivated on Alfisols in the Semi-arid region of South India.

Source: Selvaraju et al., 2004.

Note: Irrigation requirement increases rapidly at 35th DAP and peaks at 60–80th DAP. It subsides later. Arrow indicates the point of time from which water requirements increase drastically. Warm = Warm climate; Cold =Cold climate; Neutral = normal climate.

Maize–wheat rotation is popular in the Northeast China. Both the crops are cultivated intensively using large supply of nutrients and irrigation. The WUE calculated based on grain yield improved until an ET of 650 mm for the entire rotation. Fang et al. (2007) state that there are several options that improve WUE and maize grain productivity. Reasonable spacing of irrigation events, optimizing frequencies and quantity of water, and matching irrigation schedules properly with rainfall will enormously improve WUE. Working out irrigation schedules for entire cropping systems has also been suggested. Excess irrigation for wheat crop enhances initial soil moisture status at the time of seeding maize. The excess residual moisture greatly hastens seedling establishment and growth. Nutrient recovery by maize seedlings is also better and it leads to higher grain productivity.

Chinese maize farmers have also been exposed to various GIS based methods that assess spatial variations in evapo-transpiration and irrigation requirements more accurately. Crop models are used to evaluate evapo-transpiration and climatic water deficits. Yield reduction due to soil moisture deficiency is also predicted. Irrigation schedules are prepared for spring maize grown in Beijing-Tianjin-Hebei region based on available data and CropWat model (Zhiming et al., 2006). Such GIS based methods allow better WUE and nutrient recovery by maize crop.

Forage maize is popular with Queensland dairy farmers. Forage maize yields 19–20 t ha^{-1} and it is better than several other fodder crops preferred in this zone. However, its productivity depends on nutrient supply, precipitation pattern and supplemental irrigation. Rainfall is highly variable and at times below that needed for optimum forage production. Hence, methods that enhance WUE and nutrient recovery by forage maize are important. Following is a comparative evaluation of maize with other fodder species grown in Queensland, Australia:

Forage Species	Forage Yield	Effective Rainfall	Irrigation	Total Water Supply
	t dry matter ha^{-1}	ML	ML	ML
Annual Rye Grass	9.2	1.8	3.4	5.3
Lucerne	13.0	5.8	5.1	10.9
Barley	8.4	1.0	1.2	2.2
Perennial Mixture	9.9	4.5	4.3	8.8
Tropical Sorghum	16.6	2.9	1.7	4.6
Sorghum/Rye Grass Mixture	25.8	4.7	5.1	9.8
Maize	**19.8**	**2.8**	**2.5**	**5.4**
Maize/Barley Mix	**28.1**	**3.8**	**3.7**	**7.6**

Source: Callow and Kenman, 2006.

Maize alone or maize/barley mixed forage crop provides greater biomass and improved WUE efficiency. Similarly, sorghum alone or sorghum/rye mix too has high biomass yield potential and excellent WUE. Regulation of irrigation and soil nutrient status seems crucial in order to achieve better WUE and forage production. In Northern Victoria, again maize forage production for dairying is popular. It is supposedly more efficient in biomass production compared with perennial rye or white clover. In New South Wales, maize production is efficient in terms of dry matter accumulated per unit rainfall received and irrigation. Measurements have shown that WUE of maize is 23–34 kg dry matter ha^{-1} mm^{-1}. Improved WUE and nutrient efficiency is essential since dairying is a costly enterprise.

Water and Nitrogen Interaction

Water deficits and N stress are common features in much of tropics and semi arid regions that support large tracts of maize production. In many locations, it is difficult to precisely identify the factor that reduces maize growth and productivity. According to White and Elings (2004) an overall assessment of impact of each factor, that is soil moisture (or irrigation) and N supply is essential. It provides a basis for preparing irrigation and N schedules. There are many strategies to ascertain extent of yield loss due to each factor. They are: evaluation by researchers in the experimental fields, survey of maize farming belts, use of simulations and models, and use of GIS and remote sensing. Models that allow us to ascertain effect of water and N stress and then recommend the farmers with appropriate agronomic procedures are available.

Application of N and water improves maize productivity. There could be significant synergistic effects due to simultaneous increase in soil-N and water supply to a maize crop. The rates and pattern of nutrient recovery may also vary at different levels of N and irrigation frequencies. Kim et al. (2008) believe that prior knowledge about synergistic effects, if any, will help us in calculating N and water requirements more accurately, especially under SSNM systems. Evidences from long term trials indicate that N inputs improve WUE. For example, Kim et al. (2008) have reported that WUE was 170 kg grain cm^{-1} at zero N input and 223 kg grain cm^{-1} at 112 kg N ha^{-1}. Simultaneous supply of water increased the ability of maize crop to garner more N and attain higher N use efficiency. Following is an example depicting N recovery from soil and fertilizer at two water regimes:

	% N derived from Soil	% N derived from Fertilizer
Moderate Water Regime (Precipitation only)	61.6	44
High Water Regime (Precipitation + irrigation)	67.7	48

Source: Kim et al., 2008.

Clearly, careful manipulation of irrigation quantity and frequency plus N supply schedule can enormously benefit the farmers. Such a positive influence on nutrient dynamics and maize crop productivity needs to be replicated in different cropping regions.

It has also been suggested that N and water stress may affect N recommendations based on remote sensing and reflectance measurements. Clay et al. (2006) conducted long term factorial experiments at Brookings in South Dakota and found that a remote sensing model based on "Yield Loss due to N Stress (YLNS)" was more accurate, while predicting N requirements than models based on yield plus "Yield Loss due to Water Stress (YLWS)". They attributed it to N and Water having additive effects on yield.

Results from Northern European locations again indicate that rainfall and N supply together have positive effects on maize grain yield. Ridging further enhances maize grain yield. Ridging, precipitation and N supply together accounted for over 90% of variation in crop growth response. The precipitation use efficiency could be increased to 6 kg ha^{-1} mm and grain yield in response to N and other inputs by 4–6 folds (Jensen et al. 2002).

In the semi-arid regions of Northern Nigeria, integration of supplemental irrigation with precipitation pattern and N supply is almost mandatory. Irrigation has a major say on maize growth, its productivity and N-use efficiency. The quantum of soil moisture supplied through precipitation and irrigation is crucial. There are indeed several field investigations that have aimed at optimizing irrigation schedules and N input to match the maize grain yield. Odunze et al. (2008) suggest that maize needs 415–540 mm for optimum forage and grain production in the Savannas of Nigeria. Soil moisture supply has direct bearing on WUE, N use efficiency and ultimate grain yield harvests. Following is an example that depicts amount of water consumed by maize crop and it influence on grain yield, WUE, and N use efficiency:

Water used	Grain Yield	Water Use Efficiency	Nitrogen Use Efficiency
(mm^3)	t ha^{-1}	kg ha^{-1} mm^{-1}	kg grain kg^{-1} N
663	2.71	4.09	74.3
505	2.23	3.91	54.1
414	1.98	5.95	60.1
276	1.67	6.04	45.7

Source: Odunze et al.,2008.

Clearly, WUE increases as total amount of water supplied depreciates. Also, the total water supplied to maize crop affects N-use efficiency directly. The N-use efficiency of maize crop increases with water supply. As a consequence, biomass accumulation and grain formation also increases. In other words, irrigation schedules could be carefully tailored to obtain better returns on N supply into cropping zones and regulate N dynamics in the agroeosystem. Such examples dealing with effect of irrigation on nutrient dynamics are available for all the maize cropping belts in different continents.

SOIL EROSION CONTROL AND ITS IMPACT ON NUTRIENT DYNAMICS

Nutrient loss due to soil erosion is a major problem in most parts of maize agroecosystem world wide. It reduces fertilizer use efficiency perceptibly. Soil erosion reduces *in situ* recycling of nutrients. Crop productivity decreases due to loss of surface and subsurface nutrients. When soil erosion and nutrient loss persists for several seasons it results in degraded land that may need elaborate measures to recover soil fertility. Preventive measures that aim at controlling soil erosion and restoring nutrient dynamics include mulches using crop residues or other farm wastes; practicing conservation or NT systems; strip or alley cropping; contour farming; growing cover crops and establishing hedges with tree crops; establishing ground cover using mixed crops; splitting fertilizer inputs; placement of fertilizers instead of surface broadcasting; use of both organic and inorganic manures etc. Often, during practical maize production a combination of several measures that control soil erosion and nutrient loss are adopted. We generally call them INM methods.

Soil erosion is a natural process that causes movement of soil from one place to another. Raindrop impact provides the force required to move soil particles. Raindrops traveling at 20 mph can splash soil particles as far as 60 cm (Wolkowski and Lowery, 2000). Raindrop impact also induces runoff and loss of nutrients in dissolved state. It reduces infiltration or percolation *in situ*. Wind erosion involves detachment of soil that later gets flown or carried to different spot where it is deposited (for details see chapter 3 pp 84–86). Wind erosion is more common in plains and in regions with large patches of sandy soils prone to face high wind speeds. For example, high winds and dust storms in sahel and savanna regions of West Africa, drier regions of South Africa and Middleast cause wind erosion. They induce large scale movement of soil particles and nutrient proportionate to carrying capacity of soil particles. Soil erosion rates are high in Asia, Africa, and South America and it averages at 30–40 t ha^{-1} yr^{-1}. In the "US Corn Belt" approximately 30% of farm land suffers from soil erosion and most of them loose at least 1–3 t soil ha^{-1} yr^{-1}. Clay (2008) has reported that soil erosion rates have declined in "Corn States of USA". Continued loss of top soil, reduction in depth of upper horizon and loss of nutrients seem to be the crux of the problem (Jagadamma et al., 2009). Simulations done in Indiana, USA, have shown that soil erosion may accentuate in Corn Belt. Relative to 1990 levels, soil erosion is said to increase from 33% to 274% by the year 2050 (O'Neal et al., 2005). Halvorson et al. (2006) have point out that conversion of traditional conventional tillage to no-till systems has generally reduced loss of soil and nutrients. Loss of SOC as CO$_2$ emissions too have reduced. Following is a summary of soil erosion rates and grain productivity recorded in the Corn Belt during past decades:

Year	Soil Erosion	Productivity
	t ha^{-1} yr^{-1}	t grain ha^{-1}
1932	37.0	2.7
1982	19.5	6.5
1998	14.0	8.6
2005	<12	10.4

In the Piedmont region, sandy soils with low SOC and loose texture are highly prone to soil erosion. The most common INM techniques adopted are contour planting, dense planting, cover crops, intercropping, and adopting NT systems. The NT systems do not disturb soil to any greater extent and allow conservation of soil moisture and nutrients. A six year trial at Greensboro in North Carolina has shown that NT effectively reduces runoff by 11% compared to conventional tillage. The soil loss rate under NT was 2.3 t ha^{-1}. It is well below the USDA stipulated soil loss rate of 6.7 t ha^{-1}. Under similar conditions, soil loss from plots under conventional tillage was 74.7 t ha^{-1} (Raczkowski et al. 2006). The NT plots produced higher grain yield.

Soil erosion is a major problem in the maize cropping zones of Brazilian Cerrados. Maize crop is most vulnerable to soil erosion when it is still at seedling stage. Maize crop grown under conventional tillage system suffers relatively higher percentage of soil loss, compared to those under NT system. Planting maize at high density reduces soil erosion by 57% and water loss by 14% compared with a bare fallow (Levein et al., 2006). In addition, farmers adopt several other agronomic procedures like contour bunding, alley cropping, cover crops and mulching to reduce soil and nutrient loss form the systems. They most often adopt INM procedures since it improves nutrient efficiency and conserves soil fertility.

Soil erosion is a malady in large tracts of Eastern European plains. Maize cropping zones may experience loss of nutrients and soil moisture due to rampant soil erosion. Farmers tend to reduce soil erosion using variety of cropping systems and agronomic procedures like contours, mulches, planting grass covers during fallow season etc. Let us consider an example. In Romania, annual soil loss was 2.01 t ha⁻¹ in fields planted with beans, 0.14 t ha⁻¹ in fields with grasses, 2.57 t ha⁻¹ in maize fields and 3.06 in sunflower plots. Clearly, crop species and planting pattern seems to affect loss of soil and nutrient. Maize fields were among those that suffered relatively higher loss of nutrients. Bacur et al. (2007) have reported that annually about 335 kg humus, 17.5 kg N, 1.1 kg P, 2.4 kg K were lost from each ha of maize planted in the plains.

Gradients and slopes affect extent of soil erosion and runoff from maize fields. For example in Nigeria, a field slope of 1% did not induce perceptible soil erosion. However, at 5% slope about 156 t soil ha⁻¹ was lost; at 10% slope soil erosion became severe and it increased to 932 t ha⁻¹; at 15% it reached a maximum of 1229 t ha⁻¹. The extent of major and micronutrients lost depends on nutrient carrying capacity of soil particles. The INM measures to control such soil erosion include conservation tillage, contour farming, fallowing with cover crops and planting alley crops. It is common for farmers in semi-arid regions of Nigeria to plant grass along hedgerows and in strips to avoid soil erosion and nutrient loss. Vetiver grass (*Vetiveria zizanioides*) seems to effectively reduce soil erosion and nutrient loss. Hence, it is frequently one of the ingredients of INM procedures. Babalola et al. (2009) examined effects of vetiver on soil loss and nutrient in the eroded soil from maize fields with 6% slope. Vetiver planting improved maize grain harvests by 30–50% over plots not planted with vetiver. The net soil loss was 19.5 kg ha⁻¹, if vetiver was planted as hedge rows and 70.3 kg ha⁻¹ in non-vetiver control plots. Analysis of eroded soil has clearly shown that nutrients are generally of higher order, if plots are not protected with vetiver. Following is an example that depicts effect of vetiver planting on soil texture and nutrients:

	Characteristics of Eroded Soil						
	Clay	Silt	Sand	pH	C	Total N	P
	%	%	%		%	%	ppm
Vetiver	3.8	4.0	92.2	7.5	2.13	0.065	6
Non Vetiver	4.6	4.2	91.2	7.5	2.18	0.074	7

Source: Babalola et al., 2009.

In Kenya, maize is often planted in strips alternating with grasses like *Panicum colaratum* or *Vetiver zizanioides* as a long term measure to control soil erosion and nutrient loss from ecosystem. Short term measures include planting of creepers like *Ipomea batatus* and recycling maize stover (Nzabi, et al., 2000). Maize residue recycling improved soil fertility and quality in addition to providing soil erosion control.

At Ntecheu in Malawi, planting maize closely with Leucana was effective in controlling soil erosion and nutrient loss, even from fields with 44% slope. Hedgerows of Leucana reduced soil erosion from maize fields to 2.0 kg ha⁻¹ compared with 80 kg ha⁻¹ in unprotected control plots. The productivity of intercropped maize was maintained at 1.5–2.0 t ha⁻¹. The maize yield progressively declined in unprotected plots to 0.8–0.5 t grain ha⁻¹. Clearly, close planting and intercropping helps in conserving soil fertility and in maintaining optimum nutrient dynamics (Banda et al., 1994).

As stated earlier, maize fields with gradients are prone to loss of soil and nutrients. In upper India, farmers overcome such soil and nutrient loss by planting *Leucana leucocephala* or *Eucalyptus hybrid* as alley crops along with main cereal maize. Field trials have shown that alley crops reduce runoff by 27% compared to usual 347 mm. The soil loss reduces from 39–40 t ha^{-1} to 12.5 t ha^{-1}. Such reductions are primarily due to hedgerow or tree barriers planted in contours (Pratap et al. 1998). Soil N conserved is also appreciable. In the Southern Indian plains, it is common to sow a cover crop or intercrop maize with legumes to achieve good ground cover and avoid soil erosion. Green manure trees and grasses grown as hedgerows reduce loss of soil to low lands.

In Northeast Thailand, maize farmers regulate soil and nutrient loss using Vetiver barriers and Leucana hedgerows. Maize grown with soil conservation systems and a cover crop Canavalla produced 5.5 t grain ha^{-1} compared to 3.8 t ha^{-1} in control plots without grass barriers/hedgerows (Pansak et al., 2008). In three years, runoff decreased by 72% from an initial level of 190–264 m^3 ha^{-1}. Soil loss decreased from 24 t ha^{-1} to 1.2 t ha^{-1} due to hedgerows. Among nutrients, N loss from maize fields is a major problem. Soil conservation with Vetiver and Leucana reduced N loss due to runoff, soil loss and leaching from 55 kg N ha^{-1} in control plots to 37–40 kg N ha^{-1}. It is interesting to note that due to soil conservation systems, pathway of N loss shifted from surface runoff to small amounts of leaching and percolation.

In Philippines, maize cropping belt encounters rampant loss of soil and nutrients due to runoff and surface erosion. Both gentle and steep gradients induce loss of surface soil. Integrated soil and nutrient management procedures involve a series of measures using indigenous resources. Techniques such as applying crop residues, retaining crop stubbles and stover in the field, growing alley crops like Leucana, applying organic mulch using tree leaf droppings, etc. Hedgerow intercropping is a common feature during maize production. It is helpful in providing ground cover and reducing soil erosion (Nelson et al., 1998). Usually five rows of maize and two rows of tree legume are planted. Retaining stubble is a useful measure in most regions because it reduces runoff and improves C sequestration in soil. Following is a standard example depicting how double row hedge planting of tree legumes like Leucana improves soil and nutrient conservation *in situ*.

	Runoff	Soil Loss	Grain Yield	Stover Yield
	mm	t ha	kg ha^{-1}	
Bare (control)	53	7.08		
Corn alone (stover removed)	33	2.77	1242	2339
Corn/Leucana (stubble retained)	13	0.81	1771	3380
Corn/Leucana (stubble removed)	16	1.03	1738	3145

Source: Comia, 1999.

Adoption of NT and mulches are important measures envisaged under integrated soil and nutrient management schemes. Conventional tillage induces relatively greater loss of surface soil and nutrient. However, no-till fields supplied with mulches performed better. Following is an example:

	Runoff	Soil Eroded	Loss of Organic Matter and Nutrients			
	mm	t ha⁻¹	Organic Matter	N	P	K
			kg ha⁻¹			
Conventional Tillage	371	141.3	5916	298	2.5	266
Tilled No Mulch	209	23.7	946	47	0.4	34
Tilled, Mulched	88	2.8	224	11	0.1	9
No tillage, Mulched	99	1.7	275	14	0.1	4

Source: Comia, 1999.

According to Comia et al. (1999), mulches like *Desmanthus* pruning add 128 kg N, 9 kg P and 75 kg K ha⁻¹; but maize stover recycling adds only 35 kg N, 5 kg P, and 55 kg K ha⁻¹. In addition, adoption of integrated measures has far reaching effect on soil physico-chemical properties, especially soil nutrient availability to maize roots and nutrient recycling rates. Crop response to such measures is easily discernable. According to Hur et al. (2004), maize farming in South Korea is mostly done on fields that experience gradients. Much of the South Korean country itself is undulated and sloppy. Maize farmers therefore adopt dense planting to avoid soil erosion and nutrient loss. Following is the extent of nutrient lost via runoff, soil erosion and drainage:

Nutrient Loss (kg ha⁻¹)		Runoff	Soil Erosion	Drainage
Nitrogen	Maize	1.2	tr	11.9
	Bare fallow	13.5	0.2	4.8
Phosphorus	Maize	0.9	tr	0.8
	Bare Fallow	2.4	1.8	0.6
Potassium	Maize	1.4	tr	13.6
	Bare Fallow	15.5	3.5	5.5

Source: Hur et al., 2004.

Bare fallows without soil conservation measures account for greater loss of major nutrients via runoff and soil erosion. Drainage loss seems to be greater in irrigated fields. Hur et al. (2004) suggest that maize crop with greater ground cover or intercrop reduces nutrient loss perceptibly. In the Yunan province of China, maize-soybean rotations are prone to soil erosion and loss of nutrients applied to maize. The INM procedures suggested include organic mulches, bunding and contour planting. Application of straw mulch reduced soil erosion rates consistently (Barton et al. 1998). Depending on angle of the slope, straw mulch reduced soil erosion by 19–71% compared to conventional tilled plots without mulches. Maize-soybean rotations performed best on NT plots given straw mulch.

The dry land farms in Northern Australia that support cereal production for fodder and grain thrive mostly on red, yellow or grey earths prone to soil erosion and nutrient loss. The extent of erosion and nutrient loss depends on soil characteristics of individual farms, as well as vegetation cover maintained. The INM procedures adopted include NT system, application of massive doses of organic matter and contouring.

Tillage induces soil erosion to a tune of 1.5–2.0 t ha^{-1} more than fields retained under no-till systems. Conventional plow induced soil loss ranging from 8.1–100 kg ha^{-1} (Dilshad et al., 1996).

WEEDS AFFECT NUTRIENT DYNAMICS AND REDUCE MAIZE PRODUCTIVITY

Weeds are among the important factors that affect nutrient dynamics and productivity of a maize field or cropping zone. Weeds reduce agronomic efficiency of fertilizers and water supplied to maize fields. Basically, weeds compete with the main cereal crop in many ways such as physical exploration of soil by roots, nutrient and water resources, competition for light interception etc. Much of the early investigations during mid 20th century were aimed at standardizing herbicide application, as well as understanding critical stages of the crop, when weeds affect soil nutrient and water acquisition by maize. The inferences on minimum number of weed-free days required to obtain optimum grain yield was indeed a useful guide to farmers. Currently, detection of weeds using far red/green radiation reflectance ratio is also possible (Rajcan and Swanton, 2001). It allows us to gauge the extent of resource, especially nutrient and moisture loss to weeds at different stages during maize crop season. We can in fact, advice farmers accordingly to reduce weed infestation. Critical time for weed control measures can be effectively decided. Intercrops planted during maize cropping effectively reduce weeds. For example, squash planted in between maize rows reduced infestation by *Amaranthus retrofelxus* and *Convolvulus arvensis*. Weed infestation was higher in maize mono-crops. Fujiyoshi et al. (2007) suggest that dense planting of maize or intercropping with squash effectively reduces light interception and biomass accumulation by weeds. In the Northeast of USA, corn is generally grown under no-till or conservation tillage systems. Weeds do infest in high numbers and divert both nutrients and water away from the main cereal. Nutrient dynamics envisaged by the farmer definitely gets altered. Maize growth and production gets constrained. Teasdale et al. (2005) examined influence of effects of dense planting and cover crops like hairy vetch on the weed infestation and maize productivity. Generally, cover crops reduced loss of water and nutrients to weeds. Cover crops suppressed weed number and their growth rate. The weed biomass reduced drastically, if maize and cover crop were sown instead of a mono-crop.

Silva et al. (2004) have reported that green maize yield is about 500 kg ha^{-1}. It is very low owing to low soil fertility and weed infestation during its cultivation. Weeding is generally achieved through hoeing. The number of hoeing is increased depending on weed density and critical phases when maize crop needs to be weed-free. In some cases, maize and weeds did not compete for light interception. This was attributed to the fact that younger leaves intercepted photosynthetic radiation. However, most of weeds below 1 m in height and well below photosynthesizing maize leave. In most farms, reduction of maize productivity was attributed to loss of nutrients and water to weeds. It was very simple to decipher loss of N to weeds. In the weed infested plots maize crop showed N deficiency compared with green pigmented leaves in weed-free leaves. Weeds depleted soil-N levels. High N input to maize fields may overcome maize yield loss. In all cases, maize cultivated along with weeds had low rooting density. It means proportionately low nutrient and water absorption by maize mainly due to weed infestation.

The crop-weed interactions are complex and are under the influence of several environmental factors. Yet, maize-weed interactions that occur in the "Corn Belt of USA" could be simulated using established computer-based models such as ALMA-NAC. McDonald and Riha (2002) have reported one such simulation study that involved maize and a weed species *Abutilon theophrasti*. Parameters used included leaf area index, plant height, rooting depth, above ground biomass and grain yield in case of maize. Maize grain yield reduced by 16–35% depending on weed density. Such models could be useful while deciding on extent of weeding, nutrient supply and yield goals.

A report from Argentinean Pampas suggests that in addition to various weed control measures, such as hand weeding, herbicide application, cover crops and intercrops, maize planting density and pattern has an impact on weed density and biomass. The spatial pattern of planting affects water and nutrient recovery from soil profile. Light interception too was better based on planting pattern adopted (Acciaresi and Chidichimo, 2002; see Plate 8.2).

Plate 8.2. A weed infested maize field in Pampas of Argentina.

Source: Alfred Bono, INTA, Buenos Aires, Argentina.

Note: Weeds divert significant amounts of soil nutrients and water away from the main crop. Timely weeding at early seedling stages is important, if not, canopy development could be hampered. A minimum of 40–60 weed free days are essential to obtain optimum maize grain/forage yield.

Maize crop gown in European plains encounters stiff competition for water and nutrient, if appropriate and timely control measures like herbicides or hand weeding are not applied. The NT systems generally allow rapid build up of weed population. Hence, many farmers adopt conservation tillage to remove weeds and volunteers. Let us consider an example. In Hungary, weed biomass and coverage in no-till fields increased from 9.6% to 28% within first 30 days. However, if treated with herbicides like Acetchlor, it reduced weed coverage to 0.6% of area. Maize biomass was higher in plots treated with herbicides. It was 1.9–2.3 times greater than untreated plots. Weeds generally reduce water use efficiency. There was also a strong competition for nutrients between maize seedlings and weeds. Weeds had recovered 16 kg N, 2.5 kg P, and 28.2 kg K ha^{-1} that otherwise would have been easily available to maize roots. Maize grain yield was reduced and nutrient recovery dropped to 38 kg N, 4.2 kg P and 29 kg K ha^{-1}. It is almost 8–15% lower than in herbicide treated plots (Eva et al., 2009). Timely weeding during early seedling stage is essential to restrict nutrient loss to weeds and maintain optimum nutrient dynamics. Alexieva and Stomenova (1998) made an effort to develop computer models and decipher maize crop stages that are most vulnerable to competition by weeds. *Amaranthus retroflexus* and *Convolvulus arvensis* were dominant weeds. The aim was to assess effect of single species of weed, mixtures of weed species and different densities of weeds on nutrient and water diversion from maize crop. The idea was to standardize timing of weeding and fertilizer input schedules, in such a way that very little soil nutrients and fertilizer-based nutrients are lost to weeds. Both simulations and real yield losses ranged from 4.7 to 12% compared to weed-free plots. In Serbia, productivity of maize inbred lines planted at relatively wider spacing is affected by weeds. Slow growth of inbred maize lines makes them vulnerable to competition by weeds for nutrients and water. Herbicide application is necessary to control loss of nutrients to weeds (Stefanovic et al. (2007). This type of problem is confined to inbreds used during maize seed production. Otherwise, in general, maize crop is densely planted to avoid weed infestation. In the Spanish plains, maize encounters competition for resources from a wide range of weed species. Planting density, fertilizer supply and growth rate of maize seems crucial. A study by Cavero et al. (2002) found that *Datura stramonium* which is a common weed could effectively compete with maize and reduce its grain yield by 14–63% compared to hand weeded control fields. The yield loss increased, if weeds stayed in field for longer duration. The competitive ability of *D. stromonium* was easily attributable to its rapid growth rate, habit, its ability to intercept light better and also garner more of soil-N. In order to improve both C fixation and N absorption ability of maize, it is necessary that weeds are effectively reduced below threshold during early seedling stage.

Based on field trials at IITA, Ibadan, Nigeria, Frank et al. (2005) pointed that cultivation of green manure crops like *Pueraria phaseoloides* has definite advantage in terms of soil-C and soil-N. In addition, rapid growth of *Peuraria* suppresses weed growth during natural fallows and maize phase of the cropping system. Further, mulching with *Peuraria* leaves reduces weed biomass. It obviously reduces diversion of nutrients to weeds. Maize crop grown in most areas of Tanzania is affected by weeds. Several types of weeds compete for water, nutrients and light interception. According to Shetto and Kwiligwa (1998), maize yield loss due to weed infestation may reach 35% compared to world average of 13% loss due to weeds. In Tanzania and other East African locations, maize yield loss has ranged from 20–70% compared to hand weeded control fields. Application of herbicides or hand weeding or combinations has generally reduced loss of nutrients and water to weeds. The weed biomass reduces drastically if fields are carefully hand weed at critical stages. Fields should be weed-free in the initial stages of crop and for at least 45–60 days. Otherwise, weeds invariably affect nutrient dynamics in individual field. They also reduce maize grain yield. Following is an example showing the effect of weeding systems adopted, maize grain yield and weed biomass accumulated in Tanzanian maize cropping zone:

Weeding System	Weed Biomass	Maize Yield
	g m^{-2}	t ha^{-1}
Unweeded	642	0.6
Hand Weeded	202	6.2
Cultivator-1	99	5.9
Cultivator-2	100	5.0
Herbicide	139	6.4

Source: Shetto and Kwiligwa, 1998.

It is clear that weeding measures minimize loss of nutrients to weeds. As a consequence, maize grain/forage yield improves. It is interesting to note that farmers in Zimbabwe tend to plant maize in relatively higher densities. This step is known to reduce weed infestation and avoid undue competition for water and nutrients (Fanadzo et al., 2007).

LONG TERM TRIALS WITH MAIZE

Long term field experiments aimed at studying soil fertility changes and maize productivity have been initiated several decades ago in different countries. They are located in almost all important maize cropping belts of the world. Many of these long term trials adopt INM strategies. They aim at understanding influence of long term fertilizer and organic manure inputs into ecosystem on soil physico-chemical properties, nutrient dynamics and grain/forage productivity.

Following are few examples (not exhaustive):

Lethbridge, Canada: Long term assessment of maize mono-cropping and maize-based rotations with blue grass or alfalfa on soil N and P dynamics and crop productivity. Also, influence of repeated cattle manure application on soil-N and P dynamics.

Chazy, New York, USA: Long term (32 years) trial on effect of stover harvest and removal versus recycling *in situ* on soil quality and SOC accumulation. Also, aims at knowing influence of tillage and no-tillage on soil fertility changes and maize productivity.

Buenos Aires, Argentina: Field trials on Arguidolls of Argentine Pampas were aimed at studying effect of continuous maize and maize-soybean, with and without tillage on soil N dynamics.

Padova, Northern Italy: Long term trial (40 years) at University of Padova Experimental Station aims at understanding effect of mineral fertilizer plus FYM supply on soil quality and maize productivity.

Pisa, Italy: Long term maize mono-cropping and rotations involving crimson clover or subterranean clover, initiated at CIRRA, in 1933. It also aimed at analyzing the effects of tillage, soil fertility changes, N supply and residue recycling on maize production trends.

Keszthaly, Hungary: Long term (1983–2008) evaluation of crop rotation, fertilizer-N plus FYM and recycling of crop residues on nutrient dynamics and maize productivity.

Samaru, Nigeria: Long term field experiments initiated in 1950, aim at understanding influence of cow dung, N, P, and K on maize and cowpea productivity. It also aims at knowing weed effects on soil nutrient dynamics.

Ibadan, Nigeria: Long term trials (1980- continuing) aim at understanding effect of fertilizer schedules on physico-chemical properties of soil and fertility changes.

Homa Bay, Kenya: Long term trials on soil fertility, fertilizer supply and weed infestation with an aim to asses their impact on maize production.

Coimbatore, Tamil Nadu: Long term (31 years) field evaluations of finger millet-maize rotation for micronutrient recovery trends, supply levels and build up in soil. Effect of major and micronutrient supply on maize grain yield.

Palampur, Western Himalayas, India: Long term experiments on maize production initiated in 1970 aims at knowing effects of fertilizer-N and lime supply on soil acidity, fertility in general and maize-wheat productivity.

Gansu, China: Long term (24 years) trials aimed at ascertaining the effects of chemical fertilizer supply, organic manure and crop residue recycling on water use efficiency and grain yield of maize.

Kyoto, Japan: A nine year long term trial to asses the effect of inorganic–N, organic manure and combinations of inorganic-N plus organic manure on N recovery pattern and maize grain yield.

A long term study extending for 32 years was conducted at Chazy in New York State. It evaluated effect of maize stover removal from fields versus reincorporation *in situ* on soil parameters, fertility status and maize production trends (Mobius-Clune et al., 2008). Among the 15 soil characteristics evaluated only eight of them were adversely affected by stover harvest and removal from fields. The deleterious effect of stover removal was accentuated under NT systems. Tillage affected almost all soil characters adversely compared to no-till systems. On a long term basis, stover removal was less adverse than repeated tillage. Partial removal of stover and no-till systems seem sustainable on a long term basis.

Halvorson et al. (2006) have pointed that in the Western region of the Corn Belt of USA, maize productivity was influenced by tillage systems and N supplied to the crop. Continuous-corn provided with low N rates produced relatively low amounts of grains under NT systems. However, sprinkler irrigated maize under conventional tillage and high N rates resulted in increased grain and residue harvest. It seems cooler soil temperature at planting under no-till affected seedling growth rate. They conducted a detailed study extending over 5 years at Fort Collins in Colorado, to optimize crop response to fertilizer-N supply under various tillage and other agronomic procedures. Six levels of N inputs ranging from nil to 230 kg N ha^{-1} were evaluated. The cob weight increased with N inputs. Maize grain and forage yield was near maximum at 268–276 kg N ha^{-1} (soil-N + fertilizer-N) depending on tillage systems. The peak crop response occurred when available-N level in soil reached 265 kg N ha^{-1}. Averaged over 5 years, conventional tillage had a 16% advantage (1700 kg grain ha^{-1}) over No tillage systems in terms of grain yield. Nitrogen removed in grain averaged at 19 kg N t^{-1} grain under NT and 20 kg N t^{-1} grain under conventional tillage system. The N-use efficiency is a crucial factor that affects fertilizer N supply schedules, grain harvest and crop residue recycling. The N-use efficiency decreased as fertilizer-N inputs are enhanced. About 43% of N supplied was recovered by the crop.

Knowledge about influence of soil texture on long term nutrient dynamics is useful. Let us consider an example. Osmond et al. (1992) have described a simple N balance based on fertilizer calculation for long term maize crops. Long term fertilizer-N requirements were predicted using N uptake trends. It was found that organic-N pools declined more rapidly in coarse texture soils than in fine textured soils. The equilibrium uptake from soil organic N pool was reached in 10 cropping cycles and it was determined to be 31 kg N ha^{-1} for coarse textured soils and 36 kg N ha^{-1} for fine textured soils. There are several advantages in simulating or calculating long term N requirements based on crop sequences and yield goals.

Now, let us consider an example dealing with long term effects of different cropping systems that include maize. Maize-based cropping systems such as continuous-maize (CC), maize-blue grass (CB), maize-alfalfa (CR) were assessed for their effect on soil-P accumulation trends, soil-P availability and fluctuations in soil-P fractions during 43 years (Zhang et al., 2006). On virgin soils, organic-P (Po) decreases as cropping progresses. Loss of Po could be severe and usual levels of P inputs may not suffice. At the bottom line, consistent cropping reduced all forms of organic-P (Po), in the order CC > CR > CB, but increased labile P in the order CB > CC > CR. Moderately stable Pi increased with fertilization. Supply of fertilizer-P in general, improved dissolved inorganic-P, dissolved organic-P and particulate-P. Over all, there is a strong need to take note of long term trends in P dynamics while preparing fertilizer schedules to suit the yield goals envisaged.

A different long term trail in the same belt has shown that high organic matter (FYM or cattle manure) supply annually may affect several soil properties. The cattle manure supply rates are sometimes so high that it takes about 24–33 years for soil NO$_3$ and soil-P to return to levels that existed before starting the long term experiment. Application of high amounts of cattle manure did not improve grain yield after a certain level. Irrigation induced NO$_3$ leaching. Soil NO$_3$ was also depleted by crop uptake. The long lasting residual effect of cattle manure application was clearly perceptible through stable grain yield (Hao and Chang, 2006). Fields kept under no-till maize or no-till maize-wheat or maize-barley or maize-soybean rotations were analyzed for changes in organic-C, N and N fractions. Spargo and Alley (2006) found that in 2–14 years, field receiving FYM or other organic amendments retained SOC and N at optimum levels. In general, changes in organic manure supply affected soil N dynamics. Elevated SOC contents meant reduced need for recycling straw. Knowledge about the rate of soil organic-C and N build up seems crucial while deciding fertilizer schedules.

Continuous no-till maize is a common feature in many locations of Argentine Pampas. Maize-soybean rotation is equally frequent. Each of these sequences may have differential effect on soil properties, especially soil-N status and N fractions, microbial load and its activity, and maize productivity in general. During a long term (14 years), inclusion of soybean in the rotation increased soil-N pool and its availability to maize roots. Microbial activity in soil was enhanced. Mineralization capacity of soil was also significantly higher under maize-soybean rotation compared to continuous maize. Maize grain productivity was higher if it was rotated with soybean. It was easily attributable to better soil-N status. In general, long term maize cropping in the Rolling Pampas has depleted soil fertility especially SOC and major nutrients. This has been attributed to intensification of maize-soybean cropping system and lack of appropriate nutrient replenishment schedules based on soil analysis (Caracova et al., 2000).

Long term trials in Netherlands have aimed at assessing the effects of FYM application on productivity of silage maize. Maize field meant for forage receives high amount of inorganic fertilizers and at periodic intervals after each harvest. In addition, farmers practice application of large quantities of organic manures-FYM. Therefore, residual effects were easily traced on rye grass that follows maize. On an average 50 t FYM ha^{-1} yr^{-1} was supplied to the field. Nitrogen released from residual FYM was equivalent to 20 kg N ha^{-1}. The residual fraction was much small if cattle manure was used. In general, residual effect was equivalent to only 10% of total N requirement of maize. Therefore, long term fertilizer N supply should consider amount of residual-N derived from FYM application (Dilz et al., 1990).

Romania is an important maize producing country in the European plains. Maize grain productivity is moderate at 3–4 t ha^{-1}. Fertilizer supply to the ecosystem has been constantly standardized and varied based on yield goals and economic advantages. Long term cultivation of maize obviously affects both soil fertility and nutrient dynamics. Therefore, long term trials were in fact conducted at Teleorman in Romania to ascertain changes in maize crop productivity. According to Negrila et al. (2007), maize cultivated on a long term for 20 years, encountered drought for 5 years. Maize grain yield was obviously immensely affected by rainfall pattern and quantity. Grain yield was > 5 t ha^{-1} during 9 years, 3–4 t ha^{-1} for 3 years and <3 t ha^{-1} for 8 years. Fertilizer-based nutrient supply to the ecosystem was most crucial input that affected productivity. Maize provided with 50 kg N and 40 kg P$_2$O$_5$ maintained best average grain yield. During rainy years, higher amounts of nutrient should be supplied. For example, maize grain yield of 5 t ha^{-1} is possible if 100–150 kg N and 80 kg P$_2$O$_5$ ha^{-1} plus FYM are supplied. Clearly, precipitation pattern and yield goals dictate the nutrient dynamics in the ecosystem.

Along with barley and wheat, maize is an important cereal in the Hungarian plains. Let us consider an example that includes maize cropping on a long run. Maize based rotation that includes barley or wheat was analyzed on a long term basis from 1983 to 2005, for 22 years, at Keszthaly in Hungary (Katalin et al., 2005). The aim was to find out influence of fertilizer-N inputs in increasing quantities; supply of FYM every 3 year once; recycling straw, stalk or green manure plus fertilizer-N on soil fertility and nutrient dynamics. Generally, the soil organic-C content increased significantly by the year 18. Maize grain yield was low in all treatments that did not envisage addition or recycling of organic matter. Obviously, recycling or replenishing organic matter from extraneous sources is crucial, if soil quality, nutrient dynamics and maize grain yield are to be sustained at optimum levels. Combinations of inorganic-N fertilizers and FYM were effective in improving soil N status and improving soil quality.

Accumulation of humic fractions was better if FYM supply was consistent. Hoffman and Nyoky (2001) have reported that long term (25 years) maize-wheat rotations practiced in Hungarian plains experience progressive depreciation of grain yield, if fields are not fertilized periodically. The method of tillage influenced grain yield of maize much less compared to wheat. maize kept under NT systems for long term produced lower amounts of grain yield compared to conventional tillage. Crop rotation influences maize grain productivity. Continuous maize produced lower amount of forage/grain. The SOM content improved marginally both in fields fertilized regularly and those kept under no-till systems. Supply of higher amounts of fertilizer-N did not affect SOC significantly. Also, soil tillage did not affect soil-N distribution in the profile. Soil N depletion was affected more by irrigation, drainage, and recovery by crop. Nitrate accumulation was high in the subsurface of well drained soils. Measures that reduce NO_3-N infiltration to ground water were necessary after a few years. Field investigations at Padova in northern Italy have shown that on a long term basis (1964–2005), organic manure application significantly increased SOC and soil quality. However, on peaty soils, repeated application of fertilizers had no perceivable effect on SOC and productivity (Morari et al., 2006).

Long term trials involving maize and various intercrops have been conducted in the West African maize cropping zone since 1920s. According to Vanlauwe et al. (2005), such long term trials have provided useful insight about the influence of various soil fertility and agronomic measures, on over all nutrient dynamics and crop productivity. Periodically, it has also helped in proper evaluation of low or high input technologies adopted by farmers. During recent decades, maize has been repeatedly intercropped or grown as alley crop with agroforestry/green manure trees. Several conclusions have been arrived from such long term trials. For example, a 15 year long term experiment has shown that growing *Leucana leucocephala* as strip crop provided 20% of N to maize crop; maize crop yield was consistently high with leucana as intercrop, but higher grain/forage productivity was obtained under maize/*Senna* system. On a long run, maize cultivation without tree intercrop depleted soil fertility. Major nutrients declined but the extent of decline was less if alley crops were grown. Selecting appropriate trees and green manure species was crucial. Preventing decline of pH and accumulation of Al in subsurface was also important. More recently, Vanlauwe et al. (2005) found that at Ibadan in Nigeria, over the years biomass produced by trees reduced. Nitrogen contribution by leucana was 200 kg ha^{-1} and that by Senna was 160 kg ha^{-1}. Maize crop yield declined over the 15 years but less rapidly under maize-senna system. Tree intercrops had positive influence on exchangeable Ca, Mg, and K. Maize crop productivity and soil fertility improved, if inorganic fertilizers and organic manures were applied in combination. Maize grain yield stabilized at 2.8 t ha^{-1} over a period of 15 years under maize/senna system.

A different set of long term trials in West African Savanna has shown that if fertilizer-N was supplied consistently, it resulted in deficiencies of P (Nziguheba et al., 2008). When all major nutrients N, P, and K were added, DRIS indices for secondary nutrients Ca and Mg were low indicating deficiencies. Zinc was also found to be deficient. Despite fertilizer-N supply, N indices remained negative indicating its deficiency. Low efficiency of fertilizer-N was one of the reasons. Therefore, it was suggested that INM procedures that envisage application of all major and micronutrient should be adopted. Further, nutrient should be supplied using both inorganic and organic manures based on soil tests and DRIS. Such approaches remove nutrient imbalances.

Maize-fallow sequences are common in the West African maize belt. Fallow lengths and types of herbaceous species grown seem to affect usefulness of fallows. Tian et al. (2005) evaluated maize and its intercrops grown for a long term of 12 years with various lengths of fallow. They concluded that fallows generally had positive influence on maize productivity. During 12 years, characteristics like soil pH, organic-C, available-P, and exchangeable-K decreased, even if fallows were extended for 3 years. The organic-C was 1.39% under alley cropping, 1.37% in natural fallow, and only 0.8% under crops. Fallows beyond 2 years did not help in improving soil fertility or maize productivity.

In the Eastern African tropics, maize is cultivated using extraneous nutrient supply. Application of FYM and other organic manures helps in improving soil quality and SOC content. Maize fields are often infested with weeds, especially striga. Statistical evaluation of long term trial initiated in 1991 have shown that a combination of fertilizer-based nutrients (80 kg N, 40 kg P ha^{-1}), residue recycling and periodic weeding restores soil fertility. Actually crop residue incorporation and weeding seem essential if the aim is to thwart soil fertility decline (Ransom and Odhiambo, 1993). Basically, weeds divert soil nutrients away from main crop and so disturb the nutrient dynamics that farmers envisage. Long term neglect can have severe impact on soil fertility and nutrient dynamics.

On Alfisol plains of South India, maize is a prominent cereal that provides both forage and grain. Farmers generally grow maize during rainy season. Farmers often supply major nutrients plus manures. Effect of curtailing micronutrient is not clearly understood. Long term study on effect of major nutrient—N, P, and K supply on Alfisols at Coimbatore in Tamil Nadu has shown that intensive cropping depletes micronutrients rapidly. Micronutrient deficiencies are easily perceptible in the surface soil. There was a steep decrease in DTPA extractable-Zn after a couple of crops. This could be avoided only if N, P, K + ZnSO$_4$ were supplied each season. Supply of Zn resulted in build up compared to fields not provided with Zn. The percent depletion of micronutrients namely Fe, Zn, Cu, and Mn ranged from 63–88% of original status at the beginning of trials in 1972. The cumulative recovery of micronutrients by maize followed the order Fe > Mn > Zn > Cu. Overall, it was clear that micronutrients replenishment was necessary, if not maize grain yields depreciated by as much as 25% compared to original levels (Selvi et al., 2006).

Long term trial (25 years) in Western Himalayas has shown that, integrated use of inorganic fertilizers and FYM improved maize grain yield to 4600 kg ha^{-1}. Repeated use of N fertilizer affected soil acidity immensely. It decreased soil pH from 5.8 to 4.7 in 25 years. The SOM increased in almost all treatments. Initially, K supply did not improve maize grain yield because soils were saturated with exchangeable-K. However, after a few years, K depleted to low levels and K inputs resulted in grain yield improvements. Fertilizer schedules affected weed flora and dominant species. Based on several such long term trials in the north Indian plains, Sharma and Subehia (2003) suggest that there is a need to periodically assess and revise fertilizer supply into the maize fields. Revisions should aim at minimizing undue alteration in physico-chemical conditions and soil fertility status, but at the same time should make moderate improvements in maize grain/forage production.

Crop rotations involving wheat and maize have been practiced consistently and for long periods in Northeast China. During past few decades, nutrient supply through fertilizers and organic manures, and quantum of irrigation has fluctuated enormously owing to several factors related to crop, soil fertility status, local environment and yield goals. On the other hand, there are also several reports that depict long term effects of maize cropping on soil physico-chemical conditions and productivity. Yang et al. (2003) have made some interesting observations regarding influence of long term mineral nutrient supply on grain production and energy efficiency of maize-wheat rotations. Under high nutrient supply regime, total grain and energy output/input ratio was lower than that noticed at low nutrient supply. In some cases, in addition to nutrient dynamics and grain productivity, farmers may have to consider long term energy efficiency. In Northwest China, maize is mostly confined to semi-arid region. Water resources and WUE has an impact on nutrient dynamics and grain yield. Long term studies show that, over a period of 24 years, maize grain production was 2.29 t ha^{-1} season^{-1} in unfertilized plot. However, it was 5.6 t ha^{-1} in field provided with N and P fertilizers. Generally, corn yield and its WUE declined as number of years elapsed. Parameters like SOM, total-N and total-P increased in fields provided with fertilizers. In case of fields not fertilized each year, total-N and total-P fluctuated marginally, but SOC and available-P decreased with time. Soil available-K decreased if straw and manure additions were curtailed. (Fan et al., 2005). Soil organic fraction increased, if fertilizer supply to field was consistent. This aspect also had positive influence on long term water holding capacity and nutrient buffering capacity of the soil.

A different long term trial in Gansu region of China aimed at knowing the effect of fertilizer application on crop yield and NO$_3$-N accumulation in soil profile. In 13 years, maize grain yield decreased by 25% compared to first year. This was attributed to depletion of soil nutrients due to incessant cropping. Maize grain yield declined least if N, P, and K plus organic manure were added compared to only chemical fertilizers. The Ymax decline was noticed in fields not supplied with fertilizers (Yang et al., 2005). The maize grain yield stabilized if N, P, K plus manure were applied repeatedly. Clearly, there is a need to match nutrient depletion with supply of major and micronutrients.

Long term fertilizer trials at Kyoto in Japan have shown that fields given chemical fertilizers plus organic manure consistently yielded higher compared to those supplied with only chemical fertilizers or organic manure. The fertilizer-N recovery ranged from 24–48% during first year, but declined with time (Li et al., 2001). Fields given fertilizers for 9 years accumulated soil-N abundantly, indicting that farmers supplied N more than that needed by maize crop.

CROPPING SYSTEMS, NUTRIENT DYNAMICS, AND PRODUCTIVITY

Selecting most suitable maize-based cropping systems, either rotations/sequences or intercrops is one of the important decisions farmers have to make under INM. Efficient utilization of land per unit area and time, in terms of photosynthetic light interception, nutrient recovery, precipitation use efficiency or WUE, avoiding weeds, but maximizing grain/forage harvest are few of the basic requirements of a useful crop rotation. Maize is grown all over the world during different seasons either as monocrop or as a mixture with other crop species. Maize/wheat or sorghum, maize/legumes, maize/vegetable, maize/oilseeds, maize/alley crops (green manure trees), maize/fiber crops like cotton are some of the most common intercropping systems encountered in different continents. Intercrops basically aim at improving land use efficiency and productivity.

Maize is a highly versatile cereal crop grown in different agro-environments. It adapts well to different types of soils, as well as variety of weather and precipitation patterns. It is compatible for rotation with a wide range of crop species within different continents. Maize based rotations aim at improving soil quality, preserving soil fertility and improving grain/forage yield of crops. Maize-based rotations also aim at efficient recycling of nutrients within the agroecosystem. Maize farmers modify or adopt different rotations to maximize profits per unit time and land area. Yet, we may realize that in "US Corn Belt", North China, Pampas of Argentina or in Southern Indian plains large patches of maize *monocrops* flourish. In the following paragraphs, let us discuss several of these maize intercrops or mixtures and rotations in greater detail regarding nutrient dynamics and productivity.

Maize-based Intercrops or Crop Mixtures: Nutrient Dynamics and Productivity

Maize is cultivated world wide in combinations with several other crops either as part of a mixture, or sequence or rotation. Multiple cropping or intercropping with maize is perhaps one of the most common methods adopted by farmers since ancient period. Crop mixtures with maize could be planted in strips or rows of pure stand at different ratios or sometimes, as in pastures, seeds of maize and other species could be broadcast in mixture. Whatever is the maize-based combination or ratio of rows or mixtures, primary aim of intercropping is to improve land-use efficiency, soil fertility status, and productivity. Intercrop combinations are chosen to avoid undue competition for photosynthetic radiation, rooting zone, nutrient, and water acquisition. Maize/legume intercrops have been practiced especially in North America to avoid NO_3-N leaching to subsurface and ground water contamination. Maize intercrops if planted closely avoid weed build up, hence, it reduces undue loss of nutrients to weeds. Let us consider a few conspicuous examples of maize-based cropping mixtures.

Maize/soybean intercropping with different row ratios or planting them in strips is most common in the North American "Corn Belt". There are indeed innumerable reports regarding nutrient dynamics and productivity of maize/soybean grown in this region. Competition for major nutrients is an important concern during intercropping. Soybean being an N fixing legume, competition for soil-N is greatly reduced. In fact, soybean adds to soil-N due to BNF and crop residue incorporation.

During the past 4–5 decades, maize/soybean intercropping has been consistently practiced in large tracts of North and South America (Brazil, Argentina), and Asia (Northeast China) (Allen and Obura, 1983). Maize is also intercropped with several other legumes such as cowpea, clover, and groundnut. Reports indicate that intercropping maize with legumes often improves total crop productivity compared to monocrops. The grain yield of monocrops may be higher than individual intercrops, but together in a field, intercrops yield higher amounts of grain/forage. Intercropped maize may yield between 60–90% of monocrop reference plot. Similarly, intercropped soybean yield may be only 56–80% of monocrop and intercropped cowpea yields only 52–54% of reference monocrop. The reduction of forage and grain yield in intercropped situation has been attributed to competition for nutrients like P and K, rooting pattern, soil moisture in profile and shading effects. The land equivalent ratio (LER) for maize/cowpea intercrop is 1.32 and that for maize-soybean is 1.22 (Allen and Obura, 1983). In the Canadian Plains, maize is intercropped with Rye grass (*Lollium perenne*). Rye grass seeds are sown 10 days after maize seeding. Zhou et al. (2000) have stated that intercropping did not affect grain/forage yield of maize. However, total above ground biomass yield increased by 2.4–3.2 t ha^{-1}. Nitrogen recovery increased by 49–71 kg N ha^{-1} due to intercropping. Intercropping induced rapid N recovery. Therefore, soil NO_3-N decreased in the profile. De-nitrification process also reduces due to intercropping. It was concluded that maize/rye grass intercrops increase N uptake, reduce loss of N to subsurface layers or ground water. Most importantly, intercropping increases biomass accumulation and C sequestration in the field. It is interesting to note that rye grass or red clover planted as intercrops did not affect the yield components of maize. However, seeding maize with large seeded legumes like soybean resulted in reduced grain/forage yield compared to monocrops. The growth and yield components of intercrops were also affected. For example, harvest index and test weight of both maize and soybean reduced due to intercropping (Carruthers et al. 2000). In case of soybean, number and weight of pods decreased due to intercropping compared to monocrops. Obviously, reduction in intercropped species is partly due to competition and its influence on growth and yield components.

Maize culture along with beans and squash as intercrops is an age old cropping system in Meso-America. The biomass production by individual plants of corn or bean is not affected much but squash plants may generally be smaller if grown as intercrop. Flowering of bean was slightly delayed but total pod yield did not get affected due to intercropping. Total grain/pod and squash yield of mixtures was always higher than monocrops. Most likely reasons for better land use efficiency and higher productivity of crop mixture was greater availability of N for maize intercropped with legume; efficient utilization of photosynthetic radiation; better rates of soil moisture and nutrient recovery; reduction in soil erosion due to close planting and reduction in diversion of nutrients to weeds (Risch and Hansen et al., 1982). Maize-based intercrop in South America often encounters acid soils that need consistent liming. There are indeed many cropping systems and agronomic procedures devised to overcome soil acidity. Computer based models have also been used to fit correct cropping systems (Horst, 2000). Reports suggest maize/legume intercrops provided with N, P, K and lime perform better in terms of nutrient recovery and productivity. In addition, maize/cowpea mixtures provided with lime performed better by 2.6 times over control in terms of WUE (Gaiser et al., 2004).

Mushagalusa et al. (2008) examined inter species competition for photosynthetic radiation and soil nutrients in a maize/potato intercrop. The interplant competition for light interception was most prominent. Photosynthetic activity and light interception were greater with potato (60%) during first 45 days and thereafter maize intercepted greater amount of light (80%). In case of maize, LAI, nutrient content of shoots and roots and grain yield deceased due to intercropping compared to sole crop. In case of potato, tuber yield decreased by 4–26% compared to sole crop. Yet, intercrops produced greater amount of biomass and tuber/grain yield if entire intercrop is considered. It improved land use efficiency.

In the European plains, maize is often grown for its quality fodder. Maize is intercropped with swards of grass/clover mixture in order to maximize biomass production. It improves land use efficiency. In order to suppress undue competition for nutrients and water, farmers generally trench or cut roots of swards. No-till systems are advantageous. Plowing immediately after sward, induces NO_3-N leaching and emission (Prins et al., 2004).

Maize–legume (cowpea, peanut, pigeonpea, field bean) rotations are most frequent in the savanna regions of West Africa. It is preferred since crop mixture provides for both carbohydrates and protein. The forage quality too improves due to crop mixture. Maize derives benefits in terms of soil-N from legumes. In Nigeria, soil erosion and nutrient loss due to it are prime factors that guide farmers to adopt different tillage systems and intercropping patterns. Intercropping and dense planting avoids loss of soil and makes use of soil fertility more efficiently. Kirchoff and Salako (2006) have pointed out that in southern Nigeria; maize/cowpea intercropping is practiced to thwart soil erosion. At the same time, cowpea being a legume adds to soil-N fertility. Cowpea residues are useful as animal feed and if recycled into soil it helps in C sequestration. Reports indicate that intercropping with cover crops like Pueraria or cowpea helps in improving maize grain/forage yield. For example, maize with Pueraria intercrop yielded 2.15 t grain ha^{-1}. Then, with cowpea as intercrop it yielded 1.92 t grains ha^{-1}. However, with macuna as intercrop, maize yielded 1.71 t grain ha^{-1}. Actually, maize grain yields reflected the extent of competition for soil moisture and nutrients between the intercrops. Hence, it was suggested that maize/cowpea intercrops are most useful in terms of nutrient dynamics. In addition, planting cowpea at higher density thwarted soil erosion effectively.

Maize/pigeonpea intercrops common to Eastern and Southern Africa yield higher than sole crops in terms of N and P recovery, total biomass and grain. The harvest indices of maize or pigeonpea did not alter due to intercropping. Residual SOC and N after sole crop or intercrop too did not differ significantly. However, pigeonpea culture did add about 60 kg N ha^{-1} to the cropping system. This is of course attributable to biological nitrogen fixation. Recycling of pigeonpea residue enhances soil-P by 6 kg P ha^{-1} (Myaka et al., 2006). Maize/cassava/groundnut crop mixtures are grown in Southern Cameroon, usually without much external nutrient supply. Crop mixture is sustained through residue recycling, slash and burn and N inputs derived from peanut. Hauser et al. (2006) examined the effects of planting density of the above crop mixture. It was found that increasing planting density did not improve maize yield, except when it was planted along with *Calliandria calothyrsus*.

In Turkey, maize is intercropped with cowpea or bean. Such intercropping improves land use efficiency, nutrient recovery and grain yield. Productivity of intercrops is generally higher than sole crops (Yilmaz et al., 2008). Planting ratio of 67:50 for maize:legume was found advantageous in terms of nutrient dynamics and productivity.

Maize intercropping systems are common all over the Southern Indian plains. Maize is intercropped with cereals like sorghum, finger millet; or with wide range of legumes like pigeonpea, cowpea, dolichos, horsegram; or with oil seeds like groundnut and sunflower. Maize/legume intercrops dominate the Vertisol regions of South India. Maize pigeonpea intercrops grown during rainy season adds to soil fertility because pigeonpea contributes to soil-N.

Maize/sorghum intercropping is beneficial compared to sole crops in terms of forage and grain yield (Reddy et al., 1980). In the dry lands of South India, planting maize/sorghum intercrop in paired rows (2:2) enhanced grain and forage yield by 44% over sole crops. Crop mixtures seem most useful due to short seasons and scanty rainfall. Short season varieties suit the precipitation pattern and nutrient paucity.

In the Alfisol plains of Tamil Nadu, maize is intercropped with wide range of crop species that includes oil seeds, legumes and vegetables. The INM methods envisage application of inorganic fertilizers (N, P, K), FYM, Azospirillum and Phosphobacteria. Among various combinations of short duration intercrops, Thavaprakash and Velayudham (2007) found that maize planted at 60 cm with coriander and radish performed better and produced Ymax. Land use efficiency and nutrient recovery were highest when maize was intercropped compared to sole crops.

Maize/legume intercrops grown in Thailand were examined for growth and yield advantages and improvement in land use efficiency. Polthanee and Trelo-ges (2003) reported that yield and yield components of maize were unaffected. It was attributed to lack of competition for photosynthetic radiation and nutrient acquisition from surface and subsurface soil. However, grain yield of legume intercrop was generally reduced. Grain yield reduction was 28% for peanut, 39% for soybean, and 51% for mungbean. The pod number per plant got reduced due to intercropping. In other wards, nutrient partitioning to pod/seeds got reduced due to intercropping. However, total productivity was higher in all intercrop combinations with legumes. The land use efficiency increased by 44–66% over sole crops.

Li and Zhang (2006) point out that maize-based intercrops have played a crucial role in food production in some parts of China. This is particularly true in areas still operating under low input systems; as well as those under high input-high productivity systems. Most common intercrops of this region are maize/wheat, maize/fababean, maize/soybean, maize/peanut, and maize/vegetable. Intercrops have been generally selected keeping in view precipitation pattern, nutrient availability in soil and absorption pattern of the intercrops. Nutrient and WUE of the intercrop compared with sole crops has been a major concern. The portion of N derived from atmosphere via biological N fixation is an advantage with maize legume intercrops. In some cases, intercrops have also been more efficient in absorbing P and Fe compared to sole crop of maize. Zhang et al. (2004) have stated that maize/groundnut intercrops perform better because of rhizosphere interactions and interspecies facilitations of nutrient absorption. Phytosiderophores released by maize affected Fe availability to peanut and maize roots. Intercropping of maize with faba bean too improved nutrient recovery and productivity.

Alley Cropping Maize with Agroforestry Tree Species

"Alley cropping" is a terminology often associated with growing food crops like maize or other cereals with rows of green manure or agroforestry tree species. Alley cropping of maize/tree combination is a preferred technology in many of the tropical and subtropical regions. It is supposed to retard the process of land degradation, improve vegetation cover, soil quality and at the same time allow nutrient recycling. Alley cropping is highly suitable in fields that are under low input technology. Alley cropping may involve several types of tree/maize combinations. Planting patterns, soil management methods, agronomic procedures and harvest timings too may vary widely depending on geographic location. The overall aim is to maintain soil fertility and improve land use efficiency. According to Kang (1997), alley cropping is a low cost technology that integrates trees and shrubs in spatial zonal arrangements with food crops. The presence of woody species is supposed to contribute towards—(1) nutrient recycling; (2) reduction of soil nutrient loss via leaching and percolation to subsurface or ground water; (3) stimulation of higher soil faunal activities; (4) soil erosion control, and (5) sustain crop production levels.

Let us consider a few more examples pertaining to alley cropping in African tropics and subtropics. Several factors related to alley crops affect the extent of nutrient recycling possible in a field. The crop species used as alley crop, its ability to produced biomass rapidly, row ratio and spacing, nutrient supply to alley and cereal crop, quality of prunings and leaf, especially its C:N ratio are some of the factors affecting usefulness of alley crop. Abunyewa et al. (2004) state that in parts of West African tropics around Northern Ghana, maize is often cultivated with *Gliricidia*. *Gliricidia* and other alley crop species grown in West Africa or South Asia may often accumulate up to 20 t biomass ha^{-1} in a crop season. Application of entire prunings increases maize grain yield by 54% over control. The extent of benefits to maize depends on Gliricidia biomass recycled. If only 10 t leaves are recycled ha^{-1}, then extent of improvement in maize grain and forage yield is negligible. Obviously, there is a threshold which has to be exceeded if crop response has to be perceived. *Gliricidia* is preferred because of its high quality leaf prunings that have low C:N ratio of 12. It is succulent and easily decomposable, rather rapidly in maize fields. It is said that nearly 70% of N released from *Gliricida* leaves are easily channeled into maize crop. However, total nutrient recovery from *Gliricidia* pruning into maize is around 20–30%. Generally hedgerows planted closely at 4 m apart provided more of biomass than rows spaced at 6 or 8 M. Pruning height also affects biomass generated and its effect on maize. Application of 40 kg P ha^{-1} enhanced biomass production compared to 20 kg P ha^{-1} or control.

Following is an example of maize crop response to application of *Gliricidia* prunings in the Savanna zones of Ghana:

	Maize Yield	
	Stover	Grain
		(t ha⁻¹)
Hedgerow	2.12	2.32
Control	1.87	1.13
CV %	6	9

Source: Abunyewa et al., 2004.

Stahr (2000) states that agroforestry systems provide substantial amounts of N to maize fields, especially if tree species are leguminous. Farmers in Benin (West Africa) adopt agroforestry to thwart nutrient loss from the maize fields. However, if tree crop density is high, competition for soil nutrients may be a clear possibility. We should note that, yet in most cases nutrients scavenged by tree crops are from layers of soil not explored by maize. Incidentally, maize crop exploits nutrients mostly from top 2 m layer. Field trial in the sub-humid region of Benin has shown that alley cropping enhances nitrogen recycling in the ecosystem. Maize/Leucana alley cropping recycles as much as 253 kg N ha⁻¹ and Maize/Cajanus recycles 131 kg N ha⁻¹ (Akonde et al., 1996). Out of the recycled N, about 22–30 kg N ha⁻¹ was recovered into maize crop. Over a period of 6 years, topsoil N accumulated in no-tree control fields was 18–37 kg N ha⁻¹, but in fields with Leucana it was 223 kg N ha⁻¹ and 110 kg N ha⁻¹ in Cajanus.

Traditional agroforestry parklands too support maize cultivation, especially as intercrop. Most commonly grown parkland tree species are *Faidherbia* and *Vitellaria* in Burkina faso. Usually, leaf and twigs littered under tree canopy decompose and release sufficient amounts of nutrients into soil. These trees actually help in recycling or pumping nutrients held in lower horizons of soil to surface layer. This helps maize or any other cereal that has its roots confined to upper layers of soil. Zacharia (2007) has stated that due to release of nutrients from leaf litter, maize crop under canopy grows better than in open field. Often, farmers use prunings and leaf litter to make composts and improve soil fertility status. Combinations of inorganic fertilizers and compost have proved better in improving maize productivity in the park lands.

Now, let us discuss a few examples form Southern African maize belt. In Zambia, farmers tend to intercrop agroforestry tree species to avoid loss of N to subsurface and ground water through leaching. Tree root systems are generally profuse and they trap the percolating nutrient effectively and recycle it via leaves, twigs, and branches that decompose on the surface soil. Again, *Gliricidia sepium, Leucana leucocephala, Accacia angustisma*, and *Sesbania sesban* are popular agroforesty trees used primarily to regulate nutrient dynamics in the field and entire farm. Chintu et al. (2003) have reported that in fields with sole maize, subsoil N accumulation was high. It means large share of fertilizer-N applied is lost to lower regions of soil that is not easily accessible to maize roots. On the contrary, fields supporting maize-agroforestry tree intercrop possessed relatively low amounts of N in subsoil. Tree roots must have retrieved sizeable amounts of nutrients that leached. Some farmers tend to plant trees in the fallows after maize harvest. It is again to retrieve nutrients already lost through leaching. In any case, tree planting helps in enhancing maize grain/forage yield because prunings, leaves, and twigs that decompose on the surface add to soil nutrient pool. Maize grain yield increase due to agroforestry procedures has reached 170% over control plots in some cases. Field trial near Lusaka in Zambia has shown that Sesbania grown in fallows after maize provides up to 56 kg N ha^{-1} through prunings. The litter and fresh leaves decompose quickly to release nutrients. Often combination of NO$_3$-N supplied via fertilizers and leaf litter has proved better than only inorganic fertilizer application. Inorganic fertilizer is supposed to act as primer by inducing decomposition of tree litter. In addition to improvement of soil nutrient status, tree leaf litter enhances soil quality and SOM content (Chirwa et al., 2004). A similar report from Malawi suggests that *Gliricida*-maize intercrops improve soil-N dynamics. Application of a combination of leaf litter, prunings and twigs of *Gilricidia* plus inorganic fertilizers improves N recovery by maize. It essentially enhances nutrient mineralization and recycling (Ikerra et al. 1999). In addition, reports by Harawa et al. (2006) suggest that *Gliricida*-maize systems improve N availability in soils that are otherwise deficient for N. Application of Sesbania prunings along with small amounts of inorganic fertilizer-N improves N mineralization and NO$_3$-N content. Intercropping also avoids undue loss of N to lower horizons of soil via leaching and percolation. Fields with agroforestry tree species like *Thephrosia* or *Gliricidia* improved soil N status by 30–60 kg N ha^{-1}. A few of the reports by World Agroforestry Center (ICRAF, 2008) suggests that in the maize cropping zones of South Africa, farmers experience nutrient mining. This is firstly attributable to removal of larger share of soil nutrients to maize crop than that replenished via fertilizers or residue recycling. On an average 20–40 kg N, 4–8 kg P, and 32 kg K ha^{-1} yr^{-1} are lost from the system. It seems a portion of nutrient loss/mining is primarily due to soil erosion. Hence, INM procedures envisaged for maize farmers in South Africa includes planting strips or hedgerows of agroforestry saplings along with cereal. It supposedly lessens soil erosion and reduces nutrient loss. An example from Malawi shows that planting *Leucana lecucocephala* in hedgerows allowed formation of terraces that stopped soil erosion to a great extent. Addition of leaf litter obviously improved soil quality. In maize plots alley cropped with Leucana, soil loss was 2 t ha^{-1}, but in control (maize alone) fields it reached 80 t ha^{-1}. The maize grain yield in intercropped fields was 1.5–2.0 t ha^{-1}, but in control fields it ranged from 0.5 to 0.8 t ha^{-1} (Banda et al.,1994).

We should note that alley tree crops may also compete for nutrient and photosynthetic radiation with maize crop. Alley crops planted closely may encounter higher intensity of competition. Shading could be lessened by frequent pruning of leaves. Roots of hedgerow tree species often grows rapidly, deeper and into wider area. It may compete for nutrients and water that otherwise was meant for cereal crop. The INM techniques envisaged to avoid competition for soil nutrients is to dig trenches periodically so that hedgerow tree roots are confined to set areas within the field.

Alley cropping of tree legumes with strips of maize is again common to cereal cropping belts in Australia. For example, *Leucana leucocephala* hedgerows are grown at 4–5 m apart with maize grown in between the groves. It is said dry leaves of leucana collected and placed at a depth of 10–15 cm depth in soil decomposes rapidly than fresh leaves strewn on surface. About 52–72% of dry leaves placed deep had decomposed in 3 months, releasing sizeable amounts of N compared to only 36% from fresh leaves placed on surface. The amount of biomass loaded into maize strips did not affect N release into soil. Mostly, entire amount (98%) of Leucana biomass applied was decomposed in 12–14 months after application. Entire quantity of N and C were mixed with soil. In a different experiment, analysis of ^{15}N labeled Leucana prunings has shown that in 52 days, about 5% of N supplied via Leucana prunings had been absorbed by maize roots, 45% stayed as residues, 25% was amalgamated into soil fraction and 25% was unaccounted. It perhaps leached or percolated. Over all, leucana prunings could substantially improve soil N status, N uptake by maize, although a supplementary N doze was needed to prime the crop response. Amount of N derived initially from the leucana prunings may be small but gradually entire quantity of biomass decomposes releasing large amounts of N, C, and other nutritional factors (Zhihong, 2002).

Maize-based Crop Rotations

Continuous maize and maize-soybean rotation are most frequent in the "Corn Belt of USA". Maize-Cotton rotation is prominent in cotton growing regions of mid-southern USA, like Arkansas, Tennessee, and Kansas. Maize-peanut, maize-sorghum, maize-legumes are frequent in other regions. These maize-based rotations in USA contribute about 40% of global maize grain/forage annually. Let us discuss, at least three major maize-based crop rotations followed in USA, namely continuous maize or maize-fallow; maize-soybean and maize-cotton. Maize cultivation in USA occurs predominantly on dry lands; as either rain fed or irrigated crop. In the Corn Belt, it is mostly grown as continuous maize or in rotation with soybean. Maize production in USA is highly intensive. Fertilizer based nutrient supply into the agroecosystem has increased enormously during past 5 decades. It has risen from a mere 60 kg N ha^{-1} in 1950s to 240–280 kg N ha^{-1} in 2005. Average grain harvest has increased steadily by 110 kg ha^{-1} yr^{-1} and total grain yield has now reached almost 11.5 t ha^{-1} (see Walters et al., 2006). Much of the increased grain productivity is derived from continuous maize or maize-soybean or maize cotton–rotations. An interesting observation is that since mid 1990s, there has been a relative stagnation in N, P, K input rate. However, this has necessitated farmers to adopt techniques that improvise on nutrient use efficiency (i.e., kg grain kg^{-1} nutrient added). Agronomic procedures that enhance fertilizer efficiency, reduce loss of soil and nutrients and induce efficient recycling of nutrients have been adopted in large scale. Further, use of nutrient-efficient maize genotypes and maximization of BNF during soybean phase of the rotation are some of the reasons for improved productivity. Most crop models and simulations of maize-based rotations suggest that potential yields are higher for both maize and legume or cotton. Yield gaps could be reduced further by devising more efficient soil/crop management procedures.

Stanger and Lauer (2008) report that fertilizer-N inputs into maize belt of USA was high at 4.5 Tg of N annually during 2007. Actually, both continuous maize and maize -soybean rotations cannot be sustained without heavy doses of fertilizer-N. This is despite N contributed by N fixing soybean or other legumes. Grain yield response of maize-soybean rotation is generally slightly higher than continuous maize or maize-fallow systems. Reports by Sawyer (2007) suggest that on an average, between years

2000 and 2007, maize-soybean rotations yielded 9.6 t ha^{-1} compared with it maize monocrops yielded 8.3 t grain ha^{-1}. The grain yield improvement during the 8 years ranged from 0.11 t ha^{-1} to 2.0 t ha^{-1} due to rotation with soybean. Tendency to supply higher amount of fertilizer N and other nutrients to ensure Ymax, it seems has reduced nutrient use efficiency. Currently only 3–36% of fertilizer-N is being garnered by maize-based rotations. However, Stanger and Lauer (2008) suggest that rotations with legumes have generally improved maize yield. It is easily attributable to extra N derived from BNF. In case of maize-soybean rotations, maize crop may derive up to 65–70 kg N ha^{-1} from the legume (Varvel and Wilhelm, 2003). In case of red clover or Alfalfa, N contribution averaged at 90–125 kg N ha^{-1}. Cultivation of maize in sequence with Alfalfa or red clover improved grain yield to 8.0–8.5 t ha^{-1}. Further, if legume residues were recycled in *toto*, it resulted in high (10.3 t grain ha^{-1}) production by the succeeding maize crop. A 35 year long term assessment of continuous maize and maize-soybean or several combinations of maize-alfalfa and maize-soybean-oats has shown that maize yield consistently improved by 79–100 kg ha^{-1} each year from 1970 to 2004 due to rotation effects. Actually, increasing fertilizer-N inputs to maize-soybean or maize-legume rotations, generally, did not improve maize grain yield after a certain level. It is believed that soybean or a legume is endowed with BNF to support N requirements of maize-based rotation. The extent of N derived from soybean crop residue ranged from 12 to 44%. It depended on fertilizer-N supply to maize. At high N inputs (>168 kg N ha^{-1}) N derived from soybean residue decreased from 44 to 19%. Rest of N required by maize was derived from inorganic fertilizers (Hesterman et al., 1987).

A long term (12 year) field experimentation has shown that inoculation of soybean results in nil to 162 kg ha^{-1} more soybean grains. In addition, a fraction of N held in stover is recycled into soil. It gets utilized efficiently by maize (Conley and Christmas, 2005).

Pikul (2005) opines that in the Corn Belt, specifically in Northern fringes of South Dakota, N and water resources limit maize crop yield. As such, water and N use efficiency are important aspects for farmers in this belt. It is believed that increased length of rotations and crop diversity added to rotations has improved both water and N use efficiency. A comparative study on Calcic Hapludolls at Brookings in South Dakota has shown that average yield was greater with corn-soybean-winter wheat (6790 kg ha^{-1}) compared to continuous corn (4000 kg ha^{-1}). The WUE was greater under corn-soybean-*Avena+Pisum* hay mix, corn-soybean-wheat and corn-soybean-alfalfa. Rotations involving alfalfa consumed higher amounts of soil moisture than corn. The interaction between water and N in different maize-based rotations was important. The comparative performance of different maize-based rotations with regard to water and N use efficiency is as fallows:

	Soil Water	Soil NO$_3$-N	Soil NO$_3$-N	Grain Yield	WUE	NUE
	1.8 m depth	1.2 m depth	difference	kg Gr ha^{-1}	kg ha^{-1} mm^{-1}	kg kg^{-1}N
Corn-Continuous	521	47	–0.7	4000	12.5	32.2
Corn-Soybean	524	49	–1.7	6000	18.1	48.3

Corn-Soybean-Mixed Hay	492	68	−20.6	6470	19.1	46.0
Corn-Soybean-Alfalfa-Alfalfa	465	50	−39.8	6060	18.1	47.8
Corn-Soybean-Winter Wheat	521	66	−13.8	6790	20.5	48.2

Source: Pikul, 2005.

Note: In the third column, difference refers to difference in NO_3-N status of soil at beginning and at end of rotation.

Maize–soybean rotations are among the most preferred cropping systems in Minnesota, Iowa, and Indiana (Lambert et al., 2006). Soil fertility could be uneven, if proper nutrient management procedures are not adopted. As such, intensive cropping of maize and soybean depletes soil nutrient rather drastically. Farmers generally apply N and P to maize and only P to soybean, since it is an N-fixing legume. Lambert et al. (2006) evaluated the spatial and temporal variations in soil fertility of a maize-soybean sequence for 5 years in a row. They found that crop response to N and P was spatially variable. Response of maize and soybean was temporally stable in parts of the fields but not entirely.

Crop residue generated and recycled during maize-soybean maize rotations has immediate effects on soil quality, extent of C sequestered, soil erosion control and restriction of nutrient loss from the fields. Wolkowski (2003) states that, if maize succeeds soybean, amount of crop residue cover obtained on the field, N and C recycled are relatively low compared to the situation immediately after maize. Soybean produces low amounts of crop residue. The tillage system followed before maize phase may also affect crop residue obtained from a maize-soybean rotation. Following is an example:

Surface residue (%) as affected by tillage and maize-based rotation:

Tillage	Continuous Corn	Soybean - Corn	Corn - Soybean
		Surface Residue %	
Chisel	36	17	34
Strip-Till	66	57	65
No-till	86	76	82

Source: Wolkowski, 2003.

In North-Central USA, planting wind break or shelter belt is a method to restrict soil erosion and loss of nutrients. According to Mize et al. (2005), extent of benefits from wind breaks especially in terms of soil fertility loss and variation in crop response that soil erosion causes could be site-specific. They used site-specific analysis and models such as CROPGRO-soybean and CERES-maize to simulate and assess the advantages of placing wind breaks in the maize belt. Long term simulations indicate that an average increase of 4.1 t grain ha^{-1} of maize and 3.1 t grain ha^{-1} of soybean was possible.

Maize-cotton-maize rotations are predominant in the Central Plains of USA. Maize-cotton rotation is preferred for several reasons such as nutrient dynamics, especially N recovery rates and recycling; crop residues and C sequestration, soil quality, pest control, grain/boll yield, and economic advantages (Boquet et al., 2009). Reports by Hons (2005) and Reddy et al. (2006) suggest that lint yield of cotton increases perceptibly, if it succeeds a maize crop. Reports from Mississippi suggest that maize was cultivated in about 0.3 m ha during 2008. Maize grown in rotation with cotton offered a clear 10–25% lint yield advantage to cotton that succeeds maize. Maize crop actually produces large amounts of crop residue that could be recycled to increase SOM and N status. Residual NO_3-N derived from maize phase of the rotation has a great value for cotton in terms of SOC, N, and other nutrients. For example, at optimal N rates of 174 kg N ha^{-1} applied to maize, NO_3-N levels in post-harvest soil was 67 kg N ha^{-1}. This fraction of N was easily available for succeeding cotton. Similarly, residual-N after cotton harvest is important, since it might affect succeeding crop and its response to fertilizer inputs. Pettigrew et al. (2006) have reported that maize-cotton rotations bring about certain changes in soil nutrient stocks. Major nutrients like N, P, and K get depleted. Soil organic matter improved from 0.61 to 0.68 % in 3 years. Micronutrient levels generally decreased as crop rotations continued.

The INM methods consider the extent of C sequestration possible as well as amount of N and C emissions under different crop rotations. For example, Haile-Mariam et al. (2008) reported that cumulative CO_2 emissions by corn and potato fields were 17–13 times higher than natural steppe vegetation (NV). Nitrous oxide losses accounted for 0.55 kg N ha^{-1} under corn and 0.59 kg N ha^{-1} under potato. The global warming potential of native vegetation was 459, that of corn fields were 7843, and potato was 6028. Clearly, modifications to maize-based cropping systems should aim at reducing loss of nutrients, especially N emissions or percolation and/or seepage. Rotations should also lessen CO_2 emission, but at the same time should improve C and N recycling into soil phase of the ecosystem.

In Central USA, maize is also rotated with vegetables. There are several variations of fertilizer-N schedules prepared and suggested to farmers who rotate vegetables like onion with maize. A few fertilizer-N schedules recommend lower levels of N to both onion and maize that succeeds. Yet, vegetables like onion are supplied with large doses of fertilizer- N to ensure high bulb yield. It leads to large pool of residual-N and other nutrients that percolate to subsurface and may even contaminate ground water. In fact, recent report in the Arkansas River Valley suggests that ground water NO_3-N is beyond the threshold of 10 µg L^{-1} in many locations that supported vegetable and maize production. In New Mexico, high rates of N supply induced leaching and reduced fertilizer-N use efficiency to 30%. In order to set right N dynamics, farmers grow maize immediately after onion, so that residual-N is efficiently used up by maize. Recent studies by Halvorson et al. (2002) suggests that fertilizer-N recovery by onion was only 11–19% and much of residual-N stayed in upper 60 cm of soil. Fertilizer-N also leached further deep from root zone. Improvements in agronomic procedures allowed higher recovery of N by maize (24%) and onion (39%). Delaying fertilizer-N supplying to onion improved recovery and fertilizer-N use efficiency. Planting maize directly on previous onion beds resulted in better N-use efficiency by maize. A few other suggestions to improve soil N dynamics are soil testing before fertilizer-N supply; changing to efficient irrigation methods like drip irrigation; periodic analysis of ground water for NO_3-N; growing cover crops that remove sizeable quantities of soil N that would otherwise be vulnerable to leaching and emissions; and use of slow-N release fertilizers (Halvorson et al., 2002).

Maize-based Rotations of Meso-America

Maize-based rotations have practically sustained cereal supply to Meso-American population, since ages. Wheat-maize rotations are most common in Mexican region (Plate 8.3). Maize-soybean, maize-soybean-green manure, maize-pastures, maize-vegetables are some of the common combinations traceable in Central America. As usual, tillage practices, their intensity, soil fertility status before sowing each crop, soil mineralization trends, and fertilizer-based nutrient supply dictate the nutrient dynamics that ensues in the entire rotation. Farmers could calculate nutrient needs based on yield goals of entire rotation. According to Basamba et al. (2006) SOM, N, and P fractions, soil moisture pattern, and inorganic nutrient supply play a vital role in deciding grain yield. Stamp (1999) has stated that both fertilizer supply and maize cultivar impart their influence on nutrient dynamics and productivity of the agroecosystem. On Andosols of Nicaragua, maize–bean rotations are preferred because it adds to soil-N fertility, in addition to improvising on land use efficiency and grain/pod yield. However, supply of organic manure along with inorganic fertilizers seems crucial. The net mineralization rates for N needs to be high, if only organic manures are supplied. Obviously, aspects like soil organic fractions, microbial component and net availability of nutrients are important during entire rotation (Miranda et al., 2008).

Plate 8.3. Maize after wheat on permanent beds in Central Mexico.
Source: Ken D. Sayre, CIMMYT, Mexico.
Note: Maize is planted in fields with standing wheat stubble on permanent raised beds.

Maize-based Rotations of Pampas of Argentina and Cerrados of Brazil

Caracova et al. (2000) have made some observations regarding crop rotations in Argentine Pampas that include maize. It seems during early 1970s small farms concentrated on wheat production, but larger farms of >200 ha consistently grew maize-based sequences like maize-maize-maize-mixed pasture-wheat/sunflower. Cattle often grazed the pastures and stubble was left after crop harvest. However, with acceptance of maize-soybean rotation, cropping got intensified. During past 3 decades, about 60% of maize belt has experienced severe depletion of SOC and nutrients in both surface and subsurface layers of soil. Loss of soil structure and decline of aggregates has been marked. Since maize is mostly grown as dry land crop with small amount of supplemental irrigation, water storage too has decreased. Expansion and intensification of maize-soybean rotations has almost reduced pastures and grazing has disappeared in many locations. Lack of fallows and incessant cropping has necessitated repeated application of fertilizer-based nutrients.

The Rolling Pampas of Argentina is agriculturally a highly productive zone. Wheat, maize and soybean are the main stay of farming enterprises. About 10 m t of wheat grain is produced in a year and sizeable fraction of it is derived from maize-wheat rotations. Other important rotations are maize-soybean-fallow, maize-sunflower (Plate 8.4), maize-sorghum-fallow. Precipitation pattern and soil fertility are the key factors that dictate maize productivity.

Plate 8.4. Maize-sunflower rotation in Pampas.

Source: Alberto Quirogo and Alfred Bono, INTA, Anguila Experimental Agricultural station, Argentina.

Note: Maize residue is applied as organic mulch and a no-till sunflower planted immediately.

Models integrating rainfall, fertilizer N and P, soil-N mineralization rates and tillage system suggest that maize productivity is highly dependent on some or all of these factors operating at different intensities. For example, wheat yield was insensitive to tillage intensity but corn yield was higher under NT conditions (Alvarez and Grigera, 2009). Farmers in this region generally hold soil fertility high by adding higher amount of N and P. Periodic soil-N tests and N release through mineralization processes guide the farmers about N supply to soil (Rozas et al., 2008). Hence, wheat and corn grain production was always high. Over all, precipitation pattern, soil N and P status seems to decide nutrient dynamics and productivity of maize-wheat rotations in Pampas.

In Brazil, many of the maize farms are integrated with live stock that requires sizeable quantities of forage. Maize-forage grass like Brachiaria rotations are practiced to serve that extra need for fodder. Maize-grass rotations are supplied with fertilizer-N, P, and K based on yield goals. Fertilizer-N supply varies widely from 0 to 200 kg N ha^{-1}. Often, high N inputs meant proportionately higher demand for P and K. Soil-K depletion is rampant. Soil-K is lost to subsurface well below the rooting zone. Hence, K inputs are based on soil tests of entire profile. Recycling crop residues restore soil-K and other nutrients (Garcia et al., 2006). In the Eastern European plains, maize-wheat sown repeatedly for long term may induce greater loss of nutrients due to erosion and runoff. As such, field sown to maize experiences greater loss of soil and nutrients due to erosion (Bacur et al., 2007). For example, under similar conditions if wheat fields suffered nutrient loss of 3.2 kg ha^{-1}, pea fields lost 6.23 kg nutrients ha^{-1} and maize fields lost 20.4 kg nutrients ha^{-1}. In Eastern Europe, it is common to change over from wheat-maize to other sequences like peas-wheat-maize, bean-wheat maize, pea-wheat, maize-grasses, etc.

Following is an example of benefits from changing to other sequences that include maize:

	Average Nutrient Loss from Cropping Systems (kg ha^{-1})					
	Humus	N	P$_2$O$_5$	K$_2$O	Ca	Mg
Wheat-Maize	195	10.2	0.62	1.38	0.66	0.19
Pea-Wheat-Maize	163	8.6	0.52	1.14	0.55	0.15
Pea-Wheat-Maize-Perennial Grass	124	6.6	0.40	0.88	0.42	0.12
Bean-Wheat-Maize-Grass	6.15	0.4	0.81	0.39	0.11	7.82

Source: Bacur et al., 2007.

Clearly, growing an assortment of crops in sequence along with maize improves soil and nutrient conservation.

Maize Cropping Systems in Africa

Maize is a staple cereal diet in West African Savanna region. Farmers adopt several variations of intercrops and rotations that include maize. Obviously, they all aim at maximizing land equivalent ratio, nutrient recovery rates and use, photosynthetic efficiency and C fixation, grain/forage yield and economic advantages. In the savanna region of Ghana, Nigeria, and other West African countries, maize is often rotated with sorghum or other millet. Maize is more frequently rotated with legumes like cowpea, groundnut, bambara nuts, beans, and vegetables (Sauerborn et al., 2001). Maize is also rotated with crops like Cassava or Banana. Such combinations improve WUE, soil fertility, restore SOC status and increase fodder/grain production. Maize grown after a fallow that was planted with green manure species is very common in savannas. It allows accumulation of biomass, conservation of moisture and improvement of soil N if green manure species is legume. Report by Tijani et al. (2008) suggests that a planted fallow basically improves soil moisture storage and precipitation use efficiency. Native fallows improved WUE by 25%, but fallows planted with *Pueraria phaseoloides* improved WUE by 19% over control. The biomass of maize was proportionately higher due to better WUE. Short fallows planted with *Pueraria phaseoloides* performed best in terms of N dynamics and WUE. Such a short fallow planted with Macuna—a cover crop is known to improve nutrient recovery and productivity of maize/cassava intercrops that succeed it. Sogbedji et al. (2006) have stated that cover cropping fallows with macuna and pigeonpea is a good method to improve N dynamics. Nitrogen budgets of maize-legume rotations showed a gain. However, under continuous maize, soil NO_3-N decreased by 57% over initial levels. Most importantly, NO_3-N loss via leaching and emissions reduced, if fallows had legume cover crops. Maize grain yield improved by 32–37% over fallows without cover crops.

Maize-legume rotations that include soybean or stylosanthes or cowpea enhance soil-N status. Field estimates suggest that 20–40 kg N ha^{-1} could be added to soil via BNF with legumes. Total N uptake by maize that succeeds such legumes is often higher by 20–25% over a control that succeeds a fallow period or cereal (Oikeh et al., 1998). Okogun et al. (2007) state that in Nigeria, farmers adopt legume–maize rotations to reap benefits from "carry over-N" that legumes provide to maize. The maize grain/forage yield often increases when it succeeds legumes. The extent of "carry over-N" may vary depending on soil type, location, legume crop, and maize genotype that succeed. For example, in case of soybean, BNF potential varies with genotype and it ranges from 57–90 kg N ha^{-1}. The N balance in a soybean-maize rotation improved by 5–17 kg N ha^{-1}. In the same area, lablab bean contributed about 22 kg N ha^{-1} through BNF. Based on a different set of field trials in Nigeria, Carsky et al. (1997) have concluded that response of maize that succeeds soybean is equivalent to that obtainable by adding 40 kg N ha^{-1}. Maize-planted (legume) fallows are practiced in parts of Kenya. The green manure species like *Crotalaria*, *Sesbania*, or *Cassia* are commonly grown during fallow season. It shifts N balance to positive due to N derived from BNF (Bunemann et al., 2004).

Maize-cotton rotations are practiced in some regions of Burkina Faso. It is believed that intensified farming that involves repeated plowing affects soil structure and induces massive loss of soil nutrients via emissions and percolation. Hence, reduced tillage is being suggested along with other INM procedures. Maize-cotton rotations currently receive nutrients both in organic and inorganic forms. This improves soil quality in addition to fertility (Ouattara, 2007).

Maize is a major cereal in the semi-arid regions of southern Africa. It is a preferred crop during rainy season. Paucity for soil moisture and its interaction with soil fertility almost decide yield goals. Since rainy season is short in many regions of Zimbabwe,

Zambia and Mozambique, farmers need to adjust all agronomic procedures and fertilizer inputs within this period. Subsistence farming supported by low nutrient turnover is most common. Fertilizer-based nutrient supply is indeed low. It ranges from 8 kg N to 20 kg N ha^{-1} in the maize cropping zones of Zimbabwe (Cimmyt, 2002). Farmers apply fertilizers to match the rainfall pattern and try to reap better grain yield. Nutrient input schedules tailored to match rainfall pattern have often fetched 21–40% more grain/forage yield. Maize-fallow rotations are affected by weed infestation. Hence, farmers need to ascertain low weed intensity prior to nutrient supply.

Maize-based Rotations in Asia

Maize is grown both for fodder and grains in North Indian Plains and Hills. It is often rotated with other cereals, legumes, oil seeds and vegetable. Maize-wheat rotations are becoming popular in the Gangetic plains. It allows rapid and high accumulation of fodder in addition to cereal grains. Nutrient demand is trifle higher. Farmers currently calculate nutrient supply per season but considering the entire rotation seems more accurate and efficient. In the Western Himalayas, maize-wheat rotations are supplied with N, P, K plus lime to correct acidity. Sometimes P is added to only one crop and residual-P available after maize is utilized by succeeding wheat. The INM procedures include major nutrients and FYM. During past 25 years, adoption of INM has consistently provided 4600 kg maize plus 3300 kg wheat grains ha^{-1} (Sharma and Subehia, 2003). In comparison, application of single major nutrient that is—N has almost reduced advantages to nil compared to control. During recent period, secondary and micro-nutrient dynamics too has got impaired due to incessant cropping. Hence, farmers are asked to supply S and micronutrients in order to achieve optimum nutrient ratios in soil.

Maize-wheat rotations in the plains of Rajasthan, Haryana, and Punjab are grown adopting INM procedures. Long term trials on Ustocrepts have shown that application of N, P, K, and FYM (10 t ha^{-1}) is essential to sustain high grain/forage yield expectations. Recovery of nutrients is substantially high. Adoption of organic farming and application of organic manures alone seems insufficient (Kanthaliya and Verma, 2006). Maize-chickpea rotations are practiced in Gangetic plains. The INM procedures adopted include application of N, P, K, biogas slurry or FYM. Maize recovers about 26 kg N, 4.5 kg P, and 25.3 kg K to produce 1.0 t grain. Chickpea recovers 46 kg N, 3.9 kg P, and 41.1 kg K to produce 1.0 t grain. The suggested average nutrient (N:P:K) input ratio is 6:1:1 for maize and 11:1:10 for chickpea (Singh et al., 2005). Periodic soil tests and revisions in grain yield goals dictate nutrient supply and recovery.

Sharma (2006) studied P dynamics in the maize-wheat cropping zones of Punjab plains. It was found that dynamics of soil P fraction gets affected markedly due to fertilizer-P and FYM inputs. Build up of inorganic P pools is crucial to productivity of entire rotation.

Maize-millet rotations practiced in Nepal generally received only organic manures. It resulted in expansion of subsistence or low input farming. However, during recent years, such rotations are supplied with major nutrients using inorganic fertilizers based on soil tests and yield goal. Further, INM procedures suggest use of 10 t FYM ha^{-1}. The FYM supplies organic-C, P, and micronutrients essential for both cereals. Nitrogen supply is staggered and split (Pilbeam et al., 2002). Splitting N input avoids excessive loss via leaching or percolation. On an average, about 58% of fertilizers applied were found in the maize crop. Similarly, about 25% of fertilizer-N supplied to millet was traceable in the shoot system. A sizeable fraction of N applied was quickly immobilized into soil organic matter, but became available to succeeding crops. Over all, it is clear that nutrient turnover rates need to be increased in order achieve higher yield goals.

Maize cultivation in rice fallows is a popular rotation in Eastern and Southern Indian coastal plains. This rotation is expanding in many regions of Southeast Asia, especially in Vietnam, Philippines, and China (Pasuquin et al., 2007). Maize is a dry land crop grown on well aerated soils held in oxidized state. However, previous crop in the sequence that is—rice is grown under flooded conditions in wet lands with stagnating water. Flooding induces anaerobic conditions and alters nutrient dynamics markedly. Nutrient transformations are definitely affected by the alteration in soil redox potential. Soil microbial pools too are affected by alternation of anaerobic and aerobic conditions that occur during rice-maize rotations.

Wheat-maize Rotations of North China Plains

Maize is an intensively cultivated crop in the North China plains. Both, wheat and maize that follows it are supplied with relatively higher levels of fertilizers. This is to augment high yield goals of 7–10 t grains plus 15 t forage ha^{-1}. Soil N is obviously a major constraint. Actually, incessant cropping removes large quantities of soil-N during each cropping season. Plant uptake, loss via erosion, leaching and emissions could be substantial leading to soil N deficiency. Repeated application of fertilizer N is recommended for both wheat and maize in order to sustain optimum soil-N levels. In the maize fields, fertilizer-N efficiency is generally reduced due to erosion, leaching and N emissions. Loss of N due to nitrification/denitrification reactions could be significant. Zou et al. (2006) have reported that on Aquic Cambisols of North China, plains about 4.7–9.7 kg N ha^{-1} could be lost due to nitrification processes. It reduces fertilizer-use efficiency and productivity of maize crop.

On calcareous soils that support wheat-maize rotation, Tsinghua (2002) found that NH$_3$ evolution resulted in perceptible N loss from the cropping system. Ammonia volatilization peaked in 3–5 days after fertilizer-N application to wheat. Loss of N through volatilization ranged from 4.4 to 13 kg N ha^{-1} depending on inherent soil-N and fertilizer-N inputs. During the summer maize phase, NH$_3$ evolution peaked rather rapidly, within 2 days after fertilizer-N supply. The NH$_3$-N lost ranged from 0.33 kg N ha^{-1} day^{-1} to 5.41 kg N ha^{-1} day^{-1} depending on fertilizer N supply. Such high rates of NH$_3$ evolution was only short lived. NH$_3$ volatilization subsided rather rapidly and became negligible by 7th day after fertilizer-N supply. Overall about 2–6% of fertilizer-N supplied to each crop could be lost via NH$_3$ volatilization. Therefore, in this wheat-maize cropping zone, farmers are advised to split fertilizer-N supply and most importantly place them deep (6–10 cm) in the soil. A light irrigation that wets the surface soil avoids NH$_3$ loss from system.

On calcerous soils of North China plains, wheat-maize rotations tend to experience higher amounts of N loss as we increase N inputs. An increase of N supply from 0 to

180 kg N ha^{-1} to winter wheat induced net loss of N as NH$_3$ and it ranged from 2.9 to 35.7 kg N ha^{-1}. Subsequent summer maize lost 0–22.7 kg N ha^{-1} for N inputs from 0 to 360 kg N ha^{-1} (Wang et al., 2004). Wang et al. (2008) state that, recycling residues and application of fertilizer N, P, and micronutrients enhances maize grain yield. Balanced nutrition is important to attain higher grain yield.

Maize-wheat rotations are a mainstay for farmers in large tracts of Northwest China, especially Loess plateau. It is predominantly a semi-arid region. Irrigation is feeble or protective. The maize-wheat belt is supposed to supply carbohydrates to about 40% of the populace in the region. Hence, its upkeep regarding soil fertility, water resources, and productivity is important. Average nutrient supply may range from 60 to 80 kg N and 40 kg P ha^{-1}. The productivity ranges from 2.2 to 5.5 t grain ha^{-1} depending on inherent soil fertility and fertilizer-based nutrient supply. Precipitation use efficiency and maintenance of SOM important aspects in the dry lands. Application of organic manure at 30 t ha^{-1} improves SOC of surface layers. Recycling crop residues is another important agronomic measure that ensures better soil quality (Fan et al. 2005).

Maize Cropping in Australia-rotations

Maize is grown in several regions of Australia. Farmers adopt wide variations in cropping systems and adopt INM methods to maximize productivity. Crop rotations that include are often adjusted to reap best WUE and nutrient recovery. Maize monocrops are also common, although it is known that monocrops yield slightly lower than crop rotations that include legumes or vegetables. In parts of Queens land, maize is being rotated with cotton. Maize farmers are advised to apply 240 kg N, 50 kg P, and 200 kg K ha^{-1} for a crop that yields 10 t grain ha^{-1}. Cotton grown after irrigated maize is said to yield 25% more than cotton monocrops sown every season. Soil fertility status especially SOM, residual-N and other nutrients are higher after maize harvest. Recycling maize residues improves SOM status during cotton phase. Holden (2005) have reported that maize after maize produced higher quantity of grain (6.4 t ha^{-1}) and fodder (8.0 t ha^{-1}) compared to maize 94.4 t grain plus 6.9 t fodder ha^{-1} after cotton. Clearly, extent of residual nutrients, fertilizer supply schedule and irrigation may all intricate to produce better grain/forage harvest in case of maize–maize sequence compared other sequences.

Maize is cultivated as a dry land crop in the Katherine region of Northern Territory. Maize cropping is actually tailored to suit the scanty precipitation pattern and is often grown as non-irrigated low input crop. Maize is easily grown in different soil types that occur in Northern region but soils prone to water logging is avoided. Soils that are less subject to nutrient leaching are preferred, since nutrient supply is itself marginal or low compared to other crops. Maize crop is provided with balanced nutrition consisting of 80–90 kg N, 30 kg P, 50 kg K, 30 kg S, and 5 kg each of Zn and Cu. Application of FYM adds to soil quality. In order to improve fertilizer-use efficiency, farmers adopt banding or foliar fertilization.

MAIZE FOR BIO-FUEL PRODUCTION: NUTRIENTS AND PRODUCTIVITY

Global energy demand is ever increasing. Currently, it is increasing at a rate of 2.2% annually. Therefore, in most areas, alternate and viable sources of biofuel (or crop residues) are being searched briskly. Harvesting ligno-cellulosic crop residue is the mainstay for biofuel production. Among them maize crop residue is an important source of biofuel.

Following is a list that compares Biofuel production and Ethanol conversion rates of different crop species:

	Biofuel Production	Ethanol Conversion Rates
	(liter ha^{-1})	(liter t^{-1})
Sugarcane	6000	81
Sugar beet	5000	103
Maize	3100	410
Sorghum	3400	402
Wheat	2500	389
Barley	1100	243
Oil Palm	4500	223
Rapeseed	1200	392
Castor	800	393
Soybean	600	183

Source: Johnston et al., 2009.

Maize is an important source of biofuel in many countries. In USA, maize contributes largest amount of crop residue that is utilized to produce ethanol. Currently, about 20% of maize grain is consumed to produce ethanol. It equates to 3.5% of total gasoline consumption in USA by volume and 2.5% by energy. It is said that maize crop residue produced in USA has still greater potential to contribute to ethanol pool (Dhugga, 2007).

The nutrient dynamics in fields that support maize crop exclusively for biofuels may be different, slightly or sometimes conspicuously, depending on agronomic procedures. Farmers adopt several variations of crop rotations in order to satisfy grain/crop residue requirements for both human consumption as well as biofuel production. Generally, it is said that, since large quantities of crop residues are lifted out of the fields for biofuel manufacture, such maize fields invariably need nutrient replenishment, especially N, P, and K. Heggenstaller et al. (2008) state that double cropping systems adopted in much of the Corn Belt has the potential to produce additional feed stock (crop residue) for bioenergy and forage for farm animals. At the same time, it allows farmers to lessen soil-N loss via NO_3-N leaching, that otherwise accentuates during fallow period (Karpenstein-Machan, 2001). The soil nutrient dynamics during a double cropping system is practically different from sole crop. The productivity of double crops is generally much higher than sole crop. For example, total dry matter accumulated by triticale-maize double crop system is at least 25% more than maize sole crop (Haggenstaller et al., 2008). Among the three double cropping systems evaluated, triticale-maize provided 1080 L ha^{-1} more ethanol than sole crop of maize. Obviously, in order to support extra biomass formation, double crops remove significantly greater amount of soil N compared to sole crop of maize. There is no doubt that farmers need to alter nutrient dynamics to suit to the requirements of a double crop. Fertilizer-N schedules and timing needs to vary depending on biomass (ethanol) yield goals, in addition to grain harvests envisaged. A double crop uses extra soil-N efficiently. High biomass production and higher concentration of nutrient in the biomass necessitates higher amount of nutrient recovery. Double crops remove 84% N, 41% P, and 177% K more than sole crops. Therefore, it reduces loss of soil-N via NO_3 leaching by 25–34% compared to sole crop.

Biofuel production has a great potential since it helps in reducing excessive use of mineral oils and petroleum. Admixture of ethanol and petroleum has been successfully used in many countries. The first generation methods of biofuel production needs use of large quantities of maize grains that otherwise is useful as animal feed or for human consumption. Second generation techniques are based on crop residue, straw and silage from maize. According to Thomsen et al. (2008) pretreatment of crop residues and en-silaging helps in improving ethanol formation. Ensilaged dry maize produces about 392 kg ethanol t^{-1} dry maize. Thomsen et al. (2003) state that devising intercropping systems that help farmer with grains for human consumption and forage/crop residue for biofuel production will be useful. Obviously, nutrient needs of intercrops should be calculated in such a way, that it satisfies grain formation plus enough forage. INM procedures should therefore calculate nutrient schedules based on grain needs, forage needs for farm animals plus that needed for biofuel production. Intensification of maize farming zones seems inevitable.

According to Lal (2009), harvesting crop residues and absence of recycling causes adverse effects on soil quality and nutrient status. On an agroecosystem basis, normally, returning crop residue partly or *in toto* helps to recycle large quantities of nutrients and improves C sequestration. Usually, 20–60 kg N, P, and K is recycled for each ton of grain crop harvested. During a year, about 100–160 kg C ha^{-1} is recycled or held back in soil through residue recycling. Such portions of soil nutrients and C are loss to the system, if it is utilized to produce biofuel. Clearly, alternate source of nutrient replenishment must be envisaged. Lal (2009) believes that such biofuel maize or any other crop should be established on specifically identified zones and soils, so that competition with areas meant for grain production is lessened. Biofuel plantation and field crops could be developed on marginal soils. Wherever, biofuel production is done, basic formula is to produce more of crop residue. Therefore, farmers have to channel extra nutrients from any number of sources. Nutrients required to produce ethanol remains the key factor and this has to be satisfied with least competition to food grain and fodder sources. Most importantly, disturbances to optimal nutrient dynamics in the ecosystem and ecosystematic functions need to be least, if not nil. Some reports suggest that high enthusiasm to produce biofuel crops like sugarcane, sugar beet, maize or sorghum may only have modest effects on food grain production levels and soil quality. However, there are reports that in countries like India and China, ambitious plans to improve biofuel crop area may really affect soil nutrient and water resources perceptibly (CGIAR, 2007). Nutrient dynamics that is usually known for food grain crops may alter significantly since crop residue recycling gets smaller and nutrient recovery from soil increases because of double cropping. Further, the fallow period shrinks. The estimates suggest that to meet biofuel needs, China has to produce about 26% more and India 16% more of maize.

SUMMARY

A great deal of field scale research on maize farming, like any other important cereal, has revolved around improving crop response to various soil and crop management techniques. Maximizing crop response to nutrient and moisture supply has been a major preoccupation. Conserving soil and its fertility status has also received considerable attention. During recent years, research efforts are essentially geared at, firstly to improve crop response, then to obtain maximum grain/forage yield. This is in addition to maintenance of ecosystematic functions.

Maize production systems have varied immensely through the ages, based on geographic location, soil, crop genotype, environment, and human preferences. At present we can identify several different farming systems. They are classified as Farmer's or Traditional Practices, State Agency Recommendations, BMPs, Maximum Yield Technology, SSNM practices or Precision Farming, etc. As explained earlier, each method has its specific advantage depending on farming conditions and yield goal. Further, each of these farming procedures has its specific influence on nutrient dynamics within the agroecosystem.

Great emphasis has been bestowed world over on use of INM procedures. The INM essentially involves most judicious, efficient and profitable use of as many nutrient resources and appropriate agronomic procedures. Adoption of INM reduces use of chemical inorganic fertilizers and enhances use of environmentally more congenial organic sources of nutrients. Several other procedures included under INM reduce soil erosion, loss of nutrients via percolation, seepage and emissions. Adoption of SSNM enhances uniformity in soil fertility and nutrient distribution. Generally, fertilizer inputs are slightly lesser than that needed under other systems like Maximum Yield Technology or State Agency Recommendation. Crop foliage and grain yield production is more uniform. The INM improves *in situ* recycling of nutrients via crop residue. Adoption of INM helps in conserving SOM. Adoption of improved fertilizer supply techniques adds to efficiency. Fertilizer-N recovery may reach up to 48–52%, if INM methods are practiced. The INM involves different cropping systems that are essentially more efficient with regard to grain/forage per unit land and/or time. Overall, INM is known to enhance maize productivity and at the same time preserve optimum ecological conditions within maize fields or expanses.

KEYWORDS

- **Best Management Practice**
- **Crop-weed interactions**
- **Fertilizer-N**
- **Integrated Nutrient Management**
- **Maize-based rotations**
- **Soil erosion**
- **Water use efficiency**

References

1

Abunyewa, A., Asiedu, E.K., Nyameku, A.L., and Cobbina, J. (2004). Alley cropping *Gliricidia sepium* with maize: 1. The effect of hedgerow spacing, pruning height and phosphorus application on maize yield. *Journal of Biological Sciences* **4**, 81–86.

Allmaras, R.R., Wilkins, D.E., Burndisde, O.C., and Mulla, D.J. (1997). Agricultural technology and adoption of conservation practices. In *The State of Site Specific Management for Agriculture*. Agronomy Association of America, Madison, Wisconsin, USA, pp. 99–158.

Amorim, P.K. and Batalha, M.A. (2009). Soil characteristics of a hyper-seasonal Cerrado compared to a seasonal Cerrado and a flood plain grassland: Implications for plant community structure. Instituto Internacional de Ecologia, Universidade Federal de Sao Carlos, Sao Paolo, Brazil, pp. 1–12.

Andrews, J. (1993) Diffusion of Mesoamerican food complex to Southeastern Europe. *Geographical Review* **83**, 194–204.

Asif, M. and Anwar, M. (2007). Phenology, leaf area and yield of spring maize as affected by levels and timings of potassium application. *World Applied Sciences Journal* **2**, 299–303.

Bahr, A.A., Zeidan, M. S., and Hozayan, M. (2006). Yield and quality of maize (*Zea mays* L.) as affected by slow-release nitrogen in newly reclaimed sandy soil. *American-Eurasian Journal of Agriculture and Environment* **1**, 239–242.

Basamba, T.A., Barrios, E., Amezquita, E., Rao, I.M., and Singh, B.R. (2006). Tillage effects on maize yield in a Columbian Savanna Oxisol: Soil organic matter and P fractions. *Soil Tillage Research* **91**, 131–142.

Beadle, G. (1939). Teosinte and the origin of maize. *The Journal of Heredity* **30**, 245–247.

Beadle, G. (1972). The mystery of maize. *Field Museum of Natural History Bulletin* **43**, 2–11.

Beck, D. (2000). Research on Tropical Highland Maize. http://www.cimmyt.org/Research/Maize/results/mzhigh99-00mrhigh99-00_res.pdf pp. 1–12.

Bennetzen, J., Buckler, E., Chandler, Y., Doebley, J., Dorweiler, J., Gaut, B., Freeling, M., Hake, S., Kellog, E., Poething R.S., Wlabot, V., and Wessler, S. (2001). Genetic evidence and the origin of maize. *Latin American Antiquity* **12**, 84–86.

Bennetzen, J.L. and Hake, S.C. (2009). *Hand Book of Maize*. Springer Verlag, New York, p. 248.

Benz, B. (1999). On the origin, evolution and dispersal of maize. In *Pacific Latin America in Prehistory: The Evolution of Archaic and Formative Cultures*. Blake, M. (Ed.). Washington State University Press, Pullman, Washington State, USA, pp. 25–38.

Beyer, L. (2002). Soil geography and sustainability of cultivation. In *Soil Fertility and Crop Production*. Krishna K.R. (Ed.) Science Publishers Inc. New Hampshire, USA, pp. 33–63.

Birch, C.J., Robertson, M.J. Humphrey, E., and Hutchins N. (2007). Agronomy of Maize in Australia-internal Review and Prospect. http://espace.library.uq.edu.au/eserve/UQ:9698/Agronomy_of Maiz.pdf pp. 1–17.

Birch, W.D. (1875). *The Commentaries of the Great Alfonso D'Albuquerque, Second Viceroy of India*. Hakluyt Society, London, United Kingdom, pp. 83–92.

Brady, N.C. (1995). US soil taxonomy. In *Nature and Properties of Soils*. Prentice Hall of India, New Delhi, p. 623.

Brandolini, A. and Brandolini, A. (2009). Maize introduction, evolution and diffusion in Italy. *Maydica* **54**, 233–242.

Brown, R. (1896). *The History and Descriptions of Africa*. Leo Africanus (Giovanni Leone). Translated by John Pory into English in 1600. Hakluyt Society, London. Reprinted by Franklin, New York, pp. 82–88.

Carcova, J., Maddonni, G.A., and Ghersa, C.M. (2000). Long-term cropping effects on maize productivity. *Agronomy Journal* **92**, 1256–1265.

CGIAR, (2008). Maize (*Zea mays* L.). Consultative Group on International Agricultural Research. Rome, Italy. http://www.cgiar.org/impact/reaserch/maize.html pp. 1–9.

Chirwa, P.W., Ong, C.K., Maghembe, J., and Black, C.R. (2007). Soil water dynamics in cropping systems containing *Gliricidia sepium*, pigeonpea and maize in Malawi. *Agroforestry Systems* **69**, 1–7.

Devereux, A.F., Fukai, S., and Hullugale, N.R. (2008). The Effects of Maize Rotation on Soil Quality and Nutrient Availability in Cotton-based Cropping. www.reginal.org.au/au/asa/2008/concurrent/plant_nutrition/5815_devereuxaf.htm pp. 1–5.

DeWet, J. and Harlan, J.R. (1972). Origin of maize. *The Tripartite Hypothesis. Euphytica* **21**, 271–279.

Diaz, R.A., Graciela, O., Maria, M., Travasso, I., Rafeal, O., and Rodgriguez, A.A. (1997). Climatic change and its impact on the properties of agricultural soils in Argentinean rolling Pampas. *Climate Research* **9**, 25–30.

Doebley, J., Renfroe, W., and Blanton, A. (1987). Restriction site variation in the *Zea* chloroplast genome. *Genetics* **117**, 139–147.

Doebley, J., Stee, A., Wendel, J., and Edwards, M. (1990). Genetic and morphological analysis of maize-teosinte F$_2$ population: Implications for the origin of maize. *Proceedings of the National Academy Science, USA.* **87**, 9888–9892.

Dowswell, C.R., Paliwal, R.L., and Cantrell, R.P. (1996). *Maize in the Third World.* Westview Press, Boulder, Colorado, USA, p. 275.

Eubanks, M.W. (2001). An interdisciplinary perspective on the origin of maize. *Latin American Antiquity* **12**, 91–98.

FAO, (2004). The Origin of Maize and its Cultivation. FAO Corporate Document Repository. www.fao.org/docrp/005/y3841e/y3841e04.htm pp. 1–6.

FAO, (2006). Zimbabwe-Remote Sensing Based Contribution to the Assessment of the 2006 Maize Harvest Prospects. www.fao.org/GIEWS/emglish/otherpub/gmfs-zwe-2006.pdf pp. 1–6.

FAO, (2008). Food and Agricultural Organization of the United Nations: Economics and Social Department-Statistics Department. www.faostat.fao.org

FAO-AGLW Water Management Group, (2002). *Maize.* Food and Agricultural Organization of the United Nations. Rome, Italy, pp. 1–7.

FAOSTAT, (2005). Maize Statistics. Food and Agriculture Organization of the United Nations. Rome, Italy, www.faostat.fao.org

Frost, A. (1993). Sir Joseph Banks and the transfer of plants to and from South Pacific-1786–1798. The Colony Press, Melbourne, Australia, p. 12.

Gallinat, W.C. (1983). The origin of maize as shown by key morphological traits of its Ancestor –Teosinte. *Maydica* **28**, 121–138.

Gallinat, W.C. (1996). The patterns of plant structures in maize. In *The Maize Hand Book.* Freeling, M. and Walbot, V. (Eds.). Springer Verlag, New York, USA, pp. 61–65.

Gallinat, W.C. (1999). Maize: Gift from Americas first peoples. In *Chiles to Chocolate: Food the Americans gave the World.* Foster, N and Cordell, L.S. (Eds.). University of Arizona Press, Tuscon, pp. 47–60.

Gerpacio, R.V. (2002). Maize Economy of Asia. International Maize and Wheat Center, Mexico. www.cimmyt.org/Research/Economics/map/.../pdfs/ImpactsAsia_Chapter1.pdf pp. 1–15.

GMO Compass, (2009). Genetically Modified Plants: Maize. www.gmo-compass.org/eng/grocery_shopping/crops/18.genetically_modified_maize_eu.html

Hair, P.E.H., Jones, A., and Law, R. (1992). Barbot on Guinea: The writings of Jean Barbot on West Africa 1678–1712. Hakluyt Society, London, United Kingdom, pp. 579–651.

Hall, A.J., Vivella, F., Trapani, N., and Chimenti, C.A. (1992). The effects of water stress and genotype on the dynamics of pollen shedding and silking in maize. *Field Crops Research* **5**, 349–363.

Harlan, J. (1992). *Crops and Man.* American Society of Agronomy, Madison, WI, USA, p. 143.

Harsono, A. and Karsono, S. (1997). Yield and irrigation-use efficiency of groundnut, with and without intercropping with maize on an Alfisol. *International Arachis News Letter* **17**, 56–57.

Hilton, H. and Gaut, B. (1998). Speciation and domestication in maize and its wild relatives: Evidence from the *Globulin-1* gene. *Genetics* **150**, 863–872.

Horst, W.J. and Hardter, R. (2006). Rotation of maize with cowpea improves yield and nutrient use maize compared to maize mono-cropping in an Alfisol in the Northern Guinea savanna of Ghana. *Plant and Soil* **160**, 171–183.

Hudson, J.C. (1994). *Making the Corn Belt: A Geographical History of Middle-Western Agriculture*. Indiana University Press, Bloomington, p. 260.

IFPRI, (2005). *IFPRI 2020—Projections*. International Food Policy Research Institute, Washington, D.C., pp. 1–73.

Ikena, J.E. and Amusa, N.A. (2004). Maize research and production in Nigeria. *African Journal of Biotechnology* **3**, 302–307.

Ikisan, (2007). Maize: Crop Technologies. http://www.ikisan.com/lomks/ap_maizeCropTechnologies.shtml pp. 1–9.

IPNI, (2008). Maize Planting and Production in the World. International Plant Nutrition Institute, Norcross, Georgia, USA http://www.ipni.net/ppiweb/nchina.nsf/$webindex/3FA09D72EC945883482573BB0030EB2A pp. 1–8.

Jean du Plessis (2008). *Maize Production*. Grain Crops Institute, Pretoria, South Africa, pp. 1–38.

Jenkins, K. (2002). Latin America: The Lore and History of Maize. http://www.mythinglinks.org/ip~ maize.html pp. 1–14.

Johannessen, C.L. and Parker, A.Z. (1989). Maize sculptured in 12th and 13th century A.D. as indicators of pre-Columbian diffusion. *Economic Botany* **43**, 164–180.

Kamara, A.Y., Ekeleme, F., Chikoye, D., and Omogui, L.O. (2009). Planting date and cultivar effects on grain yield in dry land corn production. *Agronomy Journal* **101**, 91–98.

Krishna K.R. (2003). *Agrosphere: Nutrient Dynamics, Ecology and Productivity*. Science Publishers Inc. Enfield, New Hampshire, USA, pp. 92–97.

Krishna K.R. (2008). *Peanut Agroecosystem: Nutrient Dynamics and Productivity*. Alpha Science International Inc. Oxford, United Kingdom, pp. 80–84.

Krishna, K.R. (2010). *Agroecosystems of South India: Nutrient Dynamics, Ecology and Productivity*. BrownWalker Press Inc, Boca Raton, Florida, USA, p. 543.

La Fleur, J.D. (2000). *Pieter Van der Broecke's Journal of voyages to Cape Verde, Guinea and Angola 1605–1612*. Hakluyt Society, London, United Kingdom, p. 90.

Larson, W.E. and Cardwell, Y.B. (1999) History of US Corn Production. http://www.mindully.org/Farm/US-Corn-Production1999.htm p. 106.

Laufer, B. (1907). The Introduction of Maize into Eastern Asia. *Proceedings of the International Congress of Americanists*. **1**, 223–257.

Lopes, S. (1996). Soils under Cerrado: A success story in soil management. *Better Crops International* **10**, 8–14.

Maize Statistical Reports, (2006). All India Area, Production and Yield of Maize. http://www.indiacommodity.com/statistic/maizestat.htm pp. 1–4.

Mangelsdorf, P. (1939). *The Origin of Indian Corn and its Relatives*. Texas Agricultural Experimental Station. College Station, USA, Bulletin 574, p. 85.

Mangelsdorf, P. (1974). *Corn: Its Origin, Evolution and Improvement*. Harvard University Press, Cambridge, Massachusetts, USA, p. 210.

McCann, J. (2000). Maize and Grace: History, Corn and Africas New landscapes, 1500–1999. http://ruafrica.rutgers.edu/events/media/0405_media/maize_and_grace_jamesmccann.pdf pp. 1–29.

McClintok, B., Kato, T.A., and Blumenschien, A. (1981). *Chromosome Constitution of Races of Maize: Its Significance in the Interpretations of Relationships Between Races and Varieties in the Americas*. Cologio de Postgraduados, Chapinga, Mexico, p. 178.

Mugendi, D.N., Nair, P.K.R., Mugwe, J.N., O'Niell, M.K., Swift, M.J., and Woomer, P.L. (1999). Alley cropping of maize with *Calliandra* and *Leucana* in the sub humid highlands of Kenya: Biomass decomposition, N mineralization and N uptake by maize. *Agroforestry Systems* **46**, 232–239.

Mwangi, T.J., Ngeny, J.M., Wekesa, F., and Mulati, J. (2000). Acidic soil amendment for maize production in Uasin Gishu district, North Rift Valley in Kenya. http://www.kari.org/Legume_Project/Legume2conf_2000/5.pdf pp. 1–14.

NAB, (2007). Grain Controlled Crops-White Maize. http://www.nab.com.na/white_maize.htm pp. 1–2.

Nyamangara, J., Bergstrom, L.F., Piha, M.L., and Giller, K.E. (2003). Fertilizer-use efficiency and nitrate leaching in a tropical sandy soil. *Journal of Environmental Quality* **32**, 599–606.

Ofori, E. and Kyei-Baffour, N. (2008). Agrometeorology and Maize Production in Africa. www.wmo.int/pages/prog/wcp/agm/gamp/documents/chap13C-draft.pdf pp. 1–27.

Ofori, E., Kyei-Baffour, N., and Agodzo, S.K. (2004). Developing effective climate information for managing rained crop production in some selected farming centers in Ghana. *Proceedings of the School of Engineering Research.* Accra, Ghana, pp. 1–18.

OKSTATE, (2009). World Wheat and Maize Production. http://nue.okstate.edu/Crop_information/World_Wheat_Producion.htm pp. 1–11.

Oliviera, P.S. and Marquis, R.J. (2002). *The Cerrados of Brazil: Ecology and Natural History of a Neotropical Savanna.* Columbia University Press, New York, USA, p. 346.

OSU, (2009). Zea: Introduction. http://www.gramene.org/species/zea/maize_intro.html pp. 1–3.

Otegui, M.E., Nicolini, M.G., Ruiz, R.A., and Dodds, P.A. (1995). Sowing date effects on grain yield components for different Maize genotypes. *Agronomy Journal* **87**, 29–33.

Perales, H.R., Benz, B.F., and Brush, S.B. (2008). Maize diversity and ethnolignuistic diversity in Chiapas, Mexico. *Proceedings of National Academy of Sciences* **2005**, 949–954.

Phiri, E., Verplancke, H., Kwesiga, F., and Mafangoya, P. (2003). Water balance and maize yield following improved Sesbania fallow in eastern Zambia. *Agroforestry Systems* **59**, 197–205.

Piperno, D.R. and Flannery, K.V. (2001). The Earliest Archeological Maize (*Zea mays* L.) from Highland Mexico: New Accelerator Mass Spectrometry Dates and Their Implications. www.pnas.org/content/98/4/2101.abstract pp. 1–2.

Pouketat, T.R. (2005). Ancient Cahokia and Mississippians. In *Civilization of North America.* Cambridge University Press, New York, http://assests.cambridge.org/97805218/17400/excerpt/9780521817400_excerpt.pdf pp. 1–16.

Pray, C., Rozelle, S., and Huang, J. (1998). Country case study in China. In *Maize Seed Industry.* Morris, M.L. (Ed.). Lynne Reinner Publishers, USA, p. 319.

Qiao, C., Wanf, Y.J., Guo, H.A., Chen, X.J., Liu, J.Y., and Li, S.Q. (1996). A review of advances in maize production in Jilin province during 1974–1993. *Field Crops Research* **47**, 65–75.

Racz, F., Illes, O., Pok, I., Szoke, C., and Zsubori, Z. (2003). Role of Sowing Time in Maize Production. www.date.hu/acta-agraria/2003–11i/racz.pdf pp. 1–6.

Ratter, J.A., Ribiero, J.F., and Bridgewater, S. (1997). The Brazilian cerrado vegetation and threats to its biodiversity. *Annals of Botany* **80**, 223–230.

Rebourg, M., Chastnet, R., Gouesnard, C., Welcker, P., Dubreuil, A., and Charcosset, S. (2003). Maize introduction into Europe: The history reviewed in the light of molecular data. *Theoretical and Applied Genetics* **106**, 895–903.

Sene, M. and Niane-Badiane, A. (2001). Effect of Manure and P and Ca source on the optimization of soil water and nutrient use in the corn/peanut rotation system in the peanut basin. Institute Senegalais de Researches Agricoles. Ministere de L'Agriculture, Republique Du Senegal, Senegal report pp. 18–31.

Soliman, M.S.M. (2006). Stability and environmental interactions of some promising yellow maize genotypes. *Research Journal of Agriculture and Biological Sciences* **2**, 249–255.

Sykes, P. (1902). *Ten Thousand Miles in Persia or Eight Years in Iran.* Scribners, New York, USA, pp. 46–112.

Tijani, F.O., Oyedele, D.J., and Aina, P.O. (2008). Soil moisture storage and water-use efficiency of maize planted in succession to different fallow treatments. *International Agrophysics* **22**, 81–87.

Turrent, A. and Serratos, J.A. (2004). Context and background on maize and its wild relatives in Mexico. In *Maize and Biodiversity: The Effects of Transgenic Maize in Mexico.* Secretariat of Commission for Environmental Cooperation of North America. www.cec.org/maize/ pp. 1–35.

USDA, (2008). Mahindi and Milho in Africa. In *Plants and Crops.* United States Department of Agriculture, Beltsville, Maryland, USA, http://www.nal.usda.gov/research/maize/chapter3.shtml pp. 1–3.

Vigouroux, Y., Glaubitz, J.C., Matsuoka, Y., Goodman, M.M., Jesus Sanchez, G., and Doebley, J. (2008). Population structure and genetic diversity of new world maize races assessed by DNA microsatellites. *American Journal of Botany* **95**, 1240–1253.

Wenguang, H., Shufen, D., and Qingwei, S. (1995). High yielding technology for groundnut. *International Arachis News Letter* **15** (suppl.), 1–22.

White S. and Doebley, J. (1999). The molecular evolution of terminal Earl—A regulatory gene in the genus *Zea. Genetics* **153**, 1455–1462.

Zambezi, B.T. and Mwambula, C. (2006). The impact of drought and low soil nitrogen on maize production in the SADC. *Proceedings of Symposium on Developing Drought and Low Soil Nitrogen Tolerant Maize Symposium.* International Maize and Wheat Center (CIMMYT). Abstract 4, 1.

Zere, T.B., Van Huysteen, C. W., and Hensley, M. (2007). Quantification of long term precipitation use efficiencies of different maize production practices on a semi-arid ecotype in the free state province. *Water in South Africa* **33**, 61–66.

2

Al-Kaisi, M. (2007). Tillage Challenges in Managing Continuous Corn. http://www.ipm.iastate.edu/ipm/icm/2007/2-12/tillagechallenge.html pp. 1–3.

Al Kaisi, M. and Kwaw Mensah, D. (2007). Effect of tillage and nitrogen on corn yield and nitrogen and phosphorus uptake in corn-soybean rotation. *Agronomy Journal* **99**, 1548–1558.

Beyer, L. (2002). Soil geography and sustainability of cultivation. In *Soil Fertility and Crop Production.* Krishna. K.R. (Ed.). Science Publishers Inc. Enfield, NH, USA, pp. 33–63.

Blanco-Canqui, H. and Lal, R. (2008). No-tillage and soil profile carbon sequestration: An on-farm assessment. *Soil Science Society of America Journal* **72**, 693–701.

Blume, P. (1998). Soil formation, soil taxonomy and soil geography. In *Text Book of Soil Science.* Schachtschable, P., Blume, P., Brummer, G., Hartge, K.H., and Schwertmann, U. (Eds.). Springer Verlag, Stuttgart, Germany, pp. 373–400.

Boeme, F.H.G., Alvrez, C.R., Cabello, M.J., Fernanedez, P.L., Bono, A., Prystupa, P., and Taboada, M.A. (2008). Phosphorus retention on soil surface of tilled and no-tilled soils. *Soil Science Society of America Journal* **72**, 1158–1162.

Bono, A., Alvarez, R., Buschiazzo, D.E., and Cantet, R.J.C. (2008). Tillage effects on soil carbon balance in a semi-arid agroecosystem. *Soil Science Society of America Journal* **72**, 1140–1149.

Brady, N.C. (1995). US Soil Taxonomy. In *Nature and Properties of Soils.* Prentice Hall of India. New Delhi, pp. 623.

Brye, K.R., Norman, J.M., Bundy, L.G., and Gower, S.T. (2001). Nitrogen and carbon leaching in agroecosystems and their role in denitrification. *Environmental Quality* **30**, 58–70.

Buerkert, A, Pieopho, H.P., and Bationo, A. (2002). Multi-site time trend analysis of soil fertility management effect on crop production in Sub-Saharan West Africa. *Experimental Agriculture* **38**, 163–183.

Carter, M.A. and Singh, B. (2004). Response of Maize and Potassium dynamics in Vertisols following Potassium fertilization. http://www.regional.org/au/au/asssi/supersoil2004/s13/oral/1603_ carterm.htm pp. 1–12.

Diaz, R.A., Graciela, O., Maria, M., Travasso, I., Rafeal, O., and Rodriguez, A.A. (1997). Climatic change and its impact on the properties of agricultural soils in Argentinean Rolling Pampas. *Climate Research* **9**, 25–30.

Duiker, S.W., Heldman, J.F., and Johnson, D.H. (2006). Tillage x maize-hybrid interactions. *Agronomy Journal* **98**, 436–442.

FAO-AGLW Water Management Group (2002). Maize. Food and Agricultural Organization of the United Nations. Rome, Italy, pp. 1–7.

Filho, S.P., Feigl, B.J., Piccolo, M.C., Fante, L. Jr., Neto, M.S., and Cerri, C.C. (2004). Root systems and soil microbial biomass under no-tillage system. *Science Agricultura (Piracicaba, Brazil)* **61**, 529–537.

Follet, R.F., Varvel, G.E., Kimble, J.M, and Vogel, K.P. (2009). No-till crop after bromegrass: Effect on soil carbon and soil aggregates. *Agronomy Journal* **101**, 261–268.

Ganesh, S.S., Dhakshinamoorthy, M., Kumaraperumal, R., Anandakumar, G., and Devrajan, S.

(2006). Assessment of Soil Carbon Turnover in the Long term fertilizer Experiment and Validation of Roth C –26.3 Model. 18th World Congress of Soil Science, Philadelphia, USA. http:// crops.confex.com/crops/wc2006/techprogram/ P12769.HTM pp. 1–2.

Garbulsky, M.F. and Deregibus, V.A. (2005). Argentina: The Country, Pasture and Forage Resource. http://www.fao.org/waicent/faoinfo/agricult/agpc/doc/counprof/argentina.html pp. 1–23.

Garcia, J.P., Wormann, C.S. Mamo, M. Drijber, R., and Tarlalson, D. (2007). One time tillage of no-till: Effects of nutrients, mycorrhizae and phosphorus uptake. *Agronomy Journal* **99**, 1093–1103.

Gerpacio, R.V. (2002). The Maize Economy of Asia. International Maize and Wheat Center, Mexico. www.cimmyt.org/Research/Economics/ map/pdfs/ImpactsAsia_Chapter1.pdf pp. 1–15.

Glatzle, A. (1999). Compendio para el manejo de pastures en el Chaco. Proyecto Estacion Experimental Chaco Central (MAG-GTZ), GTZ, El lector, p. 188.

Gwenzi, W., Taru, M., Mutema, Z., Gotosa, J., and Munshir, S.M. (2008). Tillage system and genotype effects on rain fed maize (*Zea mays*) productivity in Semi-arid Zimbabwe. *African Journal of Agricultural Research* **3**, 101–110.

Halvorson, A.D., Mosier, A.R., Reule, C.A., and Bausch, W.C. (2006). Nitrogen and Tillage effects on Irrigated Continuous Corn yields. *Agronomy Journal* **98**, 63–70

Halvorson, A.D. and Reule, C.A. (2007). Irrigated no-till corn and barley response to nitrogen in Northern Colorado. *Agronomy Journal* **99**, 1521–1529.

Higgins, D.R., Allmaras, R.R., Lamb, G.W., and Randall, G.W. (2007). Corn-soybean sequence and tillage effects on soil carbon dynamics and storage. *Soil Science Society of America Journal* **71**, 145–154.

Hoette, G.D. (2009). Missouri No-Till planting Systems manual. http://extension.missouri.edu/ xplor/ manuals/m00164.htm pp. 1–25.

Hood, R. (2002). The use of stable isotopes in soil fertility research. In *Soil Fertility and Crop Production*. Krishna, K.R. (Ed.). Science Publishers Inc. Enfield, New Hampshire, USA, pp. 313–336.

Ikisan, (2007). Maize: Soils and Land Preparation. http://www.ikisan.com/lnks/ap_maizeLand%20 preparation.shtml pp. 1–2.

Iqbal, M., Hasssan, A., and Lal, R. (2006). Nutrient content of maize and soil organic matter status under various tillage methods and farmyard manure levels. *Acta Agriculture Scandinavica: Plant-Soil Science* **57**, 349–356.

Jean du Plessis (2008). *Maize Production*. Grain Crops Institute, Pretoria, South Africa, pp. 1–38.

Kaspar, T.C., Logsdon, S.D., and Prieksat, M.A. (1995). Traffic pattern and tillage effects on corn root and shoot growth. *Agronomy Journal* **87**, 1046–1051.

Krishna, K.R. (2003). Agrosphere: Nutrient Dynamics, Ecology and Productivity. Science Publishers Inc. New Hampshire, USA, pp. 68–104.

Krishna, K.R. (2008). Peanut Agroecosystem: Nutrient Dynamics and Productivity. Alpha Science International Ltd. Oxford, England, p. 293.

Krishna, K.R. and Rosen, C. (2002). Nitrogen in soil: Transformations and influence on crop productivity. *In Soil Fertility and Crop Production*. Krishna. K.R. (Ed.). Science Publishers Inc. Enfield, New Hampshire, USA, pp. 91–108.

Lopes, S. (1996). Soils under Brazilean Cerrado: A success story in soil management. *Better Crops International* **10**, 8–14.

Maly, S., Sarapalka, B., and Krskova, M. (2002). Seasonal variability in soil-N mineralization and nitrification as influenced by N fertilization. *Rostlinna Vyroba* **48**, 389–396.

Mclaughlin, N.B., Lapen, D.R., Kroetch, D., Wang, X., Gregorich, B.L., and Li, Y.X. (2004). Soil compaction and roots. http://www.farmwest.com/index.cfm.cfm?method=library.showpage& librarypageid=145.htm pp. 1–5.

Moebius-Clune, B.N., Van Es, H.M., Idowu, O.J., Schindelbeck, R.R., Moebius-Clune, D.J., Wolfe, D.W., Abawi, G.S., Thies, J.E., Gugino, B.K., and Lucey, R. (2008). Long term effects of harvesting Maize Stover and Tillage on soil quality. *Soil Science Society of America Journal* **72**, 960–969.

Mozafar, A., Anken, T., Ruh, R., and Frossard, E. (2000). Tillage intensity, mycorrhizal and non-mycorrhizal fungi and nutrient concentrations in maize, wheat and canola. *Agronomy Journal* **92**, 1117–1124.

Nakamoto, T. and Suzuki, K. (2001). Influence of soybean and maize roots on the seasonal change in soil aggregate size and stability. *Plant Production Science* **4**, 317–319.

Oorts, K., Garnier, P., Findeling, A. Mary, B., Richard, G., and Nicolardot, B. (2007). Modeling soil carbon and nitrogen dynamics in no-till and conventional tillage using PASTIS model. *Soil Science Society of America Journal* **7**, 336–346.

Pedelini, R. (2002). Peanut production in Argentina. *Proceedings of American Peanut Research and Education Society* **30**, 60.

Perry, M.C. (1992). Cereal and Fallow/Pasture System in Australia. In *Field Crop Ecosystem*. Pearson, C.J. (Ed.). Elsevier, Amsterdam, Netherlands, pp. 451–481.

Poirer, V., Angers, D.A., Rochette, P., Chantigny, M.H., Zaidi, N., Tremblay, G., and Josee Fortin. (2009). Interactive effects of tillage and mineral fertilization on soil carbon profiles. *Soil Science Society of America Journal* **73**, 255–261.

Qin, R., Stamp, P., and Richner, W. (2005). Impact of tillage and banded starter fertilizer on maize root growth in the top 25 cm of the soil. *Agronomy Journal* **97**, 674–683.

Quarles, D. (1994). Effects of 10 years of continuous conservation tillage crop production and infiltration for Missouri Claypan soils. Vol. 12, pp 1–28.

Riveros, F. (2005). *The Gran Chaco*. Technical paper on Gran Chaco. Crop and Grassland Service. Food and Agricultural Organization, Rome, Italy, pp. 1–42.

Sanchez, P.A. (1978). *Properties and Management of Soils in the Tropics*. Wiley, New York, p. 372.

Sharifi, M., Zebrath, B.J., Burton, D.L., Grant, C.A., Bittman, S. Drury, C.F., McConkey, B.G., and Zaidi, N. (2008). Response of potentially mineralizable soil nitrogen and indices of nitrogen availability to tillage system. *Soil Science Society of America Journal* **72**, 1124–1131.

Singh, A.K. (2006). Effect of tillage, water and nutrient management on soil quality parameters under rice-wheat and maize-wheat cropping zones. 18th World Congress of Soil Science, Pennsylvania, USA. http://www.Idd.

go.th/18wcss/techprogram/ P19367.HTM pp. 1–2.

Sridevi, T.K., Shirefaw, B., and Wni, S.T. (2004). Adharsha water shed at Kothapally: Understanding drivers of higher impact. *Global Theme on Agroecossytems*. International Crops Research Institute for the Semi-arid Tropics. Patancheru, A.P. India Report No. 10 pp. 1–24.

Staley, T.E. and Perry, H.D. (1995). Maize silage utilization of fertilizer and soil nitrogen on a hillland ultisol relative to tillage method. *Agronomy Journal* **87**, 835–842.

Subba Rao, A. and Sami Reddy, K. (2005). Emerging strategies for sustaining higher productivity and ensuring soil quality under intensive agriculture. *Indian Journal of Fertilizer* **1**, 61–76.

Taboada, M.A. and Alvarez, C.R. (2008). Root abundance of maize in conventionally-tilled and zero-tilled soils of Argentina. *Research Brasiliera Cienco Solo* **32**,769–779.

Walters, D. (2001). Is Conservation Tillage the answer to Global Warming. http://agecon.ok-state. edu/isct/labranza/walters/answer.doc pp. 1–7.

Wang, X. (2006). *Conservation tillage and Nutrient Management in dry land farming in China*. Wageningen University, The Netherlands. Doctoral Thesis, p. 187.

Wolkowski, D. (2006). *Tillage considerations for first year Corn after Soybean*. New Horizons in Soil Science, University of Wisconsin, USA, Vol 1, pp. 1–9.

Yoo, G. and Wander, M.M. (2008). Tillage effects on aggregate turnover and sequestration of particulate and humified soil organic carbon. *Soil Science Society of America* **72**, 670–676.

Yost, R. Doumbia, M., and Berthe, A. (2002). Systems diagnosis for technology adaptation and transfer. http://www.oired.vt.edu/projects/current/projectsynthesis.pdf. pp. 26–34.

Zengjia, L. and Tangyuan, N. (2009). Prospects and present situation of conservation tillage in shandong province. http://www.unapcaem.org/Activities%20Files/A0710/proceedings/21.pdf pp. 1–12.

Zougmore, R., Nagumo, F., and Hosikawa, A. (2006). Nutrient uptakes and maize productivity as affected by tillage and cover crops in a sub-tropical climate at Ishigaki, Okinawa, Japan. *Soil and Plant Nutrition* **52**, 509–518.

3

Abdelrahman, M., Bakar, R.A., Zauyah, S., and Rahim, A.A. (2001). Balance of applied ^{15}N ammonium sulphate in maize (*Zea mays*) field. *Japanese Journal of Tropical Agriculture* **45**, 176–180.

Abendroth, L. and Elmore, R. (2007). Demand for more corn following Corn. http://www.agronext. iastate.edu/corn/production/management/cropping/demand.html. pp. 1–2.

Acharya, G.P. (1999). *Review on soil and soil fertility loses from the cultivated hill regions of Nepal and their conservation.* Agricultural Research Station, Lumle, Kaski, Nepal. Lumle Review Paper No. **99**(1), 1–78.

Adetunji, M.T. (1994). Phosphorus requirement of a maize-cowpea sequential cropping on a Paleudult. *Nutrient Cycling in Agroecosystems* **39**, 1385–1414.

Ahmed, O.H., Hussain, A., Ahmad, H.M.H., Mohamad, U., Rahim, A.A., and Majid, N.M.A. (2009). Enhancing the urea-N use efficiency in maize (*Zea mays*) cultivation on acid soils using urea amended with Zeolite and TSP. *American Journal of Applied Sciences* **6**, 716–720.

Alvarez, R. and Steinbach, H. (2006). Ammonia volatilization from nitrogen fertilizers in pampean agroecosystems of Argentina. *18th World Congress of Soil Science*, **July 9–15**, Philadelphia, Pennsylvania, USA. http://a-c-s.confex. com/crops/wc2006/techprogram/P14749.HTM pp. 1–2.

Atreya, K., Sharma, S., and Bajracharya, R.M. (2005). Minimization of soil and nutrient losses in maize-based cropping systems in the Mid-Hills of Central Nepal. *Katmandu University Journal of Science, Engineering and Technology* **1**, 1–15.

Bala, S.P. (2006). Split application and levels of K on the yield of maize and different fractions of K in Soil. *18th World Congress of Soil Science*, Philadelphia, Pennsylvania, USA. http://a-c-s.confex.com/crops/wc2006/techprogram/ P12218.HTM pp. 1–2.

Basamba, T.A., Barrios, E., Amezquita, E., Rao, I.M., and Singh B.R. (2006). Tillage effects on maize yield in Columbian savanna Oxisol: Soil organic matter and P fractions. *Soil and Tillage Research* **91**, 131–142.

Benbi, D.K., Biswas, C.R., and Kalkat, J.S. (1991). Nitrate distribution and accumulation in an Ustochrept soil profile in a long-term fertilizer experiment. *Nutrient Cycling Agroecosystems* **26**, 173–177.

Beuchamp, E.G. (1997). Nitrous oxide emissions from agricultural soils. *Canadian Journal of Soil Science* **77**, 113–123.

Boem, F., Alvarez, C.R., Cabello, M.J., Fernandez, P.L., Bono, A., Prystupa, P., and Taboada, M.G. (2008). Phosphorus retention on soil surface of tilled and no-tilled soils. *Soil Science Society of America Journal* **72**, 1158–1162.

Bonzi, M., Ganry, F., Oliver, R., and Sedogo, M. (2003). Developing indicators for soil and fertilizer nitrogen-use efficiency for maize using on-farm experiments in Burkina-Faso. *Arid Land Research and Management* **17**, 127–128.

Bromfield, A.R. (1969). Uptake of phosphorus and other nutrients by maize in Western Nigeria. *Experimental Agriculture* **5**, 91–100.

Bruns, H.A. and Ebelhar, M.W. (2006). Nutrient uptake of maize affected by nitrogen and potassium fertility in a humid subtropical environment. *Communications in Soil Science and Plant Sciences* **37**, 275–293.

Brye, K.R., Norman, J.M., Gower, S.T., and Bundy, L.G. (2003). Effect of management practices on annual net nitrogen-mineralization in restored prairie and maize agroecosystems. *Biogeochemistry* **63**, 135–160.

Bundy, L.G. and Meisinger, J.J. (1994). Nitrogen availability indices. In *Methods of Soil Analysis. Part 2: Microbiological and Biochemical Properties*. Weaver, R.W. (Ed.). Soil Science Society of America Book Series 5, Madison, WI, USA, pp. 951–984.

Calderon, F.J., McCarty, G.W., and Reeves J.B. (2005). Nitrapyrin delays denitrification on manured soils. *Soil Science* **170**, 350–359.

Carter, M.A. and Singh, B. (2004). Response of maize and potassium dynamics in vertisols following potassium fertilization. http://www.

regional.org/au/au/asssi/supersoil2004/s13/ oral/1603carterm.htm pp. 1–12.

Cavero, J., Beltran, A., and Argues, R. (2003). Nitrate exported in drainage water of two sprinkle-irrigated watersheds. *Journal of Environmental Quality* **32**, 916–926.

Chikowo, R., Mapfumo, P. Nyammagafata, P., and Giller, K.E. (2004). Maize productivity and mineral N dynamics following different soil fertility management practices on a depleted sandy soil in Zimbabwe. *Agriculture, Ecosystems and Environment* **102**, 143–149.

Christenson, D.R. (2008). *Potassium for Crop Production*. Michigan State University Extension Bulletin, pp.1–2.

Dass, S., Jat, M.L., Singh, K.P., and Rai, H.K. (2008). Agro-economic analysis of maize-based cropping systems in India. *Indian Journal of Fertilizers* **4**, 49–62.

Davarede, L.C., Kravachenko, A.N., Hoeft, R.G., Nafziger, E.D., Bullock, D.G., Warren, J.J., and Gonzini, L.C. (2003). Phosphorus runoff: Effect of tillage and soil phosphorus levels. *Journal of Environmental Quality* **32**, 1436–1444.

Del Grosso, S.J., Halvorson, A.D., and Parton, W.J. (2008). Testing DAYCENT model simulations of corn yields and nitrous oxide emissions in irrigated tillage systems in Colorado. *Journal of Environmental Quality* **37**, 1383–1389.

Denmead, O.T., Freney, J.R., and Simpson, J.R. (1982). Dynamics of ammonia volatilization during furrow irrigation of maize. *Soil Science Society of America Journal* **46**, 149–155.

Diez, J.A., Roman, R., Cabelloro, R., and Cabellero, A. (1997). Nitrate leaching from soils under maize-wheat-maize sequence, two irrigation schedules and three types of fertilizers. *Agriculture, Ecosystems, and Environment* **65**, 189–199.

Dirk, Van Der, Janssen, B.H., and Oenema, O. (2006). Initial and residual effects of fertilizer phosphorus on soil phosphorus and maize yields on soil phosphorus fixing soils: A case study in South west Kenya. *Agriculture, Ecosystems, and Environment* **11**, 104–120.

Dobermann, A. (2001). Crop potassium nutrition: Implications for fertilizer recommendations. *Proceedings of the 31st North-Central Extension-Industry Soil fertility Conference.*

Potash and Phosphate Institute, Brookings South Dakota, pp. 1–12.

Ebeling, A., Kelling K., and Bundy, L. (2002). Phosphorus management on high phosphorus soils. New horizons in soil science. *Department of Soil Science, University of Wisconsin News letter* **12**, 1–11.

Edis, R., Chen, D., Turner, D., Wand, G., Meyer, M., and Kirkby, C. (2006). How are soil nitrogen dynamics in an irrigated maize system impacted on by nitrogen and stubble management. 18th World Congress of Soil Science, Philadelphia, Pennsylvania, USA. http://a-c-s.confex. com/crops/wc2006/techprogram/P18731.HTM pp.1–2.

Fillery, I.R.P., Simpson, J.R., and DeDatta, S.K. (2005). Contribution of ammonia volatilization to total nitrogen loss after applications of urea to wetland fields. *Nutrient Cycling in Agroecosystems* **8**, 193–202.

Fox, R.H., Piekielek,W.P., and Macneal, K.M. (2001). Comparison of late-season diagnostic tests for predicting nitrogen status of corn. *Agronomy Journal* **93**, 590–597.

Francis, D.D. and Schepers, J.S. (2005). Nitrogen uptake efficiency in Maize production using Irrigation water high in Nitrate. *Nutrient Cycling in Agroecosystems* **39**, 239–244.

Francis, D.D., Schepers, J.S., and Vigil, M.F. (1993). Post anthesis nitrogen loss from corn. *Agronomy Journal* **85**, 659–663.

Garcia, J.P., Wortmann, C.S., Mamo, M., Drijber, R.A., Quincke, J.A., and Tarkalson, D. (2007). One-time tillage of No-till: Effects of nutrients, mycorrhizae and phosphorus uptake. *Agronomy Journal* **99**, 1093–1103.

Gonzalez-Montaner, J.H., Madonni, G.A., and Dinapoli, M.R. (1997). Modeling grain yield response to N in the Argentinean Southern Pampas. *Field Crops Research* **51**, 241–252.

Gregorich, E.G., Rochete, P., St-Georges, P., McKim, U.F., and Chan, C. (2008). Tillage effects on N_2O emissions from soils under corn and Soybean in Eastern Canada. *Canadian Journal of Soil Science* **88**, 234–256.

Griganani, C., Zavattaro, L., Sacco, D., and Monaco, S. (2007). Production, nitrogen, and carbon balance of maize-based forage systems. *European Journal of Agronomy* **26**, 442–453.

Halvorson, A.D., Del Grosso, S.J., and Reule, C.A. (2008). Nitrogen, tillage, and crop rotation effects on nitrous oxide emissions from irrigated cropping systems. *Journal of Environmental Quality* **37**, 1337–1344.

Halvorson, A.D., Mosier, A.R., Reule, C.A., and Bausch, W.C. (2006). Nitrogen and tillage effects on Irrigated Continuous Corn yields. *Agronomy Journal* **98**, 63–71.

Harper, L.A. and Sharpe, R.R. (1995). Nitrogen dynamics in irrigated corn: Soil-plant nitrogen and atmospheric ammonia transport. *Agronomy Journal* **87**, 669–675.

Hawkins, J.A., Sawyer, J.E., Barker, D.W., and Lundvall, J.P. (2007). Using relative chlorophyll meter values to determine nitrogen application rates for corn. *Agronomy Journal* **99**, 1034–1040.

Heckman, J.R., Jokela, W., Morris, T., Beegle, D.B., Sims, J.T., Coale, F.J., Herbert, S., Griffin, T., Hoskins, B., Jemison, J., Sullivan, W.M., Bhumbla, D., Estes, G., and Reid, W.S. (2006). Soil test calibration for predicting corn response to phosphorus in the Northeast USA. *Agronomy Journal* **98**, 280–288.

Hedley, M.J., Stewart, J.W.B., and Chauhan, B.S. (1982). Changes in inorganic and organic soil phosphorus fractions induced by cultivation practices and laboratory incubations. *Soil Science Society of America Journal* **46**, 970–976.

Hermann, A. and Taube, F. (2005). Nitrogen concentration at maturity—An indicator of nitrogen status of forage maize. *Agronomy Journal* **97**, 201–210.

Hermiyanto, B., Zoebisch, M.A., Singh, G., Ranamukhaarachchi, S.L., and Angus, F. (2007). Comparing runoff, soil, and nutrient losses from three small watersheds in Indonesia. http://www.cigr-ejournal.tamu.edu/submissions/volume6/LW%2004%2007%20Singh%20final%20 22Sept2004.pdf pp. 1–18.

Hestermann, O. B., Russell, M.P., Shaefer, C.C., and Heichel, G.H. (1987). Nitrogen utilization from fertilizer and legume residues in legume—Corn rotations. *Agronomy Journal* **79**, 726–731.

Hong, N., Scharf, P.C., Davis, J.G., Kitchen, N., and Sudduth, K.A. (2007) Economically optimum nitrogen rate reduces soil residual nitrate. *Journal of Environmental Quality* **36**, 354–362.

Hu, K., Li, B., Chen, D., Zhang, Y., and Edis, R. (2008). Simulation of Nitrate leaching in irrigated maize on sandy soil in desert oasis in Inner Mongolia, China. *Agriculture and Water Management* **95**, 1180–1188.

ICRISAT (2004). Fertilizer needs of some crops. http://www.icrisat.org/vasat/learning_resources/ organicFAQs/fertilizer_need.htm pp. 1–8.

Ikisan, (2000). Maize: Yield Maximization. http:// www.ikisan.com/links/ap_maizeYield%20 Maximisation.shtml. pp. 1–6.

Ikisan, (2007). Forage Crop: Maize—*Zea mays*. http://www.ikisan.com/links/ap_fc_maize.shtml pp. 1–3.

IPI, (2000). Potassium—An essential nutrient. *International Potash Institute: Coordination Reports*. Potash Research Institute of India, Gurgaon, India, http://www.ipipotash.org/pdf/publications/Pfacts.pdf pp. 1–13.

Iqbal, Z., Latif, A., Sikander, A., and Iqbal, M. (2003) Effect of fertigated phosphorus on P use efficiency and yield of wheat and maize. *Songkalakarin Journal of Science and Technology* **25**, 697–702.

Jambert, C., Serca, D., and Delmas, R. (1997). Quantification of N loses as NH_3, NO, N_2O, and N_2 from fertilized maize fields in Southern France. *Nutrient Cycling in Agroecosystems* **48**, 91–104.

Jarecki, M.K., Parkin, T.R., Chan, A.S.K., Hatfield, J.L., and Jones R. (2008). Comparison of DAYCENT-simulated and measured Nitrous Oxide emissions from a Corn field. *Journal of Environmental Quality* **37**, 1685–1690.

Jaynes, D.B. and Colvin, T.S. (2006). Corn yield and Nitrate loss in subsurface drainage from Mid-season Nitrogen Fertilizer Application. *Agronomy Journal* **98**, 1479–1487.

Jaynes, D.B., Colvin, T.S., Karlen, D.L., Cambardella, C.A., and Meck, D.W. (2001). Nitrate loss in subsurface drainage as affected by nitrogen fertilizer rate. *Journal of Environmental Quality* **30**, 1305–1314.

Jean du Plessis (2008). *Maize Production*. Grain Crops Institute, Pretoria, South Africa, pp. 1–38.

Jensen, L.S., McQueen, D.J., Ross, D.J., and Tate, K.R. (1996). Effects of Soil Compaction on N-Mineralization and Microbial-C and -N.

2. Laboratory Simulation. *Soil and Tillage Research* **38**, 189–202.

Johnston, A.M. and Dowbenko, R. (2009). Essential elements in Corn. www.farmwest.com/indexcfm?method=library.showPage&librarypageid=`124 pp. 1–7.

Jones, C.A. and Kiniry, J.R. (1986). CERES–Maize, A simulation model of Maize Growth and Development. Texas A and M University, College Station, USA, pp. 1–28.

Joshi, P.K., Singh, N.N., Singh, R.V., and Gerpacio, R.L. (2001). *Maize in India*. www.cimmyt.org/english/docs/maize-producsys/india.pdf pp. 1–75.

Kamukondiwa, W. and Bergstrom, L. (1994). Nitrate leaching in field lysimeters at an agricultural site in Zimbabwe. *Soil Use and Management* **10**, 118–124.

Kamukondiwa, W., Frost, P.G.H., and Bergstrom, L. (1998). Crop and soil recovery of Nitrogen from Nitrogen-15 labeled Maize residues. In *Carbon and Nutrient Dynamics in Natural and Agricultural Tropical Ecosystems*. Bergstrom, L. and Kirchmann, H. (Eds.). CAB International, Oxford, United Kingdom, pp. 103–112.

Kaore, S.V. (2006). An approach to Crop-wise Plant Nutrient Prescription. *Indian Journal of Fertilizers* **1**, 57–62.

Karlen, D.L., Hunt, P.G., and Matheny, T.A. (1996). Fertilizer [15]Nitrogen recovery by corn, wheat, and cotton grown with and without preplant tillage on Norfolk Loamy sand. *Crop Science* **36**, 975–981.

Kastori, R.R. (2004). Nitrogen volatilization from plants. *Proceedings of National Science Matica Sprska Novi Sad.* **107**, 111–118.

Keller, G.D. and Mengel, D.B. (1986). Ammonia volatilization from nitrogen fertilizers surface applied to No-till Corn. *Soil Science Society of America Journal* **50**, 1069–1063.

Kemmler, G. (1980). Potassium deficiency in soils of the tropics as a constraint to food production. In *Soil Related Constraints to Food production the Tropics*. International Rice Research Institute, Manila, Philippines, pp. 29–32.

Kent Keller, C., Butler, C.N., Smith, J.L., and Allen-King, R.M. (2008). Nitrate in Tile drainage of the Semi-arid Palouse Basin. *Journal of Environmental Quality* **37**, 353–361.

Khalil, K., Mary, B., and Renault, P. (2004). Nitrous oxide production by nitrification and denitrification in soil aggregates as affected by oxygen concentration. *Soil Biology and Biochemistry* **36**, 687–689.

Kimetu, J.M., Mugendi, D.N., Palm, C.M., Mutuo, P.K., Gachengo, C.N., Nandwa, S., and Kungu, J.B. (2001). Nitrogen Fertilizer Equivalency values for different Organic materials based on Maize performance at Kabeta, Kenya. http://flar.org/webciat/tsbf_institute/managing_nutrient_cycles/AfNetCh15.pdf pp. 207–224.

King, K.W. and Torbert, H.A. (2007). Nitrate and ammonium loss from surface-applied organic and inorganic fertilizer. *The Journal of Agricultural Sciences* **145**, 345–356.

Kogbe, J.O.S. and Adediran, J.A. (2003). Influence of nitrogen, phosphorus, and potassium application on the yield of maize in the Savanna zone of Nigeria. *African Journal of Biotechnology* **2**, 345–349.

Krauss, A. (1997). Potassium—The Forgotten Nutrient in West Asia and North Africa. In *Accomplishments and Future Challenges in Dry Land Soil fertility Research in the Mediterranean Area*. Ryan, J. (Ed.). International Center for Agricultural Research in Dry areas, Aleppo, Syria, pp. 9–21.

Krishiworld (2002). *The Pulse of Indian Agriculture*. Maintenance of Soil Fertility. http://www.krishiworld.com/html/soil_ferti3.html pp. 1–9.

Krishna, K.R. (2002). *Soil Fertility and Crop Production*. Science Publishers Inc. Enfield, New Hampshire, USA, p. 463.

Krishna, K. R. (2003). Agrosphere: Nutrient Dynamics, Ecology and Productivity. Science Publishers Inc. Enfield, New Hampshire, USA, pp. 208–240.

Krishna, K. R. (2008). *Peanut Agroecosystem: Nutrient Dynamics and Productivity*. Alpha Science International Pvt Ltd, Oxford, England, pp. 145–216.

Krishna, K.R. (2010). *Agroecosystems of South India: Nutrient Dynamics and Productivity*. BrownWalker Press Inc, Boca Raton, Florida, USA, p. 543.

Kumar Rao, J.V.D.K., Wni, S.P., and Lee, K.K. (1996). Biological nitrogen fixation through grain legumes in different cropping systems of the semi-arid tropics. In *Roots and Nitrogen in cropping systems of the Semi-arid Tropics*. Ito, O. Johansen, S., Adu-Gyamfi, Katayama, K., Kumar Rao, J.V.D.K., and Rego, T.J. (Eds.). Japan International Research Center for Agricultural Sciences, Tsukuba, Japan, Series 3, pp. 323–334.

Kumwenda, J.D.T., Waddington, S.S., Snapp, R.B., and Blackie, M.J. (1996). *Soil fertility management research for the maize cropping systems of small holders in Southern Africa*. International Maize and Wheat Center (CIMMYT), Mexico, NRG Paper: 96–102.

Laboski, C. (2009). Potential for Nitrogen Loss from Heavy Rainfalls. *University of Wisconsin-Extension Services, Madison USA, News Letter* **86**, 1–4.

Lawrence, J.R., Ketterings, Q.M., and Cherney, J.H. (2008). Effect of nitrogen application on yield and quality of silage corn after forage legume-grass. *Agronomy Journal* **100**, 73–79.

Lehman, J., Lilenfien, J., Rebel, K., Do Carmo Lima, S., and Wilcke, W. (2006). Subsoil retention of organic and inorganic nitrogen in a Braziliean savanna oxisol. *Soil Use and Management* **20**, 163–172.

Lemaire, G. and Gastal, F. (1997). Nitrogen uptake and distribution on plant canopies. In *Diagnosis of Nitrogen Status in Crops*. Lemaire, G. (Ed.). Springer Verlag, Berlin, pp. 3–43.

Li, K., Tatsuhiko, S., Tatsuya, I., Kakuzhi, S., and Takeshi, K. (2001). *Recovery of ^{15}N-labelled Ammonia by Barley and Maize grown on the Soils with Long term application of Chemical and Organic Fertilizers* **4**, 29–35.

Li, Z., Dong, S.T., Wang, K.J., Liu, P., Wang, W.Q., and Liu, C.X. (2007). *In situ* study on influence of different fertilization strategies for summer maize on soil nitrogen leaching and volatilization. *Plant Nutrition and Fertilizer Science* **13**, 998–1005.

Li, Z., Dong, S.T., Wang, K.J., Liu, P., Zhang, J.W., Wang, W.Q., and Liu, C.X. (2008). Soil Nutrient leaching patterns in maize field under different fertilization: An *in situ* study. *Ying Yong Sheng Tai Xue Bao* **19**, 65–70.

Lopez, R.C. and Vlek, P.G. (2006). Potassium (K): Principal Constraint to Maize Production in Imperrata-infested fields at Central Sulawest,

Indonesia. Proceedings of Conference on International Agricultural Research for Development. University of Bonn, Germany, pp.1–8.

Ma, Y., Li, J., Li, X., Tang, X., Liang, Y., Huang, S., Wang, B. Liu, H., and Yang, X. (2009). Phosphorus accumulation and depletion in soils in wheat-maize cropping systems: Modeling and validation. *Field Crops Research* **110**, 207–212.

Maly, S., Sarapalka, B., and Krskova, M. (2002). Seasonal variability in Soil N Mineralization and Nitrification as influenced by N fertilization. *Rostlinna Vyroba* **48**, 389–396.

McDonagh, J.F., Tomsan, B., Limpunantana, V., and Giller, K. (1993). Estimates of the Residual N benefit of Groundnut to Maize in Northeast Thailand. *Plant and Soil* **154**, 267–277.

Meena, S., Senthilvalavan, P., Malarkodi, M., and Kaleeswari, R.K. (2007). Residual effect of phosphorus from organic manures in sunflower—Assessment using radio tracer technique. *Research Journal of Agriculture and Biological Sciences* **3**, 377–379.

Merbauch, A. (1998). Uptake of Weed-borne ^{15}N by maize in field experiments. *Isotopes in Environmental and Health Studies* **34**, 45–52.

Mogge, B., Kaiser, E.A., and Munch J.C. (1999). Nitrous oxide emissions and denitrification N losses from Agricultural soils: Influence of Organic fertilizers and land use. *Soil Biology and Biochemistry* **31**, 1245–1252.

Moreno, F., Cayuela, A., Fernandez, E., Fernandez-Boy, Murillo, J.M., and Cabrera, F. (1996). Water balance and Nitrate leaching in irrigated maize crop in SW Spain. *Agricultural Water Management* **32**, 71–83.

Mtambanengwe, F. and Mapfumo, R. (2006). Effects of organic source quality on soil profile N dynamics and maize yields on sandy soils of Zimbabwe. *Plant and Soil* **28**, 234–240.

Mubarak, A.R., Roseneni, A.B. Anaur, A.R., and Zauyah, D.S. (2003). Recovery of Nitrogen from Maize residue and inorganic fertilizer in a Maize-Groundnut rotation in Humid Malaysia. *Communications in Soil Science and Plant Analysis* **34**, 17–18.

Mulvaney, R.L., Khan, S.A., and Ellsworth, T.R. (2006). Need for a soil-based approach in managing nitrogen fertilizer for profitable corn production. *Soil Science Society of America Journal* **70**, 172–182.

Mulvaney, R.L., Khan, S.A., Hoeft, R.G., and Brown, H.M. (2001) A soil organic nitrogen fraction that reduces the need for nitrogen fertilization. *Soil Science Society of America Journal* **65**, 1164–1172.

Murthy, K.V.S., Sahrawat, K.L., and Pardhasaradhi, D. (2000). Plant nutrient contribution by rainfall in highly industrialized and polluted Patancheru area in Andhra Pradesh. *Journal of Indian Society of Soil Science* **48**, 174–177.

Nakamura, K., Harter, T., Hirono, Y., and Mitsuno, T. (2004). Assessment of Root zone Nitrogen Leaching as affected by Irrigation and Nutrient management Practices. *Vadose Zone Journal* **3**, 1353–1366.

Niaz, A., Ibrahim, M., and Ishaq, M. (2007). Assessment of nitrate leaching in wheat-maize cropping systems: A Lysimeter study. *Pakistan Journal of Water Resources* **7**, 1–6.

Nilawonk, W., Attanandana, T., Phonphoem, A., Yost, R., and Shuai, X. (2008). Potassium release in representative maize-producing soils of Thailand. *Soil Science Society of America Journal* **72**, 791–797.

Nkonya, C. Kaizzi, C., and Pender, J. (2004). Determinants of nutrient balances in a maize farming system in Eastern Uganda. *Agricultural Systems.* **85**, 155–182.

Nyamangara, J. (2007). Mineral N distribution in the soil profile of a maize field amended with cattle manure and mineral-N under humid subtropical conditions. In *Advances in Integrated Soil Fertility Management in sub-Saharan Africa: Challenges and Opportunities.* Bationo, A., Waswa, B., Kihara, J., and Kimetu, J. (Eds.). Springer Netherlands, pp. 737–748.

Nyamangara, J., Bergstrom, L.F., Piha, M.L., and Giller, K.E. (2003). Fertilizer—Use efficiency and nitrate leaching in a tropical sandy soil. *Journal of Environmental Quality* **32**, 599–606.

Nziguheba, G., Merckx, R., and Palm, C.A. (2004). Soil phosphorus dynamics and maize response to different rates of phosphorus fertilizer applied to an acrisol in Western Kenya. *Plant and Soil* **243**, 1–10.

Oberle, S.L. and Keeney, D. R. (2006). Factors influencing corn fertilizer N requirements in the Northern US Corn Belt. *Journal of Production Agriculture* **3**, 527–534.

Oberson, A., Bunemann, E.K., Friessen, D.K., Rao, I.M., Smithson, P.C., Turner, B.I., and Frossard, E. (2006). Improving Phosphorus fertility in Tropical Soils through biological interventions. http://ciat-library.ciat.cgiar.org/Articulos_Ciat/DK3724_C037.pdf pp. 1–19.

Obogun, I.Z.J.A., Sanginga, N., and Abaidoo, C. (2007). Evaluation of maize yield in an on-farm maize-soybean and maize-lablab rotation system in the Northern Guinea Savanna of Nigeria. *Pakistan Journal of Biological Sciences* **10**, 3905–3090.

Obogun, I.Z.J.A., Sanginga, N., and Abaidoo, C., Dashiell, K.E., and Diels, J. (2005). On-farm evaluation of biological nitrogen fixation potential and grain yield of lablab and two soybean varieties in the Northern Guinea of Nigeria. *Nutrient Cycling in Agroecosystems* **73**, 2267–275.

Ogoke, I. (2006). Residual effects of phosphorus and soybean crop on maize in the Guinea Savanna. 18th World Congress of Soil Science, Philadelphia, USA. http://www.ldd.go.th/ 18wcss/techprogram/P15797.HTM pp. 1–2.

Okonkwo, C.I. (2008). Nitrogen mineralization from prunings of three multipurpose legume and maize uptake in alley cropping system. *AgroScience* **7**, 143–148.

Pacholski, A., Cia, G.C., Fan, X., Chen, D., Nieder, R., and Roelcke, M. (2008). Comparison of different methods for the measurement of ammonia volatilization after urea application in Henan Province, China. *Journal of Plant Nutrition and Soil Science* **171**, 361–369.

Parentoni, S.N. and Lopes de Souza, C. (2008). Phosphorus acquisition and internal utilization efficiency in tropical maize genotypes. http://www.scielo.br/pdf/pab/v43n7/14.pdf pp. 1–13.

Patzek, T.W. (2008). Thermodynamics of agricultural sustainability: The case of US maize agriculture. http://petroleum.berkeley.edu/papers/Biofuels/816patzek4-8-08.pdf pp.1–50.

Pettigrew, W.T. and Schorring, J.K. (2008). Potassium influences on yield and quality production for maize, wheat, soybean and cotton. *Physiologia Plantarum* **133**, 670–681.

Picone, L.L., Cabrera, M.L., and Franzleubber, A.J. (2002). A rapid method to estimate potentially mineralizable nitrogen in soil. *Soil Science Society of America Journal* **66**, 1843–1847.

Piekielek, W.P. and Fox, R.H. (1992). Use of chlorophyll meter to predict side-dress nitrogen requirements for maize. *Agronomy Journal* **84**, 59–65.

Pierce, F.J. and Fortin, M.C. (1997). Long term tillage and periodic plowing of a No-Tilled soil in Michigan: Impacts, yield ad soil organic matter. Paul, E.A. (Ed.). In *Soil Organic Matter in Temperate Agroecosystems*. CRC Press, Boca Raton, Florida, USA, pp. 220–245.

Pikul, J.L., Hammack, L., and Riedell, W.E. (2005). Corn yield, nitrogen use and corn root worm infestation of rotations in the northern corn belt. *Agronomy Journal* **97**, 854–863.

Pilbeam, C.J. and Warren, G.P. (1995). Use of ^{15}N for fertilizer N recovery and N mineralization studies in Kenya. *Fertilizer Research* **42**, 123–128.

Planet, D. and Cruz, P. (1997). Maize and Sorghum. In *Diagnosis of the Nitrogen Status in Crops*. Lemaire, G. (Ed.). Springer Verlag, Berlin, Germany, pp. 132–167.

Poss, R., Fardeau, J.C., and Saragoni, H. (2004). Sustainable agriculture in the tropics: The case of potassium under maize cropping in Togo. *Nutrient Cycling in Agreocosystems* **46**, 205–213.

Quincke, J.A., Wortmann, C.S., Mamo, M., Franti, T., Drijber, R.A., and Garcia, J.P. (2007). One-time tillage of No-till systems: Soil physical properties, phosphorus runoff and crop yield. *Agronomy Journal* **99**, 1104–1110.

Prasad, R., Kumar, D., Sharma, S.N., Gautam, R.C., and Dwivedi, M.K. (2004). Current status and strategies for balanced fertilization. *Fertilizer News* **49**, 73–80.

Redersma, S., Lusiana, B., and Van Noordwijk, M. (2005). Simulation of soil drying induced phosphorus deficiency and phosphorus mobilization as determinants of maize growth near tree lines on a ferralsol. *Field Crops Research* **91**, 171–184.

Reidell, W. E., Beck, D. L., and Schumacher, T.E. (2000). Corn response to fertilizer placement treatments in an irrigated No-Till system. *Agronomy Journal* **92**, 316–320.

Rhoades, C. (2004). Seasonal pattern of nitrogen mineralization and soil moisture beneath *Faidherbia albidia* in Central Malawi. *Agroforestry Systems* **29**, 133–145.

Rochette, P. (2008a). No-tillage only increases N_2O emissions in Poorly-aerated soils. *Soil and Tillage Research* **101**, 97–100.

Rochette, P., Andgers, D.A., Chantigny, M.H., and Bertrand, N. (2008b). Nitrous oxide emissions respond differently to No-till in a loam and a heavy clay soil. *Soil Science Society of America Journal* **72**, 1363–1369.

Roose, T. and Fowler, A.C. (2004). *A Mathematical Model for Water and Nutrient Uptake by Plant Root Systems* **228**, 173–184.

Rozas, H.S., Calvino, P.A., Echeverria, H.E., Barbieri, P.A., and Redolatti, N. (2008). Contribution of Anaerobically mineralized nitrogen to the reliability of planting or pre-side dress soil nitrogen test in maize. *Agronomy Journal* **100**, 1020–1025.

Rubio, G., Cabello, M.J., and Boem, F.G. (2006). Models estimating fertilizer requirements to increase available soil phosphorus. 18th World Congress of Soil Science, Philadelphia, Pennsylvania, USA. http://www.Idd.go.th/18wcss/techprogram/P15968.HTM pp. 1–2.

Ruiz Diaz, D.A., Hawkins, J.A., Sawyer, J.E., and Lundvall, J.P. (2008). Evaluation of in-season nitrogen management strategies for corn production. *Agronomy Journal* **100**, 1711–1719.

Saavedra, C., Velasco, J., Pajuelo, P., Perea, F., and Delgado, A. (2007). Effects of tillage on phosphorus release potential in a Spanish Vertisol. *Soil Science Society of America Journal* **71**, 56–63.

Sainz Rozas, H.R., Echevarria, H.E., and Barbieri, P.A. (2004). Nitrogen balance as affected by application time and nitrogen fertilizer rate in irrigated No-tillage maize. *Agronomy Journal* **96**, 1622–1631.

Sainz Rozas, H.R., Echevarria, H.E., and Picone, L.I. (2001). Denitrification in maize under No-tillage. *Soil Science Society of America Journal* **65**, 1314–1323.

Sainz Rozas, H.R., Echeverria, H.E., Studdert, G.A., and Andrade, F.H. (1999). No-till maize nitrogen uptake and yield: Effect of urease inhibitor and application time. *Agronomy Journal* **91**, 950–955.

Sanginga, N. (2003). Role of biological nitrogen fixation in legume-based cropping systems: A case study of West Africa Farming Systems. *Plant and Soil* **252**, 25–39.

Sawyer, J. (2007) Nitrogen Fertilization for Corn following Corn. http://www.ipm.iastate.edu/ipm/icm/2007/2-12/nitrogen.html pp. 1–4.

Sawyer, J.E., Mallarino, A.P., Killorn,R., and Barnhart, S.K. (2008). *A general guide for crop nutrient and limestone recommendations in Iowa*. Iowa State University Extension Bulletin pp. 1–18.

Scharf, P.C., Brouder, S.M., and Hoeft, R.G. (1998). Chlorophyll meter readings can predict Nitrogen need and yield response of corn in the North-Central USA. *Agronomy Journal* **98**, 655–665.

Scheppers, J.S., Francis, D.D. Vigil, M., and Below, R.E. (1992). Comparison of corn leaf nitrogen concentration and chlorophyll meter readings. *Communications in Soil Science and Plant Analysis* **23**, 2173–2187.

Schmidt, J.D., Dellinger, A.E., and Beegle, D.B. (2009). Nitrogen recommendation for corn: An on the go sensor compared with current recommendation methods. *Agronomy Journal* **101**, 916–924.

Selvi, D., Malavizhi, P.P., and Santhy, P. (2006). Effects of long-term fertilizer application and intensive cropping on dynamics of soil micronutrients under tropical agro ecosystem. 18th World Congress of Soil Science, Philadelphia, USA. http://crops. confex.com/cropswc2006/techprogam pp. 1–6.

Sharma, K.N. (2006). Soil phosphorus fractionation dynamics and phosphorus sorption in a continuous maize-wheat cropping system. 18th World Congress of Soil Science, Philadelphia, Pennsylvania, USA http://www.Idd.go.th/18wcss/techprogram/P12763.HTM pp. 1–3.

Shepherd, G, Buresh, R.J., and Gregory, P.J. (2001). Inorganic soil nitrogen distribution in relation to soil properties in small holder maize fields in the Kenyan Highlands. *Geoderma* **101**, 97–103.

Sholtanyuk, V.V. and Nadtochaev, N.F. (2004). Kukuruza: Timing and methods of fertilizer input in maize cultivation. National Academy of Sciences,

Belarus, http://www.cababstractsplus.org/abstracts/Abstract.aspx?AcNo=20043148094 pp. 1–2.

Singh, M.V. (2001). Evaluation of Micronutrient stocks in different Agroecological zones of India for sustainable Crop production. *Fertilizer News* **46**, 13–35.

Sipasueth, N., Attanandana, T., Vichan, Vichukit, V., and Yost, R.S. (2007). Subsoil Nitrate and maize root distribution in two important maize soils. *Soil Science* **172**, 861–875.

Smaling, E.M.A. and Janssen, B.H. (1993). Calibration of QUEFTS, a model predicting Nutrient Uptake and Yields from Chemical Fertility Indices. *Geoderma* **59**, 21–44.

Smaling, E.M.A, Toure, M., Ridder, N.D., Sanginga, N., and Brenan, N. (2006). *Fertilizer use and the environment in Africa: Friends or Foes*. Background paper on African Fertilizer Summit. NAPED-IFDC, Abuja, Nigeria, pp. 1–26.

Smith, W.C. (1995). Crop Production: Evolution, History and Technology. John Wiley Inc. New York, USA, p. 342.

Solari, F, Shanahan, J, Ferguson, R., Schepers, J., and Gitelson, A. (2008). Active Sensor Reflectance measurement of Corn Nitrogen Status and Yield potential. *Agronomy Journal* **100**, 571–579.

Staley, T.E. and Perry, H.D. (1995). Maize Silage Utilization of fertilizer and Soil Nitrogen on a Hill-land Ultisol relative to tillage method. *Agronomy Journal* **87**, 835–842.

Stanger, T.F. and Lauer, J.G. (2008). Corn grain yield response to Crop Rotation and Nitrogen over 35 years. *Agronomy Journal* **100**, 643–650.

Sterk, G., Stroosnidjer, I., and Raats, P.A.C. (1998).Wind erosion process and control techniques in Sahelian zone of Niger. http://www.wu.ksu.edu/symposium/prodceedings/sterk.pdf pp. 1–14.

Surendran, U. Murugappan, Y., Bhaskaran, A., and Jagadeeswaran, R. (2005). Nutrient Budgeting Using Nutmon-toolbox in an Irrigated Form of Semi-arid Tropical region in India—A micro and Meso level modeling study. *World Journal of Agricultural Sciences* **1**, 89–97.

Suwanarit, A. and Sestapukdee, M. (1989). Stimulating effects of foliar K fertilizer applied at the appropriate stage of development of Maize: A new way to increase Yield and improve Quality. *Plant and Soil* **120**, 111–124.

Szolokine, I.Z. and Szaloki, S. (2002) Relationships of water and nutrient supply yield and evapo-transpiration. *Idojaras* **106**, 197–213.

Tang, X., Li, J., Ma, Y., Hao, X., and Li, X. (2008). Phosphorus efficiency in long term (15 years) wheat-maize cropping systems with various soil and climate conditions. *Field Crops Research* **108**, 231–237.

Tan,Y.S., Van Es, H.M., Duxbury, J.M., Melonian, J.J., Schindelbeck, R.R., Geohring, L.D., Hively, D., and Moebius, B.N. (2009). Single event Nitrous oxide losses under maize production as affected by soil types, tillage rotation and fertilization. *Soil and Tillage Research* **102**, 19–26.

Taylor, K., Singh, B., and Schwenke G. (2006). Potassium Dynamics in Vertisols following Potassium fertilization and Plant Uptake. 18th World Congress of Soil Science, Philadelphia, USA. http://a-c-s.confex.com/crops/wc2006/techprogram/P19116.HTM pp. 1–2.

Tiwari, K.N. (2001). The changing phase of balanced fertilizer use in India. *Better Crops* 15, 24–27.

Tsinghua, L. (2002). *In Situ* Determination of Ammonia Volatilization from Wheat-Maize Rotation System in North China. Acta Ecologica Sinica. http://www.shvoog.com/exact-sciecnes/agronomy-agriculture/1603323-situ-determination-ammnia-volatilization-wheat pp. 1–2.

Twomlow, S.J. (1994). Field moisture characteristics for two ferrallitic soils in Zimbabwe. *Soil Use Management* 9, 53–58.

University of Illinois at Urbana-Champaign. (2004). The Illinois Soil Nitrogen Test for Amino sugar-N: Estimation of potentially mineralizable soil-N and [15]N. Department of Natural Resources. University of Illinois Urbana-Champaign Technical Note 02-01, pp. 1–23.

Vaje, P.I., Singh, B. R., and Lal, R. (1999). Erosional effects on Nitrogen balance in Maize grown on a Volcanic ash soil in Tanzania. *Nutrient Cycling in Agroecosystems* **54**, 113–123.

Van Der Kruijs, A.C.B.M., Wong, M.T. F., Juo, A.S. R., and Wild, A. (1988). Recovery of [15]N—Labeled fertilizer in crops, drainage water and soil using monolith lysimeters in Southeast

Nigeria. *European Journal of Soil Science* **39**, 483–492.

Van Der Pol, F. and Traore, B. (1993). Soil nutrient depletion by agricultural production in Southern Mali. *Nutrient Cycling Agroecosystem* **36**, 1385–1394.

Van Dijk, W. and Brouwer, G. (1998). Nitrogen recovery and Dry matter Production of silage (*Zea mays*) as affected by subsurface band application of Mineral Nitrogen fertilizer. *Netherlands Journal of Agricultural Sciences* **46**, 139–155.

Venkateswarulu, J. (2004). *Rain fed Agriculture in India*. Indian Council of Agriculture, New Delhi, p. 566.

Venterea, R.T. and Stanenas, A.J. (2008). Profile analysis and modeling of reduced tillage effects on soil Nitrous Oxide flux. *Journal of Environmental Quality* 37, 1360–1367.

Wang, H., He, P., Wang, B, Zhao, P., and Guo, H. (2005). Nutrient Management within Wheat-Maize Rotation system in China. http://www.ppi-ppic.org/ppiweb/bcrops.nsf/$webindex/3FDC22FAD896A69852574AA0070641C/$file/BC08-3p12.pdf pp. 1–5.

Wang, Z.H., Liu, X.J., Ju, X.T., Zhang, F., and Malhi, S.S. (2008). Ammonia Volatilization loss from Surface-broadcasted Urea: Comparison of Vented and Closed Chamber methods and Loss in Winter Wheat-Summer Maize rotation in North China plain. *Communications in Soil Science and Plant Analysis* **35**, 2917–2939.

Weight, D. and Kelly, V. (2009). Restoring Soil fertility in Sub-Saharan Africa: Technical and Economic Issues. http://www.aec.msu.edu/fs/polsyn/No37.htm pp. 1–6.

Weinhold, B.J. and Halvorson, A.D. (1999). Nitrogen mineralization responses to Cropping, Tillage and Nitrogen rate in the Great Plains. *Soil Science Society of America Journal* **63**, 192–196.

White, J. (2000). Potassium in Agriculture. Agrow Australia Pvt Ltd, Sydney Australia, pp. 1–12.

Williams, J.D., Crozier, C.R., White, J.G., Sripada, R.P., and Crouse, D.A. (2007a). Comparison of Soil Nitrogen Tests for Corn Fertilizer Recommendations in the Humid Southeastern USA. *Soil Science Society of America Journal* **71**, 171–180.

Williams, J.D., Crozier, C.R., White, J.G., Heiniger, R.W., Sripada, R.P., and Crouse, D.A. (2007b). Illinois Soil Nitrogen Test predicts Southeastern US corn Economic Optimum Nitrogen rates. *Soil Science Society of America Journal* **71**, 735–744.

Wolkowski, R.P. (2000). Soil Management and Potassium availability. New Horizons in Soil Science. *University of Wisconsin-Madison, Agricultural Extension Bulletin* **1**, 1–16.

Wortmann, C.S., Dobermann, A.R., Ferguson, R.B., Hergert, G.W., Shapiro, C.A., Tarkalson, D.D., and Walters, D.T. (2009). High yielding Corn response to applied Phosphorus, Potassium and Sulphur in Nebraska. *Agronomy Journal* **101**, 546–555.

Yerokun, O.A. (1997). Ammonia Volatilization from Ammonium nitrate, Urea and Urea-Phosphate Fertilizers applied to alkaline soils. *South African Journal of Plant and Soil* **14**, 248.

Yamoah, C., Ngueguim, M., Ngong, C., and Dias, D.K.W. (1996). Reduction of fertilizer requirement using lime and macuna on high P-sorption soils of North West Cameroon. *African Crop Science Journal* **4**, 441–451.

Zaidi, N., Brassard, M., Belangerie, G., Claessens, A., Tremblay, N., Cambouris, A.N., Nolin, M.C., and Parent, L.E. (2008). Chlorophyll measurements and Nitrogen Nutrition Index for the evaluation of Corn Nitrogen Status. *Agronomy Journal* **100**, 1264–1273.

Zhang, J., Blackmer, A.M., Ellsworth, J.W., and Koehler, K.J. (2008a). Sensitivity of chlorophyll meters for diagnosing nitrogen deficiencies of corn in production agriculture. *Agronomy Journal* **100**, 543–550.

Zhang, J., Blackmer, A.M., Ellsworth, J.W., Kyveryga, P.M., and Blackmer, T.M. (2008b). Luxury production of leaf chlorophyll and mid-season recovery from nitrogen deficiencies in corn. *Agronomy Journal* **100**, 658–664.

Zhang, S.L., Cai, G.X., Wang, X.Z, Xu, Y.H., Zhu, Z.L., and Freney, J.R. (1992). Losses of urea-nitrogen applied to maize grown on a calcareous Fluvo-aquic soil in North China Plain. *Pedosphere* **2**, 171–178.

Zhou, J.B., Xi, J.G., Chen, Z.J., and Li, S.X. (2006). Leaching and Transformation of Nitrogen fertilizers in Soil after application of N with

irrigation: A soil column method. *Pedosphere* **16**, 245–252.

Zingore, S., Mafangoya, P., Nyammugafata, F. Giller, K.E. (2003). Nitrogen mineralization and maize yields following application of tree prunings to a sandy soil in Zimbabwe. *Agroforestry Systems* **57**, 199–211.

Zou, G., Zhang, F.S., Ju, X., Chen, X., and Liu, X. (2006). Study on Soil De-nitrification in Wheat-Maize Rotation system. *Agricultural Sciences in China* **5**, 45–49.

Zucker, L.A. and Brown, L.C. (1998). Agricultural Drainage: Water quality impacts and subsurface drainage studies in the Midwest. *Ohio State University Extension Bulletin* **871**, 1–38.

4

Abunyewa, A.A. and Mercer-Quarshie, H. (2004). Response of maize to magnesium and zinc application in the semi-arid zone of West Africa. *Asian Journal of Plant Sciences* **3**, 1–5.

Bharathi, C. and Poongthai, S. (2008). Direct and residual effect of sulphur on growth, nutrient uptake, yield and its use efficiency in maize and subsequent green gram. *Research Journal of Agriculture and Biological Sciences* **4**, 368–372.

Chen, L., Kost, D., and Dick, W.A. (2008). Flue gas desulfurization products as sulfur sources for corn. *Soil Science Society of America Journal* **72**, 1464–1470.

Christenson, D.R., Warnecke, D.D., and Leep, R. (1993). *Lime for Michigan Soils*. Michigan State University Extension Bulletin e-471, pp. 1–6.

Farina, M.P.W. and Channon, P. (2006). A field comparison of lime requirement indices for maize. *Plant and Soil* **134**, 127–135.

Hermiyanto, B., Zoebisch, M.A., Singh, G., Ranamukhaarachchi, S.L., and Angus, F. (2007). Comparing Runoff, Soil and Nutrient Losses from Three Small Watersheds Located in Indonesia. http://www.cigrejournal.tamu.edu/submissions/volume6/LW%2004%2007%20Singh%20final%2022Sept2004.pdf pp. 1–18.

Hoeft, R. (2004). *Do you Need Micronutrient Soil Tests*. Crop Sciences Department, University of Illinois-Urbana-Champaign, Urbana Illinois, USA. http://www.cropsci.uiuc.edu/classic/2004/Article2/ pp. 1–9.

IPI, (2000). *Potassium—An Essential Nutrient.* International Potash Institute: Co-ordination Reports. Potash Research Institute of India, Gurgaon, India. http://www.ipipotash.org/pdf/publications/Pfacts.pdf pp. 1–13.

IPNI, (2007). *Nutrient Removal Rates in Maize and Wheat.* International Plant Nutrition Institute, Norcross, Georgia, USA. http://www.ppi-far.org/ppiweb/usanc.nsf/$webindex/E71D-7CA9BD24A18D86257060007A8EB3 pp. 1–3.

Johnston, A.M. and Dowbenk, R. (2009). *Essential Elements in Corn.* www.farmwest.com/index.cfm?method=library. showPage&librarypageid=`124 pp. 1–7.

Kayode, G.O. (1984). Effect of iron on maize yields in forest and Savanna zones of Nigeria. *Experimental Agriculture* **20**, 335–337.

Lisuma, J.B., Semoka, J.M.R., and Semu, E. (2006). Maize yield response and nutrient uptake after micronutrient application on a volcanic soil. *Agronomy Journal* **98**, 402–406.

Lopez-Valdevia, L.M., Fernandez, M.D., Obrador, A., and Alvarez, J.M. (2002). Zinc transformations in acidic soil and zinc efficiency on maize by adding six organic zinc complexes. *Journal of Agricultural and Food Chemistry* **50**, 1455–1460.

McKenzie, R.H. (2001). *Micronutrient Requirements of Crops.* http://www1.agric.gov.ab.ca/$department/deptdocs.nsf/all/agdex713 pp. 1–8.

Mengel, D.B. (2009). *Role of Micronutrients in Efficient Crop Production.* http://www.ces.purdue.edu/extmedia/AY/AY-239.html pp. 1–5.

Omafra, S. (2002). Corn: Secondary and micronutrients. In *Agronomy Guide for Field Crops.* Omafra Publication, Ministry of Agriculture, Canada, Vol. 811, pp. 1–4.

Patil, P.L., Rader, B.M., Patil, S.G., Aladkatti, Y.R., Meti, C.B., and Khot, A.B. (2006). Effect of moisture regimes and micronutrients on yield, water use efficiency and nutrient uptake by maize on Vertisols of Malaprabha Command, Karnataka. *Journal of the Indian Society of Soil Science* **54**, 445–452.

Sawyer, J.E., Mallarino, A.P., Killorn, R., and Barnhart, S.K. (2008). *A General Guide for Crop Nutrient and Limestone Recommendations in Iowa.* Iowa State University Extension Bulletin pp. 1–18.

Selvi, D., Malavizhi, P.P., and Santhy, P. (2006). *Effects of Long-term Fertilizer Application and Intensive Cropping on Dynamics of Soil Micronutrients Under Tropical Agro Ecosystem.* 18th World Congress of Soil Science, Philadelphia, Pennsylvania, USA. http://crops.confex.com/cropswc2006/techprogam pp. 1–6.

Shabaan, M.M. (2001). Effect of Trace-nutrient foliar fertilizer on nutrient balance, growth, yield and yield components of cereals. *Pakistan Journal of Biological Sciences* **4**, 770–774.

Sherchan, D.P., Upreti, R., and Maskey, S. (2004). Effect of micronutrients of maize (*Zea mays*) in the acid soils of Chitwan valley. In *Proceedings of an International Workshop on Micronutrients in South and Southeast Asia.* International Center for Integrated Mountain Development (ICIMOD). pp. 169–180.

Singh, M.V. (2001). Evaluation of micronutrient stocks in different agroecological zones of India for sustainable crop production. *Fertilizer News* **46**, 13–35.

Szulc, P., Waligora, H., and Skrzypczak, W. (2008). Better effectiveness of maize fertilization with nitrogen through additional application of magnesium and sulphur. *Nauka Przyroda Technologie* **23**, 1–9.

Wolkowski, R.P. (2003). *Using Recycled Wallboard for Crop Production.* University of Wisconsin-Extension Bulletin. http://www.soils.wisc.edu/extension/publications/ pp. 1–8.

Wortmann, C.S., Ferguson, R.B., Hergert, G.W., and Shapiro, C.A. (2008). *Use and Management of Micronutrient Fertilizers in Nebraska.* University of Nebraska-Lincoln, Extension Bulletin G 180, pp. 1–4.

ZARI, (2008). *Soil Fertility Research-critical Levels of Zinc in Intensive Agriculture.* http://www.zari.gov.zm/soils_zinc.php pp. 1–2.

5

Abunyewa, A., Asiedu, E.K., Nyameku, A.L., and Cobbina, J. (2004). Alley cropping *Gliricidia sepium* with maize: 1. The effect of hedgerow spacing, pruning height and phosphorus application on maize yield. *Journal of Biological Sciences* **4**, 81–86.

Adriano, P., Santos, R., Urquiaga, S.S., Guerra, J.G.M., and Freitas, G. (2006). Sun hemp and millet as green manure for tropical maize production. http://biblioteca.universia.net/ficha.do?id=10255750.htm pp. 1–3.

Al-Bakeir, H.M. (2003). Yield, growth rate and nutrient content of corn (*Zea mays L.*) hybrids. *Hebron University Research Journal* **1**.

Alexieva, S. and Stoimenova, I. (1998). A model conserning to the yield loss of maize from weed density or dry biomass. http://www.toprak.org.tr/isd/isd_82.htm pp. 1–5.

Atreya, K., Sharma, S., and Bajracharya, R.M. (2005). Minimization of soil and NUTRIENT losses in maize-based cropping systems in the Mid-Hills of Central Nepal. *Kathmandu University Journal of Science, Engineering and Technology* **1**, 1–15.

Azeez, J.O., Adetunji, M.T., and Adebusuyi, M. (2007). Effect of residue burning and fertilizer application on soil nutrient dynamics and dry grain yield of maize (*Zea mays*) in an Alfisol. *Nigerian Journal of Soil Science*. **17**, 71–80.

Babalola, O., Jimba, S.C., Maduakolam, O., and Dada, O.A. (2009). Use of Vetiver Grass for Soil and Water Conservation in Nigeria. http://www.vetiver.org/NIG_SWC.pdf pp. 1–10.

Basamba, T.A., Barrios, E., Amexquita, E., Rao, I.M., and Singh, B.R. (2006). Tillage effects on maize yield in a Colombian Savanna Oxisol: Soil organic matter and P fractions. *Soil and Tillage Research* **91**, 131–142.

Bauchmann, N. (2000). Biotic and abiotic factors controlling soil respiration rates in *Picea abies* stands. *Soil Biology and Biochemistry* **32**, 1625–1635.

Belfield, S. and Brown, C. (2008). *Field Crop Manuals: Maize—A Guide to Upland Production in Cambodia.* Australian Center for International Agricultural Research, Canberra, Australia. http://www.aciar.gov.au/publication/CoP10 pp. 1–75.

Bertora, C., Zavatarro, L., Saco, D., Monaco, S., and Grignani, C. (2009). Soil organic matter dynamics and losses in manured maize-based forage systems. *European Journal of Agronomy* **30**, 177–186.

Beyer, L., Pingpank, K., and Sieling, K. (2002). Soil organic matter in temperate arable land and its relationship to soil fertility and crop production. In *Soil fertility and Crop Production.* Krishna K.R. (Ed.). Science Publishers Inc., New Hampshire, USA, pp. 189–212.

Blagodatskaya, E.V., Blagodatsky, S., Adnerson, T.H., and Kuzyakov, Y. (2008). Substrate availability and microbial growth in soil amended with glucose, root exudates and maize straw. *Geophysical Research Abstacts* **10**, 1–2.

Bundy, L.G. (2001). Nitrogen application and residue decomposition area fertilizer dealer meeting. University of Wisconsin-Madison, Wisconsin. *Soil Science Extension papers* 1–2.

Carter, M.A. and Singh, B. (2004). Response of Maize and Potassium dynamics in Vertisols following Potassium Fertilization. http://www.regional.org/au/au/asssi/supersoil2004/s13/oral/1603_carterm.htm pp. 1–12.

Cassel, D.K. and Wagger, M.G. (1996). Residue management for irrigated maize grain and silage production. *Soil and Tillage Research* **39**, 101–114.

Chirwa, P.W., Ong, C.K., Maghembe, J., and Black, C.R. (2007). Soil water dynamics in cropping systems containing *Gliricida sepium*, pigeonpea and maize intercropping in Malawi. *Agroforestry Systems* **69**, 123–138.

Chung, H., Grove, J.H., and Six, J. (2008). Indications for soil carbon saturation in a temperate agroecosystem. *Soil Science Society of America Journal* **72**, 1132–1139.

Comia, R.A. (1999). Soil and Nutrient Conservation Oriented Practices in the Philippines. http://www.agnet.org/library/eb/472a/.htm pp. 1–13.

Copel, E.S., Fernando, A.M.D., Silva, A., Orbels, M.C., and Francois, A. (2004). Modeling crop residue mulching effects on water-use and productivity of maize under semi-arid and humid tropical conditions. *Agronomie* **24**, 383–395.

Corbeels, M., Scopel, E., Cardoso, A., Douzet, J.M., Neto, M.S., and Bernoux, M. (2004). Soil carbon sequestration and mulch-based cropping in the Cerrado region of Brazil. *Proceedings of the 4th International Crop Science Conference.* Brisbane, Australia. www.cropscience.org pp. 1–8.

Coulter, J.A. and Nafziger, E.D. (2008). Continuous corn response to residue management and nitrogen fertilization. *Agronomy Journal* **100**, 1774–1780.

David, M.B., McIsac, G.F., Darmody, R.G., and Omonade, R.A. (2009). Long term changes in Mollisols organic carbon and nitrogen. *Journal of Environmental Quality* **38**, 200–211.

Dhaliwal, S.S., Sadana, U.S., Khurana, M.P.S., Dhadli, H.S., and Manchanda, J.S. (2010). Enrichment of rice grains with zinc and iron through ferti-fortification. *Indian Journal of Fertilizers* **6**, 28–35.

Ding, W., Cai, Y., Cai, Z., Yagi, K., and Zheng, X. (2007). Soil respiration under maize crops: Effects of water, temperature and nitrogen fertilization. *Soil Science Society of America Journal* **7**, 944–951.

Duxbury, J.M (2005). Reducing greenhouse warming potential by carbon sequestration opportunities, limits and tradeoffs. In *Climate Change and Global Food security*. Lal R. (Ed.). Taylor and Francis, Boca Raton, Florida, USA, pp. 435–450.

Evers, G. and Agostini, A. (2001). No-tillage farming for sustainable land management: Lessons from the 2000 Brazil Study tour. *TCI Occasional Paper Series No* 12, pp. 1–26.

Fan, T., Stewart, B.A., Payne, W.A., Yong, W., Luo, J., and Gao, Y. (2005). Long term fertilizer and water availability effects on cereal yield and soil chemical properties in Northwest China. *Soil Science Society of America Journal* **69**, 842–855.

Follet, R.F., Varvel, G.E., Kimble, J.M., and Vogel, K.P. (2009). No-till corn after brome grass: Effect on soil carbon and soil aggregates. *Agronomy Journal* **101**, 261–268.

Fosu, M., Kuhne, R.F., and Vlek, P.L.G. (2004). Improving maize yield in the Guinea Savanna Zone of Ghana with legume crops and P, K fertilizers. *Journal of Agronomy* **3**, 115–121.

Grignani, C., Zavatarro, L., Saco, D., and Stefano, M. (2007). Production, nitrogen and carbon balance of maize-based forage systems. *European Journal of Agronomy* **26**, 442–453.

Haas, H.J., Evans, C.E., and Miles, E.A. (1957). Nitrogen and carbon changes in the Great Plains soils as influenced by cropping and soil treatments.

United States Department of Agriculture, USDA Technical Bulletin No 1164, pp. 1–87.

Habtesellassie, M.Y., Miller, B.E., Thacker, S.G., Stark, J.M., and Norton, J.M. (2006). Soil nitrogen and nutrient dynamics after repeated application of treated dairy-waste. *Soil Science Society of America Journal* **70**, 1328–1337.

Hairiah, K. and Van Noordwijk, M. (1989). Root distribution of leguminous cover crops in the humid tropics and effects on a subsequent maize crop. In *Nutrient Management for Food Crop Production in Tropical Farming Systems*. Institute for Soil Fertility, Haren, The Netherlands, pp. 157–170.

Halvorson, A.D., Mosier, A.R., Reule, C.A., and Bausch W.C. (2006). Nitrogen and tillage effects on irrigated continuous corn yields. *Agronomy Journal* **98**, 63–71.

Heidi, J. and Hairiah, K. (1989). The role of green manures in rain fed farming systems in the humid tropics. http://www.metafro.be/leisa/1989/5-2-11.pdf pp. 1–4.

Hountin, J.A., Couillard, D., and Karam, A. (1997). Soil carbon, nitrogen and phosphorus contents in maize plots after 14 years of pig slurry applications. *The Journal of Agricultural Science* **129**, 187–191.

Hur, S., Kim, W.T., Jung, K.H., and Ha, S.K. (2004). Study on soil and nutrients loss with soil textures and two crops during rainfall. *Proceedings of International Crop Science Conference*. http://www.cropscience.org.au/icsc2004/poster/1/6/1298_huros.htm pp. 1–4.

Ikpe, F.N., Ndegewe, N.A., Gbaraneh, L.D., Torunana, J.M.A., Williams, T.O., and Larbi, A. (2003). Effects of sheep browse diet on fecal matter decomposition and N and P cycling in the humid lowlands of West Africa. *Soil Science* **168**, 646–659.

Iqbal, Z., Latif, A., Sikander A., and Iqbal, M. (2003). Effect of fertigated phosphorus on P use efficiency and yield of wheat and maize. *Songkalakarin Journal of Science and Technology* **25**, 697–702.

Jenny, H. (1941). *Factors of Soil formation*. McGraw Hill, New York, USA, p. 345.

Johnson, J.M.F., Barbour, N.W., and Weyers, S.L. (2007). Chemical composition of crop biomass impacts its decomposition. *Soil Science Society America Journal* **7**, 155–162.

Juan, H., Walter, I., Undurraga, P., and Cartagena, M. (2007). Residual effects of poultry litter on silage maize (*Zea mays*) growth and soil properties derived from volcanic ash: Fertilizers and soil amendments. *Soil Science and Plant Nutrition* **53**, 480–488.

Kamara, A.Y., Ekeleme, F., Chikoye, D., and Omoigui, L.O. (2009). Planting date and cultivar effects on grain yield in dry land corn production. *Agronomy Journal* **101**, 91–98.

Karlen, D.L., Hurley, E.G., Andrews, S.S., Camberdella, C.A., Meek, D.W., Duffy, M.D., and Mallarino, A.P. (2006). Crop rotation effects on soil quality at three northern corn/soybean belt locations. *Agronomy Journal* **98**, 484–495.

Kimetu, J.M., Mugendi, D.N., Palm, C.A., Mutuo, P.K., Gachengo, C.N., Nandwa, S., and Kungu, J.B. (2002). Nitrogen fertilizer equivalency values for different organic materials based on maize performance at Kabeta, Kenya. http://fiar.org/webciat/tsbf_institute/managing_nutrient_cycles/AfNetCH51.pdf

Khattari, S. (2000). Phosphorus levels in soils of different vegetable crops in Jordan Valley. In *Plant Nutrient Management Under Pressurized Irrigation Systems in the Mediterranean Region.* Ryan, J. (Ed.). World Phosphate Institute, Casablanca, Morocco, pp. 223–228.

King, J.A., Bradley, R.I., Harrison, R., and Carter A.D. (2004). Carbon sequestration and saving potential associated with changes to the management of agricultural soils in England. *Soil Use and Management* **20**, 394–402.

Kogbe, J.D.S and Adedira, J.A. (2003). Influence of nitrogen, phosphorus and potassium application on the yield of maize in the Savanna zone of Nigeria. *African Journal of Biotechnology* **2**, 345–349.

Krishiworld, (2002). The Pulse of the Indian Agriculture: Maintenance of Soil Fertility. http://www.krishiworld.com/html/soil_ferti3.html pp. 1–9.

Lal, R. (2004). Soil carbon S sequestration impacts on global climate change and food security. *Science* **304**, 1623–1627.

Li, K., Shiraiwa, T., Saitoh, K., and Takeshi, H. (2002). Water use and growth of maize under water stress on the long term application of chemical and/or organic fertilizers. *Plant Production Science* **5**, 58–64.

Liang, Y., Gollany, H.T., Rickman, R.W., Albrecht, S.L., Follet, R.F., Wilhelm, W.W., Noak, J.M., and Douglas, C.L. (2008). CQUESTER simulation of management practice effects on long term soil organic matter. *Soil Science Society of America Journal* **72**, 1486–1492.

Lindquist, J.L., Arkebauer, T.J., Walters, D.T., Cassman, K.G., and Dobermann, A. (2005). Maize radiation use efficiency under optimal growth condition. *Agronomy Journal* **97**, 72–78.

Lisuma, J.B., Semoka, J.M.R., and Semu, E. (2006). Maize yield response and nutrient uptake after micronutrient application on a volcanaic soil. *Agronomy Journal* **98**, 402–406.

Machinet, G.E., Bertrand, I., Chabbbert, B., and Recous, S. (2006). Role of cell wall components on the decomposition of maize roots in soil: Impact on carbon mineralization. *Proceeding of 18th World Congress of Soil Science*, Session 2.2A Philadelphia, Pennsylvania, USA. Vol. 138, pp. 1–2.

Mann, L.K. (1986). Changes in soil carbon storage after cultivation. *Soil Science* **142**, 279–287.

Mapfumo, P. and Mtambanengwe, F. (2008). Base nutrient dynamics and productivity of sandy soils under maize-pigeonpea rotational systems in Zimbabwe. www.ciat.cgiar.org/tsbf_institute/managing_nutrient_cycles/Afnetch16.pdf pp. 1–19.

Mikhailova, E.A., Bryant, R., Vassenev, J.I., Schwager, S.J., and Post, C.J. (2000). Cultivation effects on soil carbon and nitrogen contents at depth in Russian Chernozems. *Soil Science Society of America Journal* **64**, 738–745.

Mtambanengwe, F., Kosina, P., and Jones, J. (2007). *Maize Crop Residue Management—Mulch Feed or Burn.* IRRI-CIMMYT Cereal knowledge bank. Knowledgebank.cimmyt.org, pp. 1–3.

Mtambanengwe, F. and Mapfumo, P. (2006). Effects of organic resource quality on soil profile N dynamics and maize yields on sandy soils of Zimbabwe. *Plant and Soil* **28**, 1132–1138.

Mtambanengwe, F., Mapfumo, P., and Vanlauwe, B. (2006). Comparative short term effects of different quality organic resources on maize productivity under two different environments in Zimbabwe. *Nutrient Cycling in Agroecosystems* **76**, 271–284.

Mubarak, A.R., Roseneni, A.B., Anuar, A.R., and Zauyah, S. (2002). Decomposition and nutrient release of maize stover and groundnut haulm under tropical field conditions of Malaysia. *Communications in Soil Science and Plant Analysis* **33**, 609–622.

Mueller, T., Magid, J., Jensen, L.S., Svendsen, H., and Nielsen, N.E. (1998). Soil C and N turnover after incorporation of chopped maize, barley straw and blue grass in the field: Evaluation of DAISY soil-organic matter sub-model. *Ecological Modeling* **111**, 1–15.

Mungai, N.W., Motavalli, P.P., Nelson, K.A., and Kremer, R.J. (2005). Differences in yields, residue composition and N mineralization dynamics of BT and Non-BT maize. *Nutrient Cycling in Agroecosystems* **73**, 101–109.

Mupangwa, W., Twomlow, S., Walker, S., and Hove, L. (2007). Effect of minimum tillage and mulching on maize (*Zea mays*) yield and water content of clayey and sandy soils. *Physics and Chemistry of the Earth* **32**, 15–18.

Nair, K.P.P. and Ghosh, G.N. (1984). Efficiency of recycled nitrogen from residues of maize, soybean and moong on wheat grain yield. *Plant and Soil* **82**, 125–134.

Oberson, A., Buneman, E.K., Friesen, D.K., Rao, I.M., Smithson, P.C., Turner, B.J., and Frossard, E. (2009). Improving phosphorus fertility in tropical soils through biological interventions. http://ciat-library.ciat.cgiar/articulos_Ciat/DK3724_CO37.pdf pp. 1–19.

Okigbo, B.N. and Lal, R. (1980). Residue mulches intercropping and agrisilviculture potential in tropical Africa. In *Basic Techniques in Ecological Farming*. Hill S. (Ed.). IFOAM Conference, Montreal, Canada, pp. 54–69.

Onim, J.F.M., Mahtuva, M., Otieno, K., and Fitzhugh, H.A. (1990). Soil fertility changes and response of maize and beans to green manures of Leucana, Sesbania and Pigeonpea. *Agroforestry Systems* **12**, 197–215.

Pare, T., Gregorich, E.G., and Nelson, S.D. (2000). Mineralization of Nitrogen from crop residues and N recovery by maize inoculated

with Vesicular arbuscular mycorrhizal fungi. *Plant and Soil* **218**, 11–20.

Pepo, P. (2001). Nitrogen-fertilization using 'Biofert' in sustainable maize production. http://www.date.hu/acta-agraria/2001-01/pepo.pdf pp. 1–8.

Pikul, J.L., Johnson, J.M.F., Schumacher, T.E., Vigil, M., and Riedell, W.E. (2008). Change in surface soil caron under rotated corn in Eastern South Dakota. *Soil Science Society of America Journal* **72**, 1738–1744.

Praharaj, C.S., Shankarnarayanan, K., Khader, S.E.S.A., and Gopalakrishnan, K. (2009). Sustaining cotton productivity and soil fertility through *in situ* management of green manure and crop residues in semi-arid irrigated condition of Tamil Nadu. *Indian Journal of Agronomy* **54**, 415–422.

Prihar, S.S., Singh, R., Singh, N., and Sandhu, K.S. (1979). Effects of mulching previous crops or fallow on dry land maize and wheat. *Experimental Agriculture* **15**, 129–134.

Purkayastha, T.J., Rudrappa, I., Singh, D., Swarup, A., and Badhraray, S. (2008). Long-term impact of fertilizers on soil organic carbon pools and sequestration rates in maize-wheat-cowpea cropping system. *Geoderma* **144**, 370–378.

Rajeshwari, R.S., Hebsur, N.S., Pradeep, H.M., and Bharamagoudar, T.D. (2007). Effect of integrated nitrogen management on growth and yield of maize. *Karnataka Journal of Agricultural Science* **20**, 399–400.

Reeder, J.D., Schuman, G.E., and Bowman, R.A. (1998). Soil C and N changes on conservation reserve program lands in the Central Great Plains. *Soil Tillage Research* **47**, 339–349.

Sainz Rozas, H.R., Echevarria, H.E., and Picone, L.I. (2001). Denitrification in maize under notillage. *Soil Science Society of America Journal* **65**, 1314–1323.

Sakala, W.D., Cadisch, G., and Giller, K.E. (2000). Interactions between residues of maize and Pigeonpea and mineral N fertilizers during decomposition and N mineralization. *Soil Biology and Biochemistry* **32**. 679–688.

Sakala, W.D., Kumweda, J.D.T., and Saka, A.R. (2009). The potential of green manures to increase soil fertility and maize yields in Malawi.

www.ciat.cgiar.org/tsbf_institute/managing_nutrient_cycles/AfnetCh26.pdf pp. 1–16.

Sangakarra, U.R., Liedgens, M., Soldati, A., and Stamp, P. (2004). Root and shoot growth of maize (*Zea mays*) as affected by incorporation of *Crotalaria juncea* and *Tithonia diversifolia* as green manures. *Journal of Agronomy and Crop Science* **190**, 339–346.

Sangoi, L., Ender, M., Guidolin, A.F., Almeida, M.L., and Konflancze, A. (2001). Nitrogen fertilization impact on agronomic traits of maize hybrids released at different decades. *Pesquisa Agroepcuaria Brasiliera* **36**, 1–14.

Schoningh, E. and Alkamper, J. (1984). Effects of different mulch materials on soil properties and yield of maize and cowpea in the eastern Amazon Oxisol. *Proc. First Symposium on Humid Tropics*, EMBRAPA, 12 17 Belem, Brazil., pp. 21–34.

Schroeder, J.J., Hilhorst, G.J., and Bruinenberg, M.H. (2007). Long term N fertilizer value of cattle slurry to maize. http://en.scientificcommons.org/14481880 pp. 1–2.

Schultz, S., Tian, G., Oyewole, B., and Bako, S. (2003). Rice mill waste as organic manure on a degraded Alfisol. *Agriculture, Ecosystems and Environment* **100**, 221–230.

Sey, B.K., Whaleu, J.K., Gregorich, E.G., Rochette, P., and Cue, R.L. (2008). Carbon dioxide and nitrous oxide content in soils under corn and soybean. *Soil Science Society of America Journal* **72**, 931–938.

Sharma, A.R. and Behera, U.K. (2008). Recycling of legume residues for nitrogen economy and higher productivity in maize (*Zea mays*)—Wheat (*Triticum aestivum*) cropping system. *Nutrient Cycling in Agroecosystems* **83**, 197–210.

Sharma, K.C. (2009). Integrated nitrogen management in fodder oats in hot arid ecosystem of Rajasthan. *Indian Journal of Agronomy* **54**, 459–464.

Silva, P.S.L., Silva, E.S., and Mesquita, S.S.X. (2004). Weed control and green ear yield in maize. *Planta Daininha* **22**, 1–9.

Smalling, E., Toure, M., Ridder, N.D., Sanginga, N., and Breman, H. (2006). Fertilizer use and the environment in Africa: Friends or foes. Background paper African fertilizer summit, NEPAD-IDFC, Abuja, Nigeria, pp. 1–26.

Sogbedji, J.M., Van Es, H.M., and Agbeko, K.L. (2006). Cover cropping and nutrient management strategies for maize production in Western Africa. *Agronomy Journal* **98**, 883–889.

Stewart, G. (2005). *Greenhouse Gas Mitigation*. Ontario Ministry of Agriculture and Food, Canada, pp. 1–3.

Tejada, M. and Gonzalez, J.L. (2006). Crushed cotton gin compost effects on soil biological properties, nutrient leaching and maize yield. *Agronomy Journal* **98**, 749–759.

Theodoro, B.L. and Ferreira, V.E. (1995). Growth and osmotic adjustment of maize plants as influenced by potassium and water stress. http://natres.psu.ac.th/link/soilCongress/bdd/symp142219-t.pdf pp. 1–10.

Tijani, F.O., Oyedele, D.J., and Aina, P.O. (2008). Soil moisture storage and water use efficiency of maize planted in succession to different fallow treatments. *International Agrophysics* **22**, 81–87.

Ultech, S. (2008). *Managing Carbon Loss*. Agronomy Society of America, Madison, Wisconsin, USA, www.agronomy.org/press/releases/2008/1201/223/ pp. 1–2.

Varvel, G.E. and Wilhelm, W.W. (2008). Soil carbon levels in irrigated western corn belt rotations. *Agronomy Journal* **100**, 1180–1184.

Walkowski, R.P. (2003). *Using recycled Wallboard for Crop Production*. University of Wisconsin-Extension Bulletin. http://www.soils.wisc.edu/extension/publications/ pp.1–8.

Wang, X.B., Cai, D.X., Hoogmooed, W.B., Perdock, U.D., and Oenema, O. (2007). Crop residue, manure and fertilizer in dry land maize under reduced tillage in Northern China 1. Grain yields and nutrient use efficiencies. *Nutrient Cycling in Agroecosystems* **79**, 1–16.

Woperies, M.C.S., Tmaelookpo, A., Ezui, K., Gnakpenou, D., Fofana, B., and Breman, H. (2005). Mineral fertilizer management of maize on farmer fields differing in organic inputs in the West African Savanna. *Field Crops Research* **96**, 355–362.

Wu, W., Chen, M., and Sun, B. (2002). Effects of land use changes on soil chemical properties of sandy soils from tropical Hainan, China. http://www.fao.org/docrep/010/ag125e/AG125E13.htm pp. 1–23.

Yamoah, C., Ngueguim, M., Ngong, C., and Dias, D.K.W. (1996). Reduction of fertilizer requirement using lime and Macuna on high P-sorption soils of North West Cameroon. *African Crop Science Journal* **4**, 441–451.

Yeboah, E., Ofori, P., Quansah, G.W., Dugan, E., and Sohi, S.P. (2009). Improving soil productivity through Biochar amendments to Soil. *African Journal of Environmental Science and Technology* **3**, 34–41.

Zingore, S., Mafogoya, P., Myammgufata, P., and Giller, K.E. (2003). Nitrogen mineralization and maize yields following application of tree prunings to a sandy soil in Zimbabwe. *Agroforestry Systems* **57**, 199–211.

Zotarelli, L., Avila, L., Scholberg, J.M.S., and Alves, B.J.R. (2009). Benefits of vetch and rye cover crops to sweet corn under no-tillage. *Agronomy Journal* **101**, 252–260.

6

Andrade, S.A.L. and Da Silviera, A.P.D. (2008). Mycorrhiza influence on maize development under Cd stress and P supply. *Brazilian Journal of Plant Physiology* **20**, 1–14.

Benjamin, J.G., Ahuja, L.R., and Allmara, R.R. (1996). Modeling corn rooting patterns and their effects on water uptake and nitrate leaching. *Plant and Soil* **179**, 223–232.

Biari, J.A., Gholami, J.A., and Rahmani, A.Z. (2008). Growth promotion and enhance nutrient uptake of maize (*Zea mays*) by application of plant growth promoting rhizobacteria in the Arid Regions of Iran. *Journal of Biological Science* **8**, 1015–1020.

Bloom, A.J., Frensch, J., and Taylor, A.R. (2005). Influence of inorganic nitrogen and pH on the elongation of maize seminal roots. *Annals of Botany* **97**, 867–873.

Cattani, I., Fragoulis, G., Bocceli, R., and Capri, E. (2006). Copper bioavailability in the rhizosphere of maize (*Zea mays*) grown in two Italian soils. *Chemosphere* **64**, 1972–1979.

Chang, C.Y., Chao, C.C., and Chao, W.L. (2007). An evaluation of the diversity of fluorescent psuedomonads in maize rhizosphere using 16S-23 rDNA intergenic spacer region restriction fragment length polymorphism and the biolog

GN plate method. *Taiwanese Journal of Agricultural Chemistry and Food Science* **45**, 67–75.

Chassot, A. and Richner, W. (2002). Root characteristics and phosphorus uptake of maize seedlings in a bilayered soil. *Agronomy Journal* **94**, 118–127.

Chassot, A., Stamp, P., and Richner, W. (2001). Root distribution and morphology of maize seedlings as affected by tillage and fertilizer placement. *Plant and Soil* **231**, 123–135.

Coelho, E.F. and Or, D. (1998). Root distribution and water uptake pattern of corn under surface and subsurface irrigation. *Plant and Soil* **206**, 123–136.

Costa, C., Dwyer, L.M., Zhou, X., Dutilleul, P., Hamel, C., Reid, L.M., and Smith, D.L. (2002). Root morphology of contrasting maize genotypes. *Agronomy Journal* **94**, 96–101.

Czarnes, S., Dexter, A.R., and Bartolli, F. (2000). Wetting and drying cycles in the maize rhizosphere under controlled conditions. Mechanics of the root-adhering soil. *Plant and Soil* **221**, 253–271.

Devereux, A.F., Fukai, S., and Hulugalle, N.R. (2008). The Effects of Maize Rotation on Soil Quality and Nutrient Availability in Cotton-based Cropping. http://www.regional.org.au/au/asa/2008/concurrent/plant_nutrition/5815_devereuxaf.htm pp. 1–5.

Dommergues, Y., Balandreau, J., Rinaudo, G., and Weinhard, P. (1972). Non-symbiotic nitrogen fixation in the rhizosphere of rice, maize and different tropical grasses. *Soil Biology and Biochemistry* **5**, 83–89.

Drinkwater, L.E. and Snapp, S.S. (2006). Understanding and Managing the Rhizosphere in Agroecosystems. http://ecommons.library.cornell.edu/bitstream/1813/3470/2/Drinkwater-Snapp%20Revised%20Chapter-3-2006-JLW-LED-Final%20version-2rtf pp. 1–16.

Engels, C. (2005). Differences between maize and wheat in growth-related nutrient demand and uptake of potassium and phosphorus at suboptimal root zone temperatures. *Plant and Soil* **150**, 129–138.

Engels, C., Mollenkopf, M., and Marschner, H. (2007). Effect of drying and rewetting the top soil on root growth of maize and rape in different soil depths. *Zeitschrift fur Pflazenernahrung und Bodenkunde* **157**, 139–144.

Evans, D.G. and Miller, M.H. (1988). Vesicular-Arbuscular Mycorrhizas and the soil-disturbance induced reduction of nutrient absorption in maize. 1. Causal relations. *New Phytologist* **110**, 67–74.

FARC, (2004). Breakdown of BT maize in soils and impacts on microorganisms. Federal Agricultural Research Center, Institute of Agroecology Braunschwig, Germany. http://www.gmosafety.eu/en/sfety_science/21.docu.html pp. 1–3.

Garcia, J.P., Wortmann, C.S., Mamo, M., Drijber, R., and Tarkalson, D. (2007). One time tillage of no-till: Effects on nutrients, mycorrhiza and phosphorus uptake. *Agronomy Journal* **99**, 1093–1103.

Goldstein, W.A. and Barber, W. (2001). Yield and root growth in a long-term trial with biodynamic preparations. www.michealfieldsaginst.org/education/biodynamic_longterm.pdf   pp. 32–41.

Gregorich, E.G., Liang, B.C., Drury, C.F., Mackenzie, A.F., and McGill, W.B. (2000). Elucidation of the source and turnover of water soluble and microbial biomass carbon in agricultural soils. *Soil Biology and Biochemistry* **32**, 581–587.

Haller, T. and Stolp, H. (1985). Quantitative estimation of root exudation of maize plants. *Plant and Soil* **86**, 207–216.

Hammer, G.L., Dong, Z., McLlean, G., Doherty, A., Messina, C., Schusler, J., Zinselmeier, C., Paszkiewicz, S., and Cooper, M. (2009). Can changes in canopy and/or root system architecture explain historical maize yield trends in the US Corn Belt. *Crop Science* **49**, 299–312.

Hess, J.L., Shiffler, A.K., and Jolley, V.D. (2005). Survey of mycorrhizal colonization in native, open pollinated and introduced hybrid maize in villages of Chiquimula, Guatemala. *Journal of Plant Nutrition* **28**, 1843–1852.

Hinsinger, P., Gobran, G.R., Gregory, P.J., and Wenzel, W.W. (2005). Rhizosphere geometry and heterogeneity arising from root mediated physical and chemical processes. *New Phytologist* **168**, 293–303.

Holanda, F.S.R., Mengel, D.B., Pula, M.B., Carvalho, J.G., and Bertoni, J.C. (1998). Influence of crop rotations and tillage systems on phosphorus and potassium stratification and root distribution in the soil profile. *Communications in the Soil Science and Plant Analysis* **29**, 2383–2394.

Inal, A. and Aydin, G. (2008). Interspecific root interactions and rhizosphere effects on soil ions and nutrient uptake between mixed grown peanut/maize and peanut barley in original saline—Sodic-boron toxic soil. *Journal of Plant Physiology* **165**, 490–503.

Inal, A., Gunes, A., Zhang, F., and Cakmak, I. (2007). Peanut/Maize intercropping induced changes in rhizopshere and nutrient concentrations in shoots. *Plant Physiology and Biochemistry* **45**, 350–356.

John, K., Kelly, J., Schroeder, P., and Wang, Z. (2005). *In situ* dynamcis of our macronutrients in the rhizosphere soil solution of maize, switch grass and cottonwood. In *Rhizopshere: Perspectives and Challenges—A Tribute to Lorez Hiltner*. Neurenberg, Germany, p. 192.

Johnson, N.C., Pfleger, F.L., Crookston, R.K., Simmons, S.R., and Copeland, P.H. (2006). Vesicular Arbuscular Mycorrhizas respond to corn and soybean cropping. *New Phytologist* **117**, 657–663.

Karasawa, T., Takebe, M., and Kasahara, Y. (2000). Arbuscular Mycorrhizal (AM) effects on maize growth and AM colonization of roots under various soil moisture conditions. *Soil Science and Plant Nutrition* **46**, 61–67.

Krishna, K.R. (2005). *Mycorrhizas: A Molecular Analysis*. Science Publishers Inc. Enfield, New Hampshire, USA, p. 343.

Kuzyakov, Y., Raskatov, A., and Kaupenjohann, M. (2004). Turnover and distribution of root exudates of *Zea mays*. *Plant and Soil* **254**, 317–327.

Laheurte, F. and Berthelin, J. (2006). Effect of a phosphate solubilizing bacteria on maize growth and root exudation over four levels of labile phosphorus. *Plant and Soil* **105**, 11–17.

Liang, B.C., Wang, X.L., and Ma, B.L. (2002). Maize root-induced change in soil organic carbon pools. *Soil Science Society of America Journal* **66**, 845–847.

Liasu, M.O. and Shosanya, O. (2007). Studies of microbial development on mycorrhizosphere and rhizosphere soils of potted maize plants and the inhibitory effect of rhizobacteria isolates on two fungi. *African Journal of Biotechnology* **6**, 504–508.

Liljeroth, E., Kulkman, P., and Van Veen, J.A. (1994). Carbon translocation to the rhizosphere of maize and wheat and influence on the turnover of native and organic matter at different soil nitrogen levels. *Plant and Soil* **16**, 233–244.

Liu, Y., Mi, G., Chen, F., Zhang, J., and Zhang, F. (2004). Rhizosphere effect and root growth of two maize (*Zea mays*) genotypes with contrasting P efficiency at low P availability. *Plant Science* **167**, 217–223.

Long, L., Li, S.M., Sun, J.H., Li, L.Z., Bao, X.G., Zhang, H.G., and Zhang, F.S. (2007). Diversity Enhances Agricultural Productivity via Rhizosphere Phosphorus Facilitation on Phosphorus Deficient Soils. http://www.pnas.org/content/104/27/11192.full p. 108.

Marcel Gomes, N.C., Fagbola, O., Costa, R., Rumjaneck, N.G., Buchner, A., Mendona-Hegler, L., and Smalla, K. (2003). Dynamics of fungal communities in bulk and maize rhizosphere soil in the tropics. *Applied and Environmental Microbiology* **69**, 3758–3766.

Martin Laurent, F., Benoit, B., Isabelle, W., Severine, P., Devers, M., Guy, S., and Phillipot, L. (2006). Impact of the maize rhizosphere on the genetic structure, the diversity and the atrazine-degrading gene composition of cultivable-degrading communities. *Plant and Soil* **282**, 99–115.

Matsumato, H., Okada, K., and Takahashi, E. (1979). Excretion products of maize roots from seedling to seed development stage. *Plant and Soil* **53**, 17–26.

McCully, M. (2005). The rhizosphere: The key functional unit in plant/soil/microbial interactions in the field. Implications for the understanding of allelopathic effects. *Journal of Agricultural Science, Cambridge* **130**, 1–7.

Merckx, R., van Ginkel, J.H., Sinaeve, J., and Cremers, A. (2006). Plant-induced changes in the rhizosphere of maize and wheat 2. Complexation of cobalt, zinc and manganese in the rhizsphere of maize and wheat. *Plant and Soil* **96**, 95–107.

Mozafar, A., Anken, T., Ruh, R., and Frossard, E. (2000). Tillage intensity, mycorrhizal and non-mycorrhizal fungi and nutrient concentrations in maize, wheat and canola. *Agronomy Journal* **92**, 1117–1124.

Nakamoto, T. (1997). The distribution of maize roots as influenced by artificial vertical macropores. *Japanese Journal of Crop Science* **66**, 331–332.

Nakamoto, T., Matsuzaki, A., and Shimado, K. (1992). Root spatial distribution of field grown maize and millets. *Japanese Journal of Crop Sciences* **61**, 304–309.

Nesci, A., Baros, G., Castillo, C., and Etcheverry, M. (2006). Soil fungal population in pre-harvest maize ecosystem in different tillage practices in Argentina. *Soil and Tillage Research* **91**, 143–149.

Neubert, S.W., Levin, I., Fischer, N., and Sonntag, C. (2000). Determination of microbial versus root-produced CO_2 in an agricultural ecosystem, by means of $^{13}CO_2$ measurements in soil air. *Tellus B* **52**, 909–918.

Neumann, G. and Romheld, V. (2000). The release of root exudates as affected by physiological status of plants. In *The Rhizosphere: Biochemistry and Organic Substances at the Plant—Soil Interface*. Pinton, R., Varanini, Z., and Nannipieri, Z. (Eds.). Marcel Dekker, New York, USA, pp. 78–110.

Odhiambo, H.O., Ong, C.K., Deans, J.D., Wilson, J., Khan, A.A.H., and Sprent, J.I. (2001). Roots, soil water and crop yield; tree crop interactions in a semi arid Agroforestry System in Kenya. *Plant and Soil* **235**, 221–233.

Oliviera, C.A., Alves, V.M.C., Mariel, I.E., Gomes, E.A., Scotti, M.R., Carneiro, N.P., Guimareas, C.T., Schafert, R.E., and Sa, N.M.H. (2009). Phosphate solubilizing microbes isolated from rhizosphere of maize cultivated in an Oxisol of the Brazilian Cerrado Biome. *Soil Biology and Biochemistry* **41**, 1782–1787.

Oliviera, M.R.G., Serralheiro, R.P., Reis, M.P.Z., and Santos, F.L. (1998). Maize root system response to furrow irrigation in a Mediterranean brown soil: Root growth related to water distribution. *Journal of Agricultural Engineering* **71**, 13–17.

Omar, S.A. (1998). The role of rock phosphates-solubilizing fungi and vesicular arbuscular mycorrhiza in growth of maize plants fertilized with rock phosphate. *World Journal of Microbiology and Biotechnology* **14**, 211–218.

Ortas, I., Kaya, Z., and Cakmak, I. (2001). Influence of Arbuscular Mycorrhizas inoculation on growth of maize and green pepper plants in phosphors and zinc deficient soil. *Plant Nutrition* **92**, 632–633.

Paskiewicz, I. and Berthelin, J. (2006). Influence of maize rhizosphere and associated microflora on weathering of Fe and Mn oxides and availability of trace elements in a New Caledonia Ferralsols. *Geophysical Research Abstracts* **8**, 762.

Payne, G.W., Ramette, A., Rose, H.L., Weightman, A.J., Jones, T.H., Tiedje, J.M., and Eswar, M. (2006). Application of a rec-A gene-based identification approach to maize rhizosphere to assess novel diversity in Burkholderia species. *FEMS Microbiology Letters* **259**, 126–132.

Pereira de Mello, I.W.M. and Mielniczuk, J. (1999). Influence of soil structure on the distribution and morphology of corn roots under three tillage methods. *Review Brazilian Cience Solo* **23**, 135–143.

Pinton, R., Varanini, Z., and Nanniperi, P. (2007). *The Rhizosphere: Biochemistry and Organic Substances at the Soil-plant Interface.* Marcel Dekker, New York, p. 348.

Pitakdanatan, R., Suwnarit, A., Nopamornbodi, O., and Sarobol, E. (2007). Comparative responses to arbuscular mycorrhizal fungi of maize cultivars different in downy mildew resistance and fertilizer requirement. *Science Asia* **33**, 329–333.

Qian, J.H., Doran, J.W., and Walters, D.T. (1997). Maize plant contributions to root zone available carbon and microbial transformations of nitrogen. *Soil Biology and Biochemistry* **29**, 1451–1462.

Qin, R., Stamp, P., and Richner, W. (2005). Impact of tillage and banded starter fertilizer on maize root growth in the top 25 centimeters of the soil. *Agronomy Journal* **97**, 674–683.

Rodriguez, M.B., Godeas, A., and Lavado, R.S. (2008) Soil acidity changes in bulk soil and maize rhizosphere in response to nitrogen fertilization. *Communications in Soil Science and Plant Analysis* **39**, 2597–2607.

Sauer, D., Kuzyakov, Y., and Stahr, K. (2006). Spatial distribution of root exudates of five plant species as assessed by ^{14}C labeling. *Journal of Plant Nutrition and Soil Science* **169**, 360–362.

Schonwitz, R. and Zeigler, H. (2007). *Interaction of Maize roots and Rhizosphere Micro-organisms* **152**, 217–222.

Schortmeyer, M., Feil, B., and Stamp, P. (1993). Root morphology and nitrogen uptake of maize simultaneously supplied with ammonium and nitrate in a split root system. *Annals of Botany* **72**, 107–115.

Sharp, R.E. and Davies W.J. (1985). Root growth and water uptake by maize plants in drying soil. *Journal of Experimental Botany* **36**, 1441–1456.

Song, Y.N., Zhang, F.S., Marschner, P., Fan, F.L., Gao, H.M., Bao, X.G., Sun, J.H., and Li, L. (2007). Effect of intercropping on crop yield and chemical and microbiological properties in rhizosphere. *Biology and Fertility of Soils* **43**, S14–15.

Strom, L., Owen, A.G., Godbold, D.L., and Jones, D.L. (2002). Organic acid mediated P mobilization in the rhizosphere and uptake by maize roots. *Soil Biology and Biochemistry* **34**, 703–710.

Tao, S., Chen, Y.J., Xu, F.L., Cao, J., and Li, B.G. (2003). Changes of copper speciation in maize rhizosphere soil. *Environmental Pollution* **122**, 447–454.

Van Beem, J., Smith, M.E., and Zobel, R.W. (1998). Estimating root mass in maize using portable capacitance meter. *Agronomy Journal* **90**, 566–570.

Walker, T.S., Bais, H.P., Grotewolde, E., and Vivinco, J.M. (2003). Root exudation and rhizosphere biology. *Plant Physiology* **132**, 44–51.

Watt, M., McCully, M.E., and Jeffree, C.E. (2004). Plant and bacterial mucilages of the maize rhizosphere: Comparison of their soil binding properties and histochemistry in a model system. *Plant and Soil* **151**, 151–165.

Wortmann, S.C., Quincke, J.A., Drijber, R.A., Mamo, M., and Franti, T. (2008). Soil microbial community change and recovery after one-time tillage of conventional tillage. *Agronomy Journal* **100**, 1681–1686.

Xu, M.G., Zhang, Y.P., and Sun, B.H. (1995). Phosphate distribution and movement in soil-root interface zone: III Dynamics. *Pedosphere* **5**, 349–355.

Yamane, Y. and Highuchi, H. (2003). Function of Arbuscular Mycorrhizal fungi associated with maize roots grown under indigenous farming systems in Tanzania. http://www. cababstractsplus.org/abstracts/Abstracts. aspx?AcNo=20036796725.htm pp. 1–2.

Yoshitomi, K.J. and Shann, J.R. (2001). Corn (*Zea mays*) root exudates and their impact on ^{14}C-pyrene mineralization. *Soil Biology and Biochemistry* **33**, 1769–1776.

7

Abendroth, L. and Elmore, R. (2007). Demand for more Corn following Corn. http://www. agronext.iastate.edu/corn/production/management/cropping/demand.html pp. 1–2.

Akintoye, H.A., Kling, J.G., and Lucas, E.O. (1999). Nitrogen Use efficiency of single, double, and synthetic maize lines grown at four N levels in three Ecological zones of West Africa. *Field Crops Research* **60**, 189–199.

Banziger, M., Betran, F.J., and Lafitte, H.R. (1997). Efficiency of High-N selection environments for improving maize for Low-N target environments. *Crop Science* **37**, 1103–1109.

Barbieri, P.A., Echeverria, H.E., SainzRosas, H.R., and Andrade, F.H. (2008). Nitrogen use efficiency in maize as affected by nitrogen availability and row spacing. *Agronomy Journal* **98**, 1094–1100.

Beem J. and Smith, M.E. (2004). Variation in Nitrogen use efficiency and root system size in temperate maize genotypes. Developing Drought and Low-Nitrogen on Maize Symposium Abstracts. http://www.cimmyt.org/Research/Maize/DLNCA/htm/DLNCp2-8.htm pp. 2–8.

Chantachume, Y., Manupeerapan, T., Grudloyma, P., Tongchuay, S., Noradechanon, S., and DeLeon, C. (1998). Selection for low nitrogen tolerance in the Thailand Maize Breeding Program. Proceedings of Symposium on Developing Drought and Low-Nitrogen on Maize Symposium. www.cimmyt.org/Resources/archive/What_is_CIMMYT/Annual_Reports/AR98/htm/AR97-98Staff p. 35.

Costa, C., Dwyer, L.M., Stewart, D.W., and Smith, D.L. (2002). Nitrogen effects on grain yield and yield components of leafy and non-leafy maize genotypes. *Crop Science* **42**, 1556–1563.

Coulter, J.A., Nafziger, E.D., Janssen, M.R., and Pedersen, P. (2010). Response of Bt and near Isoline Corn Hybrids to Planting Density. *Agronomy Journal* **102**, 103–111.

Debreczeni, K. (1999). Response of two maize hybrids to different fertilizer-N forms (NH_4-N and NO_3-N). *Communications in Soil Science and Plant Analysis* **31**, 2251–2264.

Dobermann, A. (2001). *Crop Potassium Nutrition: Implications for fertilizer recommendations.* Proceedings of the 31st North-Central Extension-Industry Soil fertility Conference. Potash and Phosphate Institute, Brookings South Dakota, pp. 1–12.

Dobermann, A., Arkebauer, T., Casman, K., Lindquist, J. Specht, D., Walters, L., and Yang, H. (2002). *Understanding and managing corn yield potential.* Proceedings of the Fertilizer Industry Round Table. Charleston, South Carolina. Fertilizer Industry Round Table, Forest Hill, Maryland, USA, pp. 260–272.

Dowswell, C.R., Paliwal, R.L., and Cantrell, R.P. (1996). *CIMMYT Mega-Environment database.* Maize in the Third World. Westview Press, Colorado, USA pp. 234–248.

Duvick, D.N. and Cassman, K.G. (1999). Post green revolution trends in yield potential of temperate maize in the North Central USA. *Crop Science* **39**, 1622–1630.

Egharbeva, P.N., Horrocks, R.D., and Zuber, M.S. (1976). Dry matter accumulation in maize in response to defoliation. *Agronomy Journal* **68**, 40–43.

Gallais, A. and Hirel, B. (2004). An approach to the genetics of nitrogen use efficiency in maize. *Journal of Experimental Botany* **55**, 295–306.

Gliessman, S.R. (1998). Agroecology: Ecological process in sustainable agriculture. Ann Arbor Press, Michigan, USA, p. 345.

Halvorson, A.D. and Johnson, J.F. (2009). Corn cob characteristics in irrigated central great plains studies. *Agronomy Journal* **101**, 390–399.

Hammer, G.L., Dong, Z., Mclean, G., Doherty, A., Messina, C., Schussler, J., Zinsclmeier, C., Paszkiewicz, S., and Cooper, M. (2009). Can changes in canopy and for root system architecture explain historical maize yield trend in the corn belt. *Crop Science* **49**, 299–312.

IASTATE, (2007a). How a Corn Plant Develops. http://www.extension.iastate.edu/hancock/info/illustrations.htm p. 1.

IASTATE (2007b). How Corn Plant Develops-V6 http://www.extension.iastate.edu/hancock/info/Corn+Develop+Stages.V6htm p. 1.

IASTATE (2007c). How Corn Plant Develops-V15. http://www.extension.iastate.edu/hancock/info/Corn+Develop+V15+Stage.htm p. 1.

IASTATE (2007d). How Corn Plant Develops-R1 Stage-Silking. http://www.extension.iastate.edu/hancock/info/Corn+Develop+R1+Silking.htm p. 1.

IASTATE (2007e). How a Corn Plant Develops-R2 Stage-blister. http://www.extension.iastate.edu/hancock/info/Corn+Develop+R2++Stage.htm p. 1.

IASTATE (2007f). How a Corn Plant Develops-R2 Stage-blister. http://www.extension.iastate.edu/hancock/info/Corn+Develop+R5+Dent+Stage.htm p. 1.

Ikisan, (2008). Maize Varieties. http://www.ikisan.com/links/ap_maizeVarieties.shtml pp. 1–10.

Johnston, A.M. and Dowbenko, R. (2009). Essential elements in Corn. www.farmwest.com/index.cfm?method=library.showPage&librarypageid=`124 pp. 1–7.

Joshi, V.N., Singh, N.N., and Sharma, G.S. (2004). Low N tolerance of maize (Zea mays) in India. Developing Drought and Low Nitrogen Tolerant Maize Symposium Abstracts. http://www.cimmyt.org/Research/Maize/DLNCA/htm/DLNCp1-23.htm pp. 1–3.

Kamara, A.Y., Ekeleme, F., Chikoye, D., and Omoigui, L.O. (2009). Planting date and cultivar effects on grain yield in dry land corn production. Agronomy Journal 101, 91–98.

Kamara, A.Y., Kling, J.G. Menkir, A., and Ibikunle, O. (2003). Agronomic performance of maize (Zea mays L.) breeding lines derived from a low nitrogen maize population. Journal of Agricultural Science 141, 221–230.

Kamara, A.Y., Menkir, A., Ajala, S.O., and Kureh, I. (2005). Nitrogen deficiency in the Northern Guinea Savanna of Nigeria. Experimental Agriculture 41, 199–212.

Kaul, J., Rakshit,S., Sain, D., Jat, M.L., Singh, R., Singh, S.B., Gupta, N.P., Sekhar, J. C., Singh, R.P., Yadav,V.K., Singh, K.P., Kumar, P., Sharma, O.P., Sekhar, M., and Singh, I. (2008). Maize Hybrid and Composites released in India-1961–2007. Directorate of Maize Research, Indian Council of Agricultural Research, New Delhi, pp. 1–12.

Kawano, K. (1990). Harvest index and evolution of major food crop cultivars in the tropics. Euphytica 46, 195–202.

Krishna, K.R. (1998). Phosphorus efficiency in Tropical and Semi-dryland crops. In Accomplishments and future challenges in dryland soil fertility research in the Mediterranean area. Ryan, J. (Ed.). International Center for Dry land Agricultural Research, Aleppo, Syria, pp. 343–363.

Krishna, K. (2002). Crop improvement towards tolerance to soil related constraints. In Soil fertility and Crop production. Krishna K.R. (Ed.) Science Publishers Inc. Enfield, New Hampshire, USA, pp. 337–370.

Krishna, K.R. (2010). Maize Agroecosystem: Nutrient Dynamics and Productivity. In Agroecosystems of South India: Nutrient Dynamics, Ecology and Productivity. Krishna, K. R. (Ed.). BrownWalker Press Inc. Boca Raton, Florida, USA, p. 553.

Kovacevic, V., Banaj, D., Antunovic, M., and Bukvic, A. (2002). Influence of genotypes and soil properties on crop potassium and magnesium status. Plant Nutrition 92, 90–91.

Kucharick, C.J. (2008). Contribution of Planting Date Trends to increased Maize yield in the Central United States. Agronomy Journal 100, 328–336.

Logrono, M.L. and Lothrop, J.E. (1998). Impact of Drought and Low Nitrogen on Maize production in South Asia. Developing Drought and Low-Nitrogen on Maize Symposium Abstracts. http://www.cimmyt.org/Research/Maize/DLNCA/htm/DLNCo-5.htm pp. 1–3.

Machado, C.T. and Furlani, A.M.C. (2004). Root phosphatase activity, plant growth and phosphorus accumulation of maize genotypes. Scientia Agricola 61, 243–244.

Minjian, C., Haiqiu, Y., Hongkui, Y., and Chunji, J. (2007). Difference in tolerance to potassium deficiency between two maize inbred lines. *Plant Production Science* **10**, 42–46.

Monneveaux, P., Zaidi, P.H., and Sanchez, C. (2005). Population density and low nitrogen affects yield associated traits in tropical maize. *Crop Science* **45**, 535 –554.

Mureithi, J.G. (2008). Maize Varieties, Soil fertility improvement and Appropriate Agronomic Practices. Kenya Agricultural Research Institute, Kitale, Kenya http://www.kari.org/Legume_Project/legume_Leaflets/MaizeVarieties020506.pdf pp. 1–12.

Ogunlela, V.B., Amoruwa, G.M., and Ologunde, O.O. (1988). Growth, Yield components and micronutrient nutrition of field grown maize (*Zea mays*) as affected by Nitrogen fertilization and plant density. *Nutrient Cycling in Agroecosystems* **67**, 1385–1414.

Oikeh, S.O., Carsky, R.J., Kling, J.G., Chude,V.O., and Horst, W.J. (2003). Differential N uptake by maize cultivars and soil N dynamics under N fertilization in West Africa. *Agriculture, Ecosystems and Environment* **100**, 181–191.

Otegui, M.E., Nicolini, M.G., Ruiz, R.A., and Dodds, P.A. (1995). Sowing date effects on grain yield components for different Maize genotypes. *Agronomy Journal* **87**, 29–33.

Paponov, I.A., Sambo, P., Erly, S.G., Presterl, T., Geiger, H. H., and Engels, S.C. (2005). Kernel set in maize genotypes in nitrogen use efficiency in response to resource availability around flowering. *Plant and Soil* **272**, 101–110.

Parentoni, S. N. and Lopes de Souza, C. (2008). Phosphorus acquisition and internal utilization efficiency in Tropical Maize genotypes. http://www.scielo.br/pdf/pab/v43n7/14.pdf pp. 1–13.

Ping, J.L., Ferguson, R.B., and Dobermann, A. (2008). Site-specific nitrogen and planting density management in irrigated maize. *Agronomy Journal* **100**, 1193–1204.

Presteri, T.G., Seitz, M., Landbeck, E.M., Thiemt, W., Schmidt, W., and Geiger, H.H. (2003). Improving N-use efficiency in European maize: Estimation of quantitative genetic parameters. *Crop Science* **43**, 1259–1265.

Racz, F., Illes, O., Pok, I., Szoke, C., and Zsubori, Z. (2003). Role of sowing time in Maize Production. www.date.hu/acta-agraria/2003-11i/racz.pdf pp. 1–6.

Sangoi, L. and Salvador, R. (1997). Dry matter production and partitioning of maize hybrids and dwarf ones at four plant populations. *Ciencia Rural* **27**, 1–11.

Sangoi, L., Ender, M., Guidolin, A.F., Almeida, M.L., and Konflanze, A. (2001). Nitrogen fertilization impact on agronomic traits of maize hybrids released at different decades. *Pesquisa Agropecuaria Brasileira* **36**, 1–14.

Simons, S., Tan, D., Belfield, S., and Martin, B. (2008). Plant population to Improve yield of Dry land Maize in northwest New South Wales. http://www.regional.orgau/au/asa/2008/poster/agronomy_landscape/5596_tan.htm pp. 1–6.

Singh, U. and Wilkins, P.W. (2001). Simulating Water and Nutrient Stress effects on phonological developments in Maize. http://www.cimmyt.org/english/docs/proceedings/gis01/simulating/simulating_singh.htm pp. 1–5.

Soliman, M.S.M. (2006). Stability and environmental interactions of some promising yellow maize genotypes. *Research Journal of Agriculture and Biological Sciences* **2**, 249–255.

Subedi, K.D. and Ma, B.L. (2007). Dry matter and Nitrogen partitioning patterns in BT and Non-Bt Near isoline maize hybrids. *Crop Science* **47**, 1186–1192.

Subedi, K.D., Ma, B.L., and Smith, D.L. (2006). Response of a Leafy and Non-leafy maize hybrid to population Densities and Fertilizer Nitrogen levels. *Crop Science* **46**, 1860–1869.

Ullah, A., Ashraf Bhatti, M., Gurmani, Z.A., and Imran, M. (2007). Studies on planting patterns of maize (*Zea mays*) facilitating legume intercropping. *Journal of Agricultural Research* **45**, 1–8.

Worku, M., Banziger, M., Erley, S.S., Freisen, D., Diallo, A. O., and Horst, W.J. (2007). Nitrogen uptake and utilization in contrasting nitrogen efficient tropical maize hybrids. *Crop Science* **47**, 519–528.

Worku, M. and Zelleke, H. (2007). Advances in improving harvest index and grain yield of maize in Ethiopia. *East African Journal of Sciences* **1**, 112–119.

8

Abunyewa, A., Asiedu, E.K., Nyameku, A.L., and Cobbina, J. (2004). Alley cropping *Gliricidia sepium* with maize: 1. The effect of hedgerow spacing, pruning height, and phosphorus application on maize yield. *Journal of Biological Sciences* **4**, 81–86.

Acciaresi, H.A. and Chidichimo, H.O. (2002). Spatial pattern effect on corn (*Zea mays*) weeds competition in the humid Pampas of Argentina. *International Journal of Pest Management* **53**, 195–206.

Akonde, T.P., Leihner, D.E., Kuhne, F., and Steinmuller, N. (1996). Alley cropping on an Ultisol in sub humid Benin. Part 3 Nutrient budget of maize cassava and trees. *Agroforestry Systems* **37**, 213–226.

Alexieva, S. and Stoimenova, I. (1998). A model concerning to the yield loss of maize from weed density or dry biomass. http://www.toprak.org.tr/isd/isd_82.htm pp. 125–138.

Allen, J.R. and Obura, R.K. (1983). Yield of corn, cowpea and soybean under different intercropping system. *American Society of Agronomy* **75**, 1005–1009.

Alvarez, R. and Grigera, S. (2009). Analysis of soil fertility and management effects on yields of wheat and corn in the Rolling Pampas of Argentina. *Journal of Agronomy and Crop Science* **191**, 321–329.

Attanandana, T., Yost, G., and Verapattananirund, P. (2004). Adapting site-specific nutrient management to small farms of the tropics. In *Proceedings of Conference on Precision Agriculture*. Minneapolis, Minnesota, USA, pp. 1–3.

Attanandana, T. and Yost, R.S. (2003). A site-specific nutrient management approach for maize. *Better Crops International* **17**, 3–7.

Babalola, O., Jimba, S.C., Maduakolam, O., and Dada, O.A. (2009). Use of vetiver grass for soil and water conservation in Nigeria. http://www.vetiver.org/NIG_SWC.pdf pp. 1–10.

Bacur, D., Jitareanu, G., Ailincai, C., Tsadilas, C., Despinas, A., and Mercus, A. (2007). Influence of soil erosion on water, soil, humus and nutrient losses in different systems in the Maldovian Plateau, Romania. *Journal of Food, Agriculture and Environment* **5**, 261–264.

Bahr, A.A., Zeidan, M.S., and Hozayan, M. (2006). Yield and quality of maize (*Zea mays L.*) as affected by slow-release nitrogen in newly reclaimed sandy soil. *American-Eurasian Journal of Agriculture and Environment* **1**, 239–242.

Banda, A.Z., Maghembe, J.A., Ngugi, D.N., and Chome, V.A. (1994). Effect of intercropping maize and closely spaced Leucana hedgerows on soil conservation and maize yield on a steep slope at Ntechu, Malawi. *Agroforestry Systems* **54**, 17–22.

Bandyopadhyay, P.K. and Mallick, S. (1996). Irrigation requirements of winter maize under shallow water table conditions in damodar valley irrigation command area. *Journal of Indian Society of Soil Science* **44**, 616–620.

Barton, A., Fullen, M.A., Mitchell, D.J., Hocking, T.J., Liu, L., Wu, B.Z., Zheng, Y., and Xia, Z.Y. (1998). Soil conservation measures on maize runoff plots on Ultisols in Yunan province, China. http://natres.psu.ac.th/Link/Soil-Congress/bddd/symp31/27-t.pdf p. 107.

Basamba, T.A., Barrios, E., Amezquita, E., Rao, I.M., and Sing B.R. (2006). Tillage effects on maize yield in Columbian savanna Oxisol: Soil organic matter and P fractions. *Soil and Tillage Research* **91**, 131–142.

Binder, J., Graeff, S., Link, J., Claupein, W., Liu, M., Dal, M., and Wang, P. (2008). Model-based approach to quantify production potentials of summer maize and spring maize in the North China Plain. *Agronomy Journal* **100**, 862–873.

Binh, D.T. (2007). Site-specific nutrient management for maize in Vietnam (Viet-Corn)—2007. International Plant Nutrition Institute, Norcross, Georgia, USA. http://www.ppi-far.org/far/farguide.nsf/$webindex/article=6AEE7B53482574200065D9C572.htm pp. 1–3.

Bouquet, D.J., Tubana, B.S., Mascagni, H.J., Holman, M., and Hague, S. (2009). Cotton yield responses to fertilizer Nitrogen rates in a cotton-corn rotation. *Agronomy Journal* **101**, 400–407.

Brye, K.R., Norman, J.M., Bundy, L.G., and Gower, S.T. (2000). Water budget evaluation of prairie and maize ecosystem. *Soil Science Society of America Journal* **64**, 715–724.

Bunemann, E.K., Smithson, P.C., Jama, B., Frossard, E., and Oberson, A. (2004). Maize productivity and nutrient in maize-fallow rotations in Western Kenya. *Plant and Soil* **264**, 195–208.

Callow, M. and Kenman, S. (2006). Water use efficiency of forages on subtropical dairy farms. http://www.dairyinfo.biz/images/content/M5/204WUE_of_Different_Forages.pdf p. 109.

Caracova, J., Maddonni, G.A., and Ghersa, C.M. (2000) Long term cropping effects on Maize: Crop evapo-transpiration and grain yield. *Agronomy Journal* **92**, 1256–1265.

Carruthers, K., Prithviraj, B., Fe, Q., Cloutier, D., Martin, R.C., and Smith, D.L. (2000). Intercropping corn with soybean, lupin and forages: Yield component responses. *European Journal of Agronomy* **12**, 103–115.

Carsky, R.J., Abaidoo, R., Dasheiell, K., and Sanginga, N. (1997). Effect of soybean on subsequent maize grain yield in the guinea savanna zone of West Africa. *African Crop Science Journal* **5**, 31–38.

Castrignano, A., Katerji, N., Karam, F., Mastrorilli M., and Hamdy, A. (1998). A modified version of CERES-Maize model for predicting crop response to salinity stress. http://ressources.cilheam.org/om/pdf/b36/05002177.pdf pp. 1–30.

Cavero, S., Zaragosa, F.N. Suso, P., and Pardo, S.P. (2002). Competition between maize and *Datura stramonium* in an irrigated field under semiarid conditions. *Weed Research* **39**, 225–240.

CGIAR, (2007). Study warns that China and India's planned biofuel boost could worsen water scarcity, compete with food production. Consultative Group on International Agricultural Research, Rome, Italy. http://iwmi.cgiar.org/news_room/Press_releases/releases/2007/iwmi_Biofuel_%20release.pdf pp. 1–5.

Chintu, R., Mafongoya, P.L., Chirwa, T.S., Mwale, M., and Matibini, J. (2003). Coppicing tree legume fallows in Eastern Zambia. http://journals.cambridge.org/acion/displayAbstract?fromPage=online&aid=229265 pp. 1–3.

Chirwa, T.S., Mafogaoya, P.L., Mbewe, D.N.M., and Chishala, B.H. (2004). Changes in soil properties and their effects on maize productivity following *Sesbania sesban* and *Cajanus cajan* improved fallow systems in Eastern Zambia. *Biology and Fertility of Soils* **40**, 20–27.

Cimmyt (2002). Exploiting interactions of soil fertility technologies with other inputs and management. http://www.cimmyt.org/Research/nrg/sfmr/htm/SFMRexploiting.htm pp. 1–2.

Clay, D.E., Kim, K.I., Chang, J., Clay, S.L., and Dalsted, K. (2006). Characterizing water and nitrogen stress in corn using remote sensing. *Agronomy Journal* **98**, 579–587.

Clay, J. (2008). Agriculture and environment: Corn. In *World Agriculture and Environment*. Island Press, USA, p. 346.

Comia, R.A. (1999). Soil and nutrient conservation oriented practices in the Philippines. http://www.agnet.org/library/eb/472a/.htm pp. 1–13.

Conley, S.P. and Christmas, E.P. (2005) Utilizing inoculants in a corn-soybean rotation. http://www.extension.purdue.edu/extmedia/SPS/SPC-100-W.pdf pp. 1–6.

CRIDA, (2008). Efficient water use: Policy for promotion of irrigated dry land crops. http://www.nrsp.org/database/documents/2265.pdf pp. 1–3.

Da Silva, P.R.G., Streider, M.L., Coser, R.B., Sangoi, L.L., argental, G., Forsthofer, E.L., and Da Silva, A.A. (2005). Grain yield and kernel crude protein content increases of maize hybrids with late nitrogen side dressing. *Scientia Agricola* **62**, 1–12.

DeJonge, K. and Kaleita, A. (2006). Simulation of Spatially Variable Precision irrigation and its Effects on Corn Growth using CERES-Maize. http://asae.frymulti.com/abstract.asp?aid=20710&t=2 pp. 1–3.

Delgrosso, S.J., Halvorson, A.D., and Parton, W.J. (2008). Testing DAYCENT model simulations of corn yield and nitrous oxide emissions in irrigated tillage systems in Colorado. *Journal of Environmental Quality* **37**, 1383–1389.

Dibb, D.W. (2000). The mysteries (myths) of nutrient use efficiency. *Better Crops* **84**, 3–5.

Dilshad, M., Motha, J.A., and Pool, L.J. (1996). Surface runoff, soil and nutrient losses from farming systems in the Australian semi-arid tropics. *Australian Journal of Experimental Agriculture* **36**, 1003–1012.

Dilz, K., Postmus, J., and Prins, W.H. (1990). Residual effect of long term applications of farm yard manure to silage maize. *Nutrient Cycling in Agroecosystems* **26**, 1385–1414.

Dhugga, K. (2007). Maize biomass yield and composition for biofuels. *Crop Science* **47**, 2211–2227.

Duvick, D.N. and Cassman, K.G. (1999). Post Green Revolution trends in yield potential of temperate maize in the north central USA. *Crop Science* **39**, 1622–1630.

Egli, D.B. (2008). Comparison of corn and soybean yields in the USA—Historical trends and future prospects. *Agronomy Journal* **100**, S78–79.

Eva, L., Nagy, P., Lencse, T., Toth, V., and Kism, A. (2009). Investigation of the damage caused by weeds competing with maize for nutrients. *Communications in Soil Science and Plant Analysis* **40**, 879–888.

Fan, T., Stewart, B.A., Payne, W.A., Yong, W., Luo, J., and Gao, Y. (2005). Long term fertilizer and water availability effects on cereal yield and soil chemical properties in Northwest China. *Soil Science Society of America Journal* **69**, 842–855.

Fanadzo, L., Mahingaidze, Z.A.B., and Nyakambe, Z.C. (2007). Narrow rows and high maize densities decrease maize grain yield but suppress weeds under dry land conditions in Zimbabwe. *Journal of Agronomy* **6**, 566–570.

Fang, Q., Chen, Y., Yu, Q., Ouyang, Z., Li, Q., and Yu, S. (2007). Much improved irrigation use efficiency in an intensive wheat-maize double cropping system in the North China Plain. *Journal of Integrative Plant Biology* **49**, 1517–1526.

FAO, (2008). NUTMON-Nutrient Monitoring for Tropical Farming. Food and Agricultural Organization, Rome, Italy. http://ww.fao.org/docrep/006/y5066e/y5066e08.htm pp. 1–30.

Ferguson, R.B., Hergert, G.W.M., Scheepers, J.S., Gotway, C.A., Cahoon, J.E., and Peterson, T.A. (2002). Site-specific nitrogen management of irrigated maize: Yield and soil residual nitrate effects. *Soil Science Society of America Journal* **66**, 544–553.

Frank, A.C., Oyewole, B.D., Abaidoo, B.D., Chickoye, R., and Schulze, S. (2005). Conservation tillage reduces weed pressure and labor demands in maize-based systems in the derived savanna of Nigeria. http://www.act.org.zw/postcongress/documents/sess5 (technChallenges)/ Franke20%20 et%20al.doc.htm pp. 1–4.

Fujiyoshi, P.T., Gliessman, S.R., and Langenheim, J. H. (2007). Factors in the suppression of weeds by squash interplanted in corn. *Weed Biology and Management* **7**, 105–114.

Gaiser, T., De Barros, I., Lange, F.M., and Williams, J.R. (2004). Water use efficiency of a maize/cowpea intercrop on a highly acidic tropical soil as affected by liming and fertilizer application. *Plant and Soil* **263**, 165–171.

Garcia, R.A., Calonego, J.O., Crusciol, C.A.C., and Rosolem, C.A. (2006). Distribution of nonexchangeable k in soil profile in a brach aria-maize cropping system. *18th World Congress on Soil Science*. Philadelphia, PA, USA. http://a-c-s.confex.com/crops/wc2006/wctechprogram/ P16603.HTM pp. 1–2.

Haile-Mariam, S., Collins, H.P., and Higgins, S.S. (2008). Green house gas fluxes from an irrigated sweet corn-potato rotation. *Journal of Environmental Quality* **37**, 759–771.

Halvorson, A.D., Follet, R.F., Bartolo, M.E., and Schwessing, F.C. (2002) Nitrogen fertilizer use efficiency of furrow-irrigated onion and corn. *Agronomy Journal* **94**, 442–449.

Halvorson, A.D., Mosier, A.R., Reule, C.A., and Bausch, W.C. (2006). Nitrogen and tillage effects on irrigated continuous corn yield. *Agronomy Journal* **98**, 63–71.

Hao, X. and Chang, C. (2006). Residual effect of Long term cattle manure application on soil nitrogen and phosphorus. *18th World Congress of Soil Science*. Philadelphia, Pennsylvania, USA. http://a-c-s.confex.com/crops/wc2006/techprogram/P15910.HTM pp. 1–3.

Harawa, R., Lehman, J., Akinnifesi, F., Fernandes, E.E., and Kanyama-Phiri, G. (2006). Nitrogen dynamics in maize based agroforestry systems as affected by landscape position in southern Malawi. *Nutrient Cycling in Agroeosystems* **75**, 271–284.

Hauser, S., Norgrove, L., and Nkem, J.N. (2006). Groundnut/maize/cassava intercrop yield response to fallow age, cropping frequency and crop plant density on an Ultisol. In Southern Cameroon. *Biological Agriculture and Horticulture* **24**, 275–292.

Heggenstaller, A.H., Amex, R.P., Liebman, M., Sundberg, D.N., and Gibson, L.R. (2008). Productivity and nutrient dynamics in bioenergy double cropping systems. *Agronomy Journal* **100**, 1740–1748.

Hendrickson, J.A. and Han, J. (2000). A reactive nitrogen management system. In *Precision Agriculture*. Robert, P.C. (Ed.). Proceedings of 5th

International Conference on Precision Farming. Minneapolis, Minnesota, USA, p. 276.

Hesterman, O.B., Russele, M.P., Sheafer, C.C., and Heichel, G.H. (1987). Nitrogen utilization from fertilizer and legume residues in legume-corn rotations. *Agronomy Journal* **79**, 726–731.

Hoffmann, S. and Nyoky, T.K. (2001). Soil fertility in a long-term fertilizer trial with different tillage systems. *Archives of Agronomy and Soil Science* **46**, 239–250.

Holden, J. (2005). Corn's benefits in cotton rotation. http://www.autralia.pioner.com/MediaRelease/Cornsbnefitticotton/tabid/163/Default.aspx pp. 1–3.

Hons, F.M. (2005). Rotation, tillage and nitrogen rate effects on cotton growth and yield. In *Proceedings of Conservation Tillage Conference-2005*. Houston, Texas, Cary, North Carolina USA, pp. 22–28.

Hoogenboom, M.L., Ahuja, G., Ascough, J.C., and Saseendran, S.A. (2006). Evaluation of the RZWQM-CERES-Maize hybrid model for maize production. *Agricultural Systems* **87**, 124–129.

Horst, W.J. (2000). Fitting maize into sustainable cropping on acid soils of the tropics of South America. http://www.ipe.uni-hannover.de/de/forschung/zyclen/horst_3.pdf pp. 1–7.

Hur, S., Kim, W.T., Jung, K.H., and Ha, S.K. (2004). Study on soil and nutrients loss with soil textures and two crops during rainfall. *Proceedings of International Crop Science Conference*. http://www.cropscience.org.au/icsc2004/poster/1/6/1298_huros.htm pp. 1–4.

ICRAF, (2008). Agroforestry for soil fertility improvement in Southern Africa. http://www.icrafsa.org/innovations/fertility.html pp. 1–3.

Ikerra, S.T., Maghembe, J.A., Smithson, P.C., and Buresh, R.J. (1999). Soil nitrogen dynamics and relationships with maize yields in a Gliricidia-maize intercrop. *Plant and Soil* **211**, 23–33.

Ikisan (2007). Maize: Crop technologies. http://www.ikisan.com/links/ap_maizeCrop Technologies.shtml pp. 1–10.

Jagadamma, S., Lal, R., and Rimal, B.K. (2009). Effects of topsoil depth and soil amendments on corn yield and properties of two Alfisols in Central Ohio. http://www.jswconline.org/content/64/1/70.refs p. 105.

Jensen, J.R., Bernhard, R.H., Hansen, S., McDonagh, J., Meberg, J.P., Nielsen, N.E., and Nordbo, E. (2003). Productivity in maize based cropping systems under various soil-water-nutrient management strategies in a semi-arid, Alfisol environment in East Africa. *Agricultural Water Management* **50**, 217–237.

Johnston, M., Foley, J.A., Holloway, T., Kucharik, C., and Monfreda, C. (2009). Resetting global expectations from agricultural bio-fuels. *Environmental Research Letters* **4**, 1–15.

Kang, B.T. (1997). Alley cropping-soil productivity and nutrient recycling. *Forest Ecology and Management* **91**, 75–82.

Kanthaliya, P.C. and Verma, A. (2006). Continuous application of fertilizer and manure on crop yield and uptake of nutrients in maize-wheat cropping system. http://a-c-s.confex.com/crops/wc2006/techprogram/P14990.HTM pp. 1–2.

Karpenstein-Machan, M. (2001). Sustainable cultivation concepts for domestic energy production and biomass. *Critical Reviews in Plant Science* **20**, 1–14.

Kartaatmadja, S. (2007). *Site-specific Nutrient Management for Maize in Indonesia: Interpretive Summaries*. International Plant Nutrition Institute. Norcross, Georgia, USA, pp. 1–3.

Katalin, B., Kismanyoky, T., and Katalin, D. (2005). Effect of organic matter recycling in long term fertilization trials and model pot experiments. *Communications in Soil Science and Plant Analysis* **36**, 191–202.

Khuong, T.Q., Tan, P.S., and Witt, C. (2008). Improving of maize yield and profitability through Site-Specific Nutrient Management (SSNM) and planting density. *Omonrice* **16**, 82–92.

Kim, K., Clay, D.E., Carlson, C.G., Clay, S.A., and Troolen, T. (2008). Do synergistic relationships between nitrogen and water influence the ability of corn to use nitrogen derived from fertilizer and soil. *Agronomy Journal* **100**, 551–556.

Kiniry, J., Williams, J.R., Vanderlip, R.L., Atwood, J.A., Reicosky, D.C., Mulliken, J., Cox, W.J., Macagani, H.J., Hollinger, S.E., and Wiebold, W.J. (1997). Evaluation of two maize models for Nine US locations. *Agronomy Journal* **89**, 421–426.

Kirchoff, G. and Salako, F.K. (2006). Residual tillage and bush fallow effects on soil properties and maize intercropped wit legumes on a tropical soil. *Soil Use and Management* **16**, 183–188.

Kitchen, N.R. Sudduth, K.A., Drummond, S.T., Schraf, P.C., Palm, C.A., Roberts, D.F., and Vories, E.D. (2010). Ground-based canopy reflectance sensing for variable rate-nitrogen corn fertilization. *Agronomy Journal* **102**, 71–84.

Krishna, K.R. (2010). *Agroecosystems of South India: Nutrient Dynamics, Ecology and Productivity.* BrownWalker Press Pres Inc. Boca Raton, Florida, USA, p. 543.

Kucharik,C., Brye, K., and Norman, J. (2008). Measurements and modeling of carbon and nitrogen cycling and crop yield in agroecosystems of Southern Wisconsin. http://sage.wisc.edu/indepth/kuchraik/CNcycling/CNcycling.html pp. 1–5.

Lal, R. (2009). Soil quality impacts of residue removal for Bioethanol production. *Soil and Tillage Research* **102**, 233–241.

Lambert, D.M., DeBoer, J.L., and Malzer, G.L. (2006). Economic analysis of spatial-temporal patterns in corn and soybean response to nitrogen and phosphorus. *Agronomy Journal* **98**, 43–54.

Levein, R., Cogo, N.P., and Rockenbach, C.A. (2006). Soil erosion and maize cropping under different prior cropping systems and methods of tillage. http://www.caabstractplus.org/absracts/Abstract.aspx?AcNo=1991195149 pp. 1–3.

Li, K., Shiraiwa, T., Inamura, T., Saitoh, K., and Horie, T. (2001). Recovery of 15N labeled ammonium by barley and maize grown on the soils with long term application of chemical and organic fertilizers. *Plant Production Science* **4**, 29–35.

Li, L. and Zhang, F. (2006). Physiological mechanism on interspecific facilitation for N, P and Fe utilization in intercropping systems. *18th World Congress of Soil Science*, Philadelphia, Pennsylvania, USA. http://a-c-s.confex.com/crops/wc2006/techprogram/P11610.HTM pp. 1–2.

Lopez, F.X., Boote, K.J., Pineiro, J., and Sau, F. (2008). Improving the CERES-Maize model ability to simulate water deficit impact on maize production and yield components. *Agronomy Journal* **100**, 296–307.

McDonald, H. and Riha, A. (2002). Model of crop; weed competition applied to maize: *Abutilon theohrasti* interactions. 1. Model description and evaluation. *Weed Research* **39**, 355–369.

Miao, Y., Mulla, D.J., Batchelor, W.D., Paz, J.O., Robert, P.C., and Wiebers, M. (2006a). Evaluating 'Management Zone' optimal nitrogen rates with a crop growth model. *Agronomy Journal* **98**, 545–553.

Miao, Y., Mulla, D.J., Hernandez, J.A., Weibers, M., and Robert, P.C. (2007). Potential impact of precision of nitrogen management on corn yield, protein content and test weight. *Soil Science Society of America Journal* **71**, 1490–1499.

Miao, Y., Mulla, D.J., Robert, P.C., and Hernandez, J.A. (2006b). Within-field variations in corn yield and grain quality responses to nitrogen fertilization and hybrid selection. *Agronomy Journal* **98**, 129–140.

Miranda, F.S., Eckersten, H., and Maria, W. (2008). Net N mineralization of an andosol influenced by chicken and cow manure applications on maize-bean rotations in Nicaragua. *Scientific Research and Essays* **3**, 280–286.

Mize, C., Egeh, M., and Batchelor, W. (2005). Predicting maize and soybean production in a sheltered field in the Corn Belt region of North Central USA. *Agroforestry Systems* **64**, 107–116.

Mobius-Clune, B.N., Van Es, H.M., Idowo, O.J., Schindelbeck, R.R., Mobius-Clune, D.J., Wolfe, D.W., Abawi, G.S., Theis, J.E., Gugino, B.K., and Lucey, R. (2008). Long term effects of harvesting maize stover and tillage on soil quality. *Soil Science Society of America Journal* **72**, 960–969.

Morari, F., Nardi, S., Berti, A., Lugato, E., Carletti, P., and Giardini, L. (2006). Soil organic matter quality after 40 years of different organic and mineral fertilization in three soils. *Proceedings of World Congress of Soil Science*, Philadelphia, Pennsylvania, USA. http://www.idd.go.th/18wcss/techprogram/p16643.HTM pp. 1–3.

Motavalli, P., Nelson, K., Kitchen, N., Anderson, S., and Scharf, P. (2008). Variable source N fertilizer application to optimize crop N use efficiency. http://aes.missouri.edu/pfcs/research/prop405.pdf pp. 1–19.

Mushagalusa, G.N., Ledent, J.F., and Draye, X. (2008). Shoot and root competition in potato/maize intercropping: Effects on growth and yield. *Environmental and Experimental Botany* **64**, 180–188.

Myaka, F.M., Sakala, W.D., Adu-Gyanfi, J.J., Kamalonga, D., Ngiwira, A., Odgaard, R., Nielsen, N.E., and High-Jensen, H. (2006). Yields and accumulations of N and P in farmer-managed intercrops of maize-pigeonpea in semi-arid Africa. http://orgprints.org/9306/ pp. 1–2.

NCSU (2009). *Soil Facts: Starter Fertilizers for Corn Production.* North Carolina State University, Raleigh, NC, USA. http://www.soil.ncsu.edu/publictions/Soilfacts/AG-439-29/ pp. 1–8.

Negrila, M., Negrila, E., and Stan, S. (2007). Maize grain yield evolution and nitrogen and phosphorus consumption in long term fertilizer experiments under non-irrigated conditions. Probleme de Agrofitotechnie Teoretica si Applicata. http://www.cababstractsplus.org/abstracts/Abstract.aspx?AcNo=20000705519 pp. 1–3.

Nelson, R., Dimes, J.P., Silburn, D.M. Paningbaen, E.P., and Cramb, R.A. (1998). Erosion/Productivity modeling of maize farming in the Philippines uplands. Simulation of alternative farming methods. *Agricultural Systems* **58**, 147–183.

Niaz, A., Ibrahim, M., and Ishaq, M. (2007). Assessment of nitrate leaching in wheat maize cropping system: A lysimeter study. *Pakistan Journal of Water Resources* **7**, 1–8.

Nzabi, A.W., Masinde, A., Gesare, M., Ngoti, B., and Mwagi, G. (2000). Soil erosion control using exotic grasses and locally available materials in Nymonyo and Kamingusa villages of Southwest Kenya. http://www.kari.org/Legume_Project/Legume2Conf_2000/30.pdf pp. 1–6.

Nziguheba, G., Tossah, B.K., Diels, J., Franke, A.C., Ajhou, K., Iwaufor, E.N.O., Nwoke, C., and Merckx, R. (2008). Assessment of nutrient deficiencies in maize in nutrient omission trials and long term field experiments in the West African Savanna. *Plant and Soil* **314**, 143–157.

Odunze, A.C., Dim, L.A., and Heng, L.K. (2008). Water supply and Rain-fed maize production in a Semi-arid zone Alfisol of Nigeria. http://tucson.ars.ag.gov/isco/isco15/pdf/Odunze%20AC_Water%20supply%20and%20rain-fed%20maize.pdf pp. 1–6.

Ofori, E. and Kyei-Baffour, K. (2008). Agrometeorology and maize production. http://wmo.int/

pages/prog/wcp/agm/gamp/documents/chap13c-draft.pdf pp. 25–32.

Oikeh, S.O., Chude, V.O., Carsky, R.J., Weber, G.K., and Horst, W.J. (1998). Legume rotation in the moist tropical savanna: Managing soil nitrogen dynamics and cereal yields in farmer's fields. *Experimental Agriculture* **34**, 73–83.

Okogun, J.A., Sanginga, N., and Abaidoo, C. (2007). Evaluation of maize yield in an On-Farm maize-soybean and maize-lablab crop rotation systems in the Northern Guinea Savanna of Nigeria. *Pakistan Journal of Biological Sciences* **10**, 3905–3999.

O'Neal, M.R., Nearing, M.A., Southworth, J., and Pfeiffer, R.A. (2005). Climate change impacts on soil erosion in Midwest United States with changes in crop management. *Catena* **61**, 165–184.

Osmond, D.L., Lathwell, D.J., and Riha, S.J. (1992). Prediction of long term fertilizer nitrogen requirements of maize in the tropics using a nitrogen balance model. *Plant and Soil* **143**, 61–70.

Ouattara, K. (2007). *Improved Soil and Water Conservatory Managements for Cotton—Maize Rotation System in the Western Cotton Area of Burkina Faso.* Swedish Agricultural University, Umea, Sweden, pp. 1–70.

Pansak, W., Hilger, T.H., Dercon, R., Kongkaew, T., and Cadish, G. (2008). Changes in the relationship between soil erosion and N loss pathways after establishing soil conservation systems in uplands of northeast Thailand. *Agriculture, Ecosystems and Environment* **128**, 167–176.

Pasuquin, J.M., Timsina, J., Witty, C., Buresh, R., Dobermann, A., and Dixon, J. (2007). *The Expansion of Rice-maize Systems in Asia: Anticipated Impact of Fertilizer Demand.* International Plant Nutrition Institute, Singapore. http://www.fertilizer.org/ifacontent/download/7320/115492/version/1/file/2007_crossroads_pasuquin.pdf pp. 1–10.

Pasuquin, J.M.C.A. and Witt, C. (2007). Yield Potential and Yield Gaps of Maize in Southeast Asia. http://www.ipipotash.org/e-ifc/2007-14/research3.php pp. 1–3.

Peinetti, H.R., Menezes, R.S.C., Tiessen, H., and Perez Marin, A.M. (2008). Simulating plant productivity under different organic fertilization practices in a maize/native pasture rotation system in semi-arid Northeast Brazil. *Computers and Electronics in Agriculture* **62**, 204–222.

Pettigrew, W.T., Meredith, W.R., Bruns, H.A., and Stetina, S.R. (2006). Effects of a short-term corn rotation on cotton dry matter partitioning lint yield and fiber quality production. *Journal of Cotton Science* **10**, 244–251.

Pikul, J.L., Hamack, L., and Riedell, W.E. (2005). Corn yield, nitrogen use and corn root worm infestation of rotations in the Northern Belt. *Agronomy Journal* **97**, 854–863.

Pilbeam, C.J., Gregory, P.J., and Tripathi, B.P. (2002). Fate of 15 nitrogen labeled fertilizer applied to maize-millet cropping systems in the mid hills of Nepal. *Biology and Fertility of Soils* **35**, 27–34.

Ping, J.l., Ferguson, R.B., and Dobermann, A. (2008). Site specific nitrogen and plant density management irrigated maize. *Agronomy Journal* **100**, 1193–1204.

Polthanee, A. and Trelo-ges, V. (2003). Growth, yield and land use efficiency of corn and legumes grown under intercropping systems. *Plant Production Science* **6**, 139–146.

Pratap, N., Singh, R., Sindhwal, N., and Joshie, P. (1998). Agroforestry for soil and water conservation in the Western Himalayan valley region of India: Runoff, soil and nutrient losses. *Agroforestry Systems* **39**, 175–189.

Prins, U., Van Eekeren, N., and Oomen, G.J.M (2004). Bi-cropping fodder maize in an existing (grass) clover sward. http://www.lousbolk.org/downloads/1762.pdf pp. 1–3.

Raczkowski, C.W., Reddy, G.B., Reyes, M.R., and Baldwin, K.R. (2006). Effectiveness of no tillage in reducing runoff and erosion in a piedmont location. *Proceedings of World Congress of Soil Science*. Philadelphia, Pennsylvania, USA. http://a-c-s.confex.com/crops/wc2006/techprogram/P144481.HTM pp. 1–3.

Rajcan, I. and Swanton, C.J. (2001). Understanding maize-weed competition: Resource competition, light quality and the whole plant. *Field Crops Research* **71**, 139–150.

Ransom, J.K. and Odhiambo, G.D. (1993). Long term effects of fertility and hand weeding on striga in maize. In *Proceedings of Biology and Management of Orobanche*. International Crops

Research Institute for the semi-arid tropics. p. 271.

Reddy, A.K., Reddy, R.K., and Reddy, D. (1980). Effects of intercropping on yield and returns in corn and sorghum. *Experimental Agriculture* **16**, 179–184.

Reddy, K.N., Locke, M.A., Koger, C.H., Zablotowicz, R.M., and Krutz, L.J. (2006). Cotton and Corn rotation under reduced tillage management: Impacts of soil properties, weed control yield and net return. *Weed Science* **54**, 768–774.

Redersma, S., Lusiana, B., and Van Noordwijsk, M. (2005). Simulation of soil drying induced phosphorus deficiency and phosphorus mobilization as determinants of maize growth near tree lines on a Ferralsol. *Field Crops Research* **91**, 171–184.

Reid, J. (2006). Land Management for Grain Maize: Recommended Best Management Practices for New Zealand. http://www.crop.cri.nz/home/products-services/crop-prodcution/rbmp/31-Maize.pdf pp. 1–58.

Risch, S.J. and Hansen, M.K. (1982). Plant growth, flowering phenologies and yield of corn, beans and squash grown in pure stands and mixtures in Costa Rica. *Journal of Applied Ecology* **19**, 901–916.

Roberts, T.L. (2008). Improving nutrient use efficiency. *Turkish Journal for Agriculture* **32**, 177–182.

Rossiter, D.G. and Riha, S.J. (1999). Modeling plant competition with the GAPS object oriented dynamics simulation model. *Agronomy Journal* **91**, 773–783.

Rozas, H.S., Calvino, P.A., Echeverria, H.E., Barbieri, P.A., and Redolatti, N. (2008). Contribution of anaerobically mineralized nitrogen to the reliability of planting or pre-side dress soil nitrogen test in maize. *Agronomy Journal* **100**, 1020–1025.

Ruffo, M. L., Weibers, M., and Below, F.W. (2006). Optimization of corn grain composition with variable rate-nitrogen fertilization. *18th World Congress on Soil Science*, Philadelphia, Pennsylvania, USA. http://a-c-s.confex.com/crops/wc2006/techprogram/P18527 pp. 1–3.

Saseendran, A.A., Ma, L., Nielsen, D.C., Vigil, M.F., and Ahuja, L.R. (2005). Simulating planting date effects on corn production using RZ-

WQM and CERES-Maize model. *Agronomy Journal* 58–71.

Sauerborn, J., Sprich, H., and Mercer-Quarshie, H. (2001). Crop rotation to improve agriculture production in sub-Saharan Africa. *Journal of Agronomy and Crop Science* **184**, 67–72.

Sawyer, J. (2007). Nitrogen fertilization for corn following corn. http://www.ipm.iastate.edu/ipm/icm/2007/2-12/nitrogen.html pp. 1–4.

Selvaraju, R., Meinke, H., and Hansen, J. (2004). Climate information contributes to better water management of irrigated cropping systems in Southern India. *Proceedings of International Crop Science Congress*, Brisbane Australia. http://www.cropscience.org.au/icsc2004/poster/2/6739-selvarajur.htm pp. 1–7.

Selvi, D., Malarvizhi, P., and Santhy, P. (2006). Effects of long term fertilizer application and intensive cropping on dynamics of soil micronutrients under tropical agroecosystem. *18th World Congress of Soil Science*, Philadelphia, Pennsylvania, USA. http://a-c-s.confex.com/crops/wc2006/techprogram/P2837.htm pp. 1–3.

Sharma, K.N. (2006). Soil phosphorus fractionation dynamics and phosphorus sorption in a continuous maize-wheat cropping system. *18th World Congress on Soil Science*. Phialdelphia, PA, SA. http://www.idd.go.th/18wcss/techprogram/P12763.HTM pp. 1–3.

Sharma, S.P. and Subehia, S.K. (2003). Effects of twenty five years of fertilizer use on maize and wheat yields and quality of an acidic soil in the Western Himalayas. *Experimental Agriculture* **39**, 55–64.

Shetto, R.M. and Kwiligwa, E.M. (1998). Weed control systems in maize based on animal drawn cultivators. http://www.fao.org/Wairdocs/ILRI/x5483B/x5483b0n.htm pp. 1–11.

Silva, P.S.L., Silva, E.S., and Mesquita, S.S.X (2004). Weed control and green ear yield in maize. *Planta Daininha* **22**, 1–9.

Singh, J., Bajaj, J.C., and Pathak, H. (2005). Quantitative estimation of fertilizer requirement for maize and chickpea in the alluvial soil of indo-gangetic plains. *Journal of Indian Society of Soil Science* **53**, 68–73.

Sipaseuth, N., Attanandana, T., and Russel, Y. (2007). Nitrogen fertilizer response of maize on some important soils from DSSAT software prediction. *Kasetsart Journal* **41**, 21–27.

Sogbedji, J.M, Van Es, H.M., and Agbeko, K.L. (2006). Cover cropping and management strategies for maize production in Western Africa. *Agronomy Journal* **98**, 883–889.

Spargo, J. and Alley, M. (2006). Duration of continuous no-tillage management and soil nitrogen status in the virginia coastal plain. *Proceedings of the 18th World Congress of Soil Science*, Philadelphia, PA, USA. http://www.idd.goth/18wcss/techprogram/P15392.HTM pp. 1–3.

Stahr, K. (2000). Soil fertility, nutrient cycling and agroforestry systems in Southern Benin. http://www.toz.uni-hohenheim.de/research/sfb308/EBcont/c1.pdf pp. 1–16.

Stamp, P. (1999). Effects of fertilizer input and cultivar on the productivity of maize in the Polochic water shed, Guatemala and in Yorito-Sulaco region of Honduras. http://.www.rdb.etz.ch/project_veiw_block.php pp. 1–7.

Stanger, T.F. and Lauer, J. G. (2008). Corn grain yield response to crop rotation and nitrogen over 35 years. *Agronomy Journal* **100**, 643–650.

Stefanovic, L., Simic, M., Rosul, J.M., Vancetovic, J., Milivojevic, M., Misovic, M., Selakovic, D., and Hojka, Z. (2007). Problems in weed control in Serbian maize seed production. *Maydica* **52**, 277–280.

Storovogel, J.J. and Smaling, E.M.A. (1990). Assessment of soil nutrient depletion in Subsaharan Africa 1983–2000. *The Winand Staring Center for Integrated Land, Soil and Water Research*. Wageningen. The Netherlands Report No 28, pp. 1–83.

Surendran, U., Murugappan, Y., Bhaskaran, A., and Jagadeeswaran, R. (2005). Nutrient budgeting using NUTMON-Toolbox in an irrigated farm of semi-arid tropical region in India-A micro and meso level modeling study. *World Journal of Agricultural Sciences* **1**, 89–97.

Suwnarit, A., Suwannarat, C., and Chotchungmaneeerat, S. (2006). Effects of land preparation and maize cultivar on efficiency of N fertilizer applied at different times and by different methods in Maize-Mungbean association using 15N. *Plant and Soil* **94**, 179–190.

Teasdale, J., Starr, J., Ali, S., and Randy, R. (2005). Corn-weed competitive effects on soil water dynamics and nutrient uptake. *Proceedings of Northeastern Weed Science Society* **59**, 12.

Thavaprakash, N. and Velayudham K. (2007). Effect of crop geometry, intercropping system and INM practices on corn yield and nutrient uptake by baby corn. *Asian Journal of Agricultural Research* **1**, 10–16.

Thomsen, M.H., Nielsen, H., Oleskowitcz-Popiel, P., and Thomsen, A.B. (2008). Pretreatment of whole crop harvested, ensiled maize for ethanol production. *Applied Biochemistry and Biotechnology* **148**, 23–33.

Thomsen, M.H., Nielsen, H., Petersson, H., Thomsen, A.B., and Jensen, E.S. (2003). http://www.risoe.dk/rispubl/reports/ris-r-1608_94-105.pdf pp. 1–17.

Tian, G., Kang, B.T., Kolawole, G.O., Idinoba, P., and Salako, F.K. (2005). Long term effects of fallow systems and lengths on crop production and soil fertility maintenance during maize production in West Africa. *Nutrient Cycling in Agroecosystems* **71**, 278–292.

Tijani, F.O., Oyedele, D.J., and Aina, P.O. (2008). Soil moisture storage and water use efficiency of maize planted in succession to different fallow treatments. *International Agrophysics* **22**, 81–87.

Tsinghua, L. (2002). *In Situ* determination of ammonia volatilization from wheat-maize rotation system in North China. Acta Ecologica Sinica. http://www.shvoog.com/exact-sciecnes/agronomy-agriculture/1603323-situ-determination-ammnia-volatilization-wheat pp. 1–2.

Van Noordwijk, M. and Lusiana, B. (2000). *WaNulCAS Version 2.0 Background on a Model of Water, Nutrient and Light Capture in Agroforestry Systems*. International Centre for Research on Agroforestry (ICRAF), Bogor, Indonesia.

Vanlauwe, B., Diels, J., Sanginga, N., and Merckx, R. (2005). Long term integrated soil fertility management in South-Western Nigeria crop performance and impact on the soil fertility status. *Plant and Soil* **273**, 337–354.

Varvel, G.E. and Wilhelm, W.W. (2003). Soybean nitrogen contribution to corn and sorghum in western corn belt rotations. *Agronomy Journal* **95**, 1220–1225.

Walters, D., Dobermann, A., Vyn, T., and Brouder S. (2006). Innovations for improving productivity and nutrient use efficiency-maize systems of North America.   http://www.Idd.go.th/19wcss/techprogram/P17102.HTM pp. 1–3.

Wang, H., He, P., Zhao, P., and Guo, H. (2008) Nutrient management with in a wheat-maize rotation system. http://www.ppi-ppic.org/ppi-web/bcrops.nsf/$webindex/3FDC722FAD896A69852547A0070641C/$fiile/BC08-3p12.pdf pp. 1–5.

Wang, Z.H., Liu, X.J., Ju, X.T., Zhang, F., and Malhi, S.S. (2004). Ammonia Volatilization loss from surface-broadcasted urea: Comparison of vented and closed chamber methods and loss in winter wheat-summer maize rotation in North China plain. *Communications in Soil Science and Plant Analysis* **35**, 2917–2939.

Whitebread, A., Muchayi, P., and Waddington, S. (2004). Modeling the effect of phosphorus on maize production and nitrogen use efficiency on small holder in sub-humid Zimbabwe. http://www.cropscience.org.au/icsc2004/poster/2/5/4/1264_whitbreada.htm pp. 1–6.

Wolkowski, R. and Lowery, B. (2000). *Soil Erosion and Productivity*. University of Wisconsin Extension Bulletin Madison, Wisconsin, USA, pp. 1–8.

Wolkowski, R.P. (2003). Wisconsin corn and soybean responses to fertilizer placement in conservation an no-till tillage systems. http://soils.wisc.edu/extension/FAPM/2003proceedings/wolkowski-2.pdf pp. 1–16.

Witt, C., Pasuquin, J.M., and Doberman, A. (2006). Towards a site specific nutrient management approach for maize in Asia. *Better Crops* **90**(2), 27–31.

Witt, C., Pasuquin, J.M.C.A., Buresh, R.J., and Dobremann, A. (2007). The principles of site specific nutrient management for maize. http://www.ipipotach.org/e-ifc/2007-14/research4.php pp. 1–6.

Xie, Y., Kiniry, J.R., Nedbalek, V., and Rosenthal, W.D. (2001). Maize and sorghum simulations with CERS-Maize, SORKAM and ALMANAC under water-limiting conditions. *Agronomy Journal* **93**, 1148.

Yang, H., Dobermann, A., Cassman, K.G., and Walters, D.T. (2004b). *Hybrid-Maize: A Simulation Model for Corn Growth and Yield*. Nebraska

Cooperative Extension Service. CD9 University of Nebraska, Lincoln, Nebraska, USA.

Yang, H., Dobermann, A., Cassman, K.G., and Walters, D.T. (2006). Features, applications and limitations of the hybrid-maize simulation model. *Agronomy Journal* **98**, 737–748.

Yang, H., Dobermann, A., Lindquist, J.L., Walters, D.T., Arekbauer, T.J., and Cassman, K.G. (2004a). Hybrid-maize: A maize simulation models that combines two crop modeling approaches. *Field Crops Research* **87**, 131–154.

Yang, S., Li, F., Suo, D., Tianwen, G., Guo, W., Sun, B., and Jin, S. (2005). Effect of long term fertilization on soil productivity and nitrate accumulation in Gansu oasis. *Sciencia Agriculture*. http://www.cababsractsplus.org/abstracts/Abstract.asps?AcNo_20053187075.htm pp. 1–3.

Yang, J., Xinping, C., Fusuo, Z., and Xingren, W. (2003). Effect of mineral fertilizer application on energy efficiency in a long term field experiment. *Journal of China Agricultural University* **8**, 31–36.

Yilmaz, F, Taka, M., and Erayman, M. (2008). Identification of advantages of maize-legume intercropping over solitary cropping through competition indices in the East Mediterranean Region. *Turkish Agricultural Journal* **32**, 111–119.

Zacharia, G. (2007). Compost and fertilizer mineralization effects on soil and harvest in parkland agroforestry systems in the south Sudanese region of Burkina Faso. http://diss-epsilon.slu.se/archive/00001655/ pp. 1–2.

Zhang, F., Shen, J., Li, L., and Liu, X. (2004). An overview of rhizosphere processes related with plant nutrition in cropping systems in China.

Plant and Soil. http://www.cababstractsplus.org/abstracts/abstaract.aspx?AcNo=20043088089.HTM pp. 1–3.

Zhang, F.S., Fan, M.S., and Zhang, W.F. (2008). Principles, dissemination and performance of fertilizer Best Management practices developed in China. In *Proceedings of International Workshop on Fertilizer Best Management Practices*. International Fertilizer Industry Association, Paris, France, pp. 193–201.

Zhang, T.Q., Tan, C.S., Drury, C.F., and Reynolds, D.W. (2006). Long Term (43 Years) Fate of Soil Phosphorus as Related to Cropping Systems and Fertilization. *Proceedings of 18th World Congress of Soil Science*. Philadelphia, Pennsylvania, USA. http://www.idd.go.th/18wcss/tech-program/P18346.HTM pp. 1–3.

Zhihong, X. (2002). Nitrogen Cycling in Leucana Alley Cropping. http://www4.gu.edu.au:8080/adt-root/public/adt-qgu20050906.155955/index.html pp. 1–2.

Zhiming, F., Dengwei, L., and Yuehong, S. (2006). Water requirements and irrigation scheduling of spring maize using GIS and CropWat model in Beijing-Tianjin-Hebei region. *Chinese Geographical Science* **17**, 56–63.

Zhou, X., Madramootoo, C.A., Mackenzie, A.F., Kaluli, J.W., and Smith, D.L. (2000). Corn yield and fertilizer N recovery in water-table controlled corn-rye grass systems. *European Journal of Agronomy* **12**, 83–92.

Zou, G., Zhang, F.S., Ju, X., Chen, X., and Liu, X. (2006). Study on soil de-nitrification in wheat-maize rotation system. *Agricultural Sciences in China* **5**, 45–49.

Index